The Polysaccharides
Volume 1

MOLECULAR BIOLOGY

An International Series of Monographs and Textbooks

Editors: BERNARD HORECKER, NATHAN O. KAPLAN, JULIUS MARMUR, AND
HAROLD A. SCHERAGA

A complete list of titles in this series appears at the end of this volume.

The Polysaccharides
Volume 1

Edited by
GERALD O. ASPINALL

Department of Chemistry
Faculty of Science
York University
Downsview, Toronto
Ontario, Canada

1982

ACADEMIC PRESS

A Subsidiary of Harcourt Brace Jovanovich, Publishers
New York   London
Paris   San Diego   San Francisco   São Paulo   Sydney   Tokyo   Toronto

ACADEMIC PRESS, INC.
111 Fifth Avenue, New York, New York 10003

United Kingdom Edition published by
ACADEMIC PRESS, INC. (LONDON) LTD.
24/28 Oval Road, London NW1   7DX

Library of Congress Cataloging in Publication Data
Main entry under title:

The Polysaccharides.

    1. Polysaccharides.   I. Aspinall, Gerald O. [DNLM:
1. Polysaccharides.   QU 83 P783]
QD321.P77   547.7'82   82-6689
ISBN 0-12-065601-9 (v. 1)   AACR2

PRINTED IN THE UNITED STATES OF AMERICA

82 83 84 85   9 8 7 6 5 4 3 2 1

Contents

Contributors   *ix*

Preface   *xi*

Contents of Other Volumes   *xv*

1 General Introduction

GERALD O. ASPINALL

   I. Polysaccharides and Other Carbohydrate-Containing
      Polymers   *1*
  II. Nomenclature and Basic Chemistry of the Constituent
       Sugars   *5*
 III. General Structure and Nomenclature of
       Polysaccharides   *14*
 IV. The Literature of Polysaccharide Chemistry   *17*
       References   *18*

2 Isolation and Fractionation of Polysaccharides

GERALD O. ASPINALL

   I. Introduction   *19*
  II. Concepts of Homogeneity and Heterogeneity   *20*
 III. Isolation and Fractionation of Polysaccharides   *26*
 IV. Criteria for Homogeneity   *31*
       References   *32*

3 Chemical Characterization and Structure Determination of Polysaccharides

GERALD O. ASPINALL

    I. Introduction   *36*
   II. Composition and Molecular Size   *37*
  III. Problems of Structure Determination   *44*
  IV. Methylation   *45*
   V. Partial Depolymerization by Hydrolysis and Related Reactions   *56*
  VI. Oxidations with Periodate and Lead Tetraacetate   *81*
 VII. Structural Modifications of Polysaccharides   *89*
VIII. Other Methods for Partial Depolymerization   *100*
     References   *124*

4 Spectroscopic Methods

ARTHUR S. PERLIN and BENITO CASU

   I. Introduction   *133*
  II. Nuclear Magnetic Resonance Spectroscopy   *135*
 III. Infrared–Raman Spectroscopy   *172*
 IV. Comparative Evaluation of Spectroscopy Methods   *184*
    References   *186*

5 Shapes and Interactions of Carbohydrate Chains

DAVID A. REES, EDWIN R. MORRIS, DAVID THOM, and JOHN K. MADDEN

   I. Conformational Principles   *196*
  II. Experimental Characterization   *204*
 III. Homopolysaccharides   *224*
 IV. Regular Copolysaccharides   *238*
  V. Disordered Chains in Solution   *255*
 VI. Hydrated Networks   *263*
 VII. Mixed Interactions   *276*
    References   *281*

6 Immunology of Polysaccharides

C. T. BISHOP and H. J. JENNINGS

  I. Introduction   *292*
 II. Molecular Architecture of Bacteria, Yeasts, and
    Fungi   *293*
III. The Immune Response to Polysaccharides   *296*
 IV. Structural Aspects of the Immune Response to
    Polysaccharides   *312*
  V. Polysaccharide Antigens in Human Disease   *320*
    References   *325*

Index   *331*

Contributors

Numbers in parentheses indicate the pages on which the authors' contributions begin.

Gerald O. Aspinall (1, 19, 35), Department of Chemistry, Faculty of Sciences, York University, Downsview, Toronto, Ontario M3J 1P3, Canada

C. T. Bishop (291), Division of Biological Sciences, National Research Council of Canada, Ottawa, Ontario K1A OR6, Canada

Benito Casu (133), Istituto di Chemica e Biochimica "G. Ronzoni," 20133 Milano, Italy

H. J. Jennings (291), Division of Biological Sciences, National Research Council of Canada, Ottawa, Ontario K1A OR6, Canada

John K. Madden (195), Unilever Research, Colworth Laboratory, Sharnbrook, Bedford MK44 1LQ, England

Edwin R. Morris (195), Unilever Research, Colworth Laboratory, Sharnbrook, Bedford MK44 1LQ, England

Arthur S. Perlin (133), Department of Chemistry, McGill University, Montreal H3A 2A7, Canada

David A. Rees (195), Unilever Research, Colworth Laboratory, Sharnbrook, Bedford MK44 1LQ, England

David Thom[1] (195), Unilever Research, Colworth Laboratory, Sharnbrook, Bedford MK44 1LQ, England

[1] Present address: Unilever UK Central Resources, Unilever House, Black Friars, London, England.

Preface

This work provides the most complete summary now available of the present knowledge of polysaccharide chemistry and related aspects of biochemistry. During the past 30 years the field has expanded so rapidly that it is no longer possible to cover even the more important areas in one volume as was done in the comprehensive "Polysaccharide Chemistry," edited by R. L. Whistler and C. L. Smart in 1953, or in a less comprehensive small textbook by the present editor, in 1970. This account is therefore divided into three volumes, and while maintaining broad coverage of polysaccharides, has been restricted in its discussion of several cognate areas, e.g., glycoproteins, glycolipids, and nucleic acids; these have been deliberately excluded as major topics.

Polysaccharides are of widespread occurrence and information on these substances is necessary for many scientists. Commercially, interest in polysaccharides is no longer limited to starch and cellulose in the food, pulp and paper, and allied industries, but is extended to include the natural gums and mucilages from higher plants and seaweeds, and the newer bio-gums from microorganisms made available through the fermentation industries. From the standpoint of nutrition, starch is no longer the only plant polysaccharide of significance now that the roles played by polysaccharides of the pectin and hemicellulose groups as constituents of dietary fiber are recognized. In animal tissues and fluids, polysaccharides are covalently linked to protein, and defects in the metabolism of proteoglycans from connective tissue are implicated in skin, ocular, and arthritic diseases. Perhaps most dramatically, many polysaccharides of microorganisms are now known to act as type-specific antigens, and large areas of immunology are incomprehensible without an appreciation of the role played by these substances in host–parasite relations.

These volumes are directed primarily to chemists and biochemists working on polysaccharides and other complex carbohydrates. In addition, one consequence of the "polysaccharide explosion" is that many scientists are encountering these materials for the first time and are in need of a reference work covering the major aspects of the subject. This treatise therefore will be of interest to chemists without specialized knowledge of the area, to those for whom polysaccharides are substances with potential for industrial exploitation, and to agricultural and medical biologists, for whom polysaccharides are of interest as functional components of living organisms. A general chemical background, especially in organic chemistry, and some familiarity with simple carbohydrates is assumed.

Volume 1 is concerned mainly with methodology. A general introduction is followed by a discussion of problems encountered in the isolation and fractionation of polysaccharides. Information on procedures used for the determination of primary or covalent structure is presented in chapters on chemical degradative and spectroscopic nondegradative methods. For an appreciation of both solid state and solution properties of polysaccharides, noncovalent aspects of secondary and tertiary structure are discussed in a chapter on shapes and interactions of carbohydrate chains. The final chapter in Volume 1 deals with medical aspects of polysaccharides resulting from their role in immunology. Biochemical methodology will be emphasized in Volume 3 in chapters on biosynthesis and on the use of enzymes in probing the detailed structure of polysaccharides. Volumes 2 and 3 deal mainly with groups of polysaccharides, individual treatment being given to cellulose, starch, and glycogen. Within other chapters the classification is based on chemical structure, but it is nevertheless convenient to treat groups of polysaccharides according to biological origin; hence the separate chapters on polysaccharides from plants, bacteria and protozoa, algae (in seaweeds), yeasts and fungi, and mammalian sources. In recognition of their commercial importance when isolated as individual chemical substances, a chapter is included on the industrial utilization of polysaccharides. In order to be reasonably comprehensive, but not unduly repetitive, the more descriptive chapters assume familiarity with chemical and spectroscopic procedures for structure determination which are dealt with in the first volume.

The range of topics in these volumes is such that no one person can attempt to deal with all of them in an equally authoritative manner. Consequently, the present multi-author work calls on the expertise of specialists in different areas. The treatise is truly international with authors now residing in Australia, Brazil, Canada, France, Italy, New Zealand, Norway, South Africa, Sweden, United Kingdom, and the United States. The editor is grateful to all collaborators who have contributed much time and effort in the preparation of manuscripts.

Gerald O. Aspinall

Contents of Other Volumes

Volume II (tentative)

Classification of Polysaccharides
Gerald O. Aspinall

Cellulose
R. H. Marchessault and P. Sundararajan

Other Plant Polysaccharides
A. M. Stephen

Algal Polysaccharides
Terence J. Painter

Bacterial Polysaccharides
Lennart Kenne and Bengt Lindberg

Fungal Polysaccharides
Philip A. J. Gorin and Eliana Barreto-Bergter

Industrial Utilization of Polysaccharides
John K. Baird and Paul A. Sandford

Volume III (tentative)

The Use of Enzymes in the Study of Polysaccharides
Norman K. Matheson and Barry V. McCleary

Biosynthesis of Polysaccharides
Douglas James, Jr., Jack Preiss, and Alan D. Elbein

Starch
Christiane Mercier and A. Guilbot

Glycogen
R. Geddes

Mammalian Polysaccharides
L. Å. Fransson

The Polysaccharides
Volume 1

General Introduction

GERALD O. ASPINALL

I. Polysaccharides and Other Carbohydrate-Containing Polymers . .   1

II. Nomenclature and Basic Chemistry of the Constituent Sugars   . .   5

III. General Structure and Nomenclature of Polysaccharides   14

IV. The Literature of Polysaccharide Chemistry   17

References   18

I. Polysaccharides and Other Carbohydrate-Containing Polymers

Natural macromolecules containing carbohydrate units are of widespread occurrence in all living organisms and include (a) polysaccharides as exclusively carbohydrate polymers; (b) glycoproteins, proteoglycans, and peptidoglycans; (c) glycolipids and lipopolysaccharides; (d) teichoic acids and related macromolecules containing phosphorodiester-linked oligosaccharide repeating units; and (e) nucleic acids. In order to keep the size of this work within reasonable bounds, these volumes are restricted to polysaccharides and their conjugates as defined in the following paragraph. Nucleic acids, although carbohydrate-containing macromolecules, display their distinctive features through their heterocyclic base constituents and obviously stand

1

in their own right as a major group of biopolymers for which there is an extensive literature. There are many excellent monographs and reviews covering different groups of glycoconjugates, and a discussion of many of these materials is excluded from the present volumes. In particular, glycoproteins have received extensive coverage in a two-volume reference work (1) as well as in other monographs (2–6) and numerous reviews (7,8). Although aspects of the chemistry of proteoglycans and their glycosaminoglycan components have been reviewed elsewhere (9–13) and are the subject of a comprehensive monograph (14), these substances are included as an important group of polysaccharide conjugates, the study of which informs and is informed by that of other classes of polysaccharides.

The term "polysaccharide" is applied in these volumes to those carbohydrate polymers that contain periodically repeating structures in which the dominant, but not necessarily exclusive interunit linkages are of the O-glycosidic type. This classification therefore includes not only classical polysaccharides consisting entirely of glycosidically linked sugar residues, but also the following classes of substances: (a) proteoglycans, in which periodic sequences of sugar residues comprise polysaccharide subunits that are linked through various types of sugar–amino acid linkages to proteins; (b) peptidoglycans, in which polysaccharide chains are covalently cross-linked through peptide bridges; (c) lipopolysaccharides, in which polysaccharide chains are covalently attached to lipid units; and (d) the teichoic acid group of substances, including the so-called capsular teichoic acids, in which there are repetitive sequences of oligosaccharide units (themselves containing glycosidically linked residues) that are mutually joined through phosphorodiester bonds. Some of the teichoic acids proper fall outside a rigorous definition of polysaccharides, but it is convenient to consider them alongside the phosphorodiester-linked oligosaccharide polymers elaborated by several bacteria and together with peptidoglycans, to which they are linked in the bacterial cell wall. In all of these classes of carbohydrate polymers the biosynthesis of major structural segments involves (1) the consecutive addition of one or more types of sugar residue in which the addition of each successive unit is governed by the specificity of the enzyme, both for the glycosyl unit being transferred and the acceptor site, and/or (2) the polymerization of preassembled "repeating units" themselves synthesized as in (1). It should be noted that for certain polysaccharide types post-polymerization modifications may take place so that the original repetitive structure is obscured.

According to the above classification the following classes of carbohydrate-containing macromolecules are omitted from detailed consideration in these volumes: (a) nucleic acids (both DNA and RNA), in which nucleosides (themselves *N*-glycosides) are joined through phosphorodiester bonds; (b) glycoproteins, in which the carbohydrate chains may carry up to 20 residues but in which no repetitive structure is apparent; and (c) glycolipids, in which oligosaccharide units, again with no obvious repetitive structure, are linked to lipid moieties. With relatively few borderline cases the vast majority of carbohydrate polymers fall clearly into the categories of macromolecules included or excluded in these volumes. However, since several of the same methods are used for the study of structure, biosynthesis, and physical properties of these glycoconjugates as for polysaccharides, references are made to them when appropriate.

The functions of all individual polysaccharides cannot be uniquely assigned, but it is evident that they may act as storage materials, as structural components, and as protective substances. Starch, glycogen, some β-D-glucans, fructans, and some galactomannans are well-known reserve polysaccharides. These substances may be rapidly metabolized and may vary markedly in amount with the state of development of the organism.

Structural polysaccharides fall into two distinct classes. The fibrous polysaccharides, cellulose in higher plants and some algae, chitin in yeasts and fungi, and less frequently 3-linked β-D-xylans and 4-linked β-D-mannans in some plants and algae, are relatively invariant in their respective structures and can adopt regular chain conformations. In contrast, the matrix polysaccharides are characterized by their gel-forming capacity, which confers flexibility on the structural assembly. These gel-forming polysaccharides can adopt regular chain conformations for substantial parts of their structures, but interruptions in regularity of structure permit modification of cell wall texture. Thus, subtle changes in structure cause marked differences in physical properties. The biosynthesis of matrix polysaccharides frequently involves structural modification after the basic skeleton has been assembled. In plant polysaccharides it is common to encounter branched structures in which linear chains of uniform linkage type carry variable proportions of rather short side chains, e.g., arabinoxylans and galactomannans, in which there is little apparent regularity of branching, suggesting that the attachment of side chains to the extent desired for the modification of properties occurs as a separate process after completion of the main chain. Another type of structurally variable polysaccharide is encountered when the initially laid down regular chain undergoes postpolymerization modification. Alterations in the configuration of individual sugar residues, e.g., of β-D-mannuronic acid to α-L-guluronic acid in alginic acid or of β-D-glucuronic acid to α-L-iduronic acid in dermatan sulfate, or in the conformation of sugar residues, e.g., of α-D-galactopyranose 6-sulfate to 3,6-anhydro-α-D-galactopyranose in the carrageenans, result in changes in polysaccharide chain conformations. Pectins, an important group of matrix polysaccharides in higher plants, show variations in structure due to the nature and frequency of side-chain attachment. In addition, physical properties of pectins may be influenced by the frequency and distribution of structural modifications occasioned by esterification of D-galacturonic acid residues, attachment of O-acetyl, and insertion of L-rhamnose residues in the galacturonan chains.

Protective polysaccharides are exemplified by the antigenic and immunogenic extracellular polysaccharides from microorganisms which are frequently highly specific for particular organisms. The exudate gums from plants appear to provide a similar protective role in sealing off injured parts of the plant from microbial infection, but specificity of the type shown by microorganisms has yet to be demonstrated.

Knowledge of the primary or covalent structure of polysaccharides forms the basis for their classification, for an understanding of their three-dimensional structures in the solid state and in solution, and for an appreciation of the ways in which polysaccharides are synthesized and broken down naturally. A large part of these volumes is therefore concerned with aspects of covalent structure. Two chapters are devoted to chemical methods for structure determination and to the use of spectroscopic techniques. Several chapters give accounts of the major groups of polysaccharides Certain polysaccharides, notably cellulose, starch, and glycogen, are of such widespread occurrence that they are discussed in separate chapters. Other types of polysaccharides are described in chapters covering those from different classes of organism, i.e., higher plants, algae, fungi, bacteria, and mammals. Within each of these chapters the polysaccharides are classified mainly by composition and/or structure.

In recognition of the importance of the highly specific immunochemical nature of many microbial polysaccharides, the immunology of polysaccharides receives separate treatment. Yet other polysaccharides of both plant and microbial origin are of increasing commercial importance, and a chapter is devoted to this aspect of polysaccharide chemistry. The physical properties and biological functions of polysaccharides are dependent on their capacity to adopt energetically favorable conformations that are stabilized either by interresidue but intrachain or by interchain interactions, and the rapid advances in this area are given separate treatment. Chapters on the biosynthesis and enzymology of polysaccharides discuss topics of obvious biochemical interest. These chapters illustrate how knowledge of covalent structure is used to obtain information on natural processes and how the reciprocal relationship of enzyme-catalyzed reactions and primary structure may advance our knowledge of the latter.

II. Nomenclature and Basic Chemistry of the Constituent Sugars

The chemistry of polysaccharides is derived from that of the constituent sugars. Most general textbooks of organic chemistry provide the classic proof of the structure of D-glucose by Emil Fischer together with the elaboration of the configurational basis for the classification of the aldose sugars in the acyclic aldehydo form according to the Fischer convention (see Fig. 1). The long established names of these aldose sugars provide the configurational prefixes for compounds with multiple centers of chirality containing chains with the same number of consecutive, but not necessarily continuous asymmetric carbon atoms. Thus, shown in the acyclic form, D-fructose with an acceptable trivial name established by usage is D-*arabino*-hexulose (1). Similarly, the commonly termed 2-deoxy-D-ribose (2) and 3-deoxy-D-glucose (3) are both incorrectly named and should be designated by the correct configurational prefixes as 2-deoxy-D-*erythro*-pentose and 3-deoxy-D-*ribo*-hexose, respectively. Commonly occurring sugars with altered functionality may retain trivial names that show the configurational relationship to aldohexoses, e.g., D-mannuronic acid (4) and D-galactosamine (5). The latter name, although in practice unambiguous for a polysaccharide constituent, is insufficient to designate structure, and the sugar should correctly have the systematic name, 2-amino-2-deoxy-D-galactose. These principles may be extended to the nomenclature of sugars containing more than six carbon atoms and, in some instances, more than four asymmetric carbons. Compounds 6–8 are examples. Rules of carbohydrate nomenclature have been adopted by international agreement (15), and the salient features are mentioned in this chapter.

The fundamental chemistry of the monosaccharides is reviewed here only to the extent necessary to appreciate the structure and stereochemistry of the simple sugars, as well as the nomenclature of their ring forms as reducing sugars and after conversion to glycosides, the universal type of intersugar linkage in polysaccharides. There are good general introductions to monosaccharide chemistry (16,17) and more extensive treatises (18,19). An excellent monograph deals with both conformational and configurational aspects of monosaccharide stereochemistry (20).

Reducing sugars exist virtually exclusively as cyclic hemiacetals. Whereas crystalline sugars exist in a single form, sugars in solution are present in both five-membered furanose and six-membered pyranose forms, each in two anomeric forms (epimeric at C-1 for aldoses or C-2 for ketoses), between which forms and the intermediate acyclic form a rapid equilibrium is established. Such an equilibrium is shown in Fig. 2 for D-glucose, for which sugar

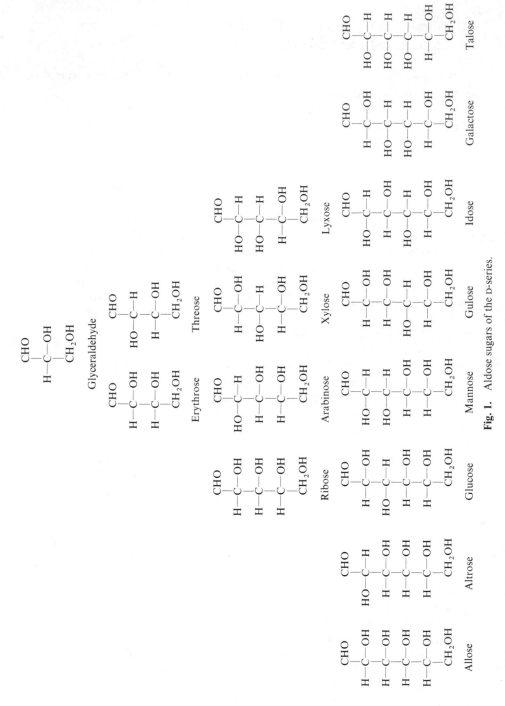

Fig. 1. Aldose sugars of the D-series.

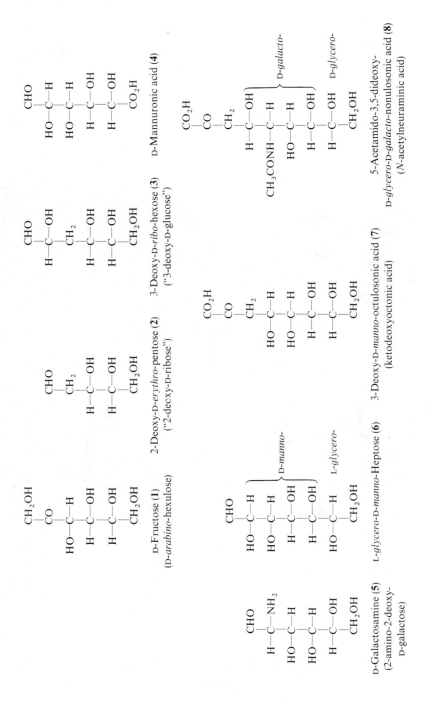

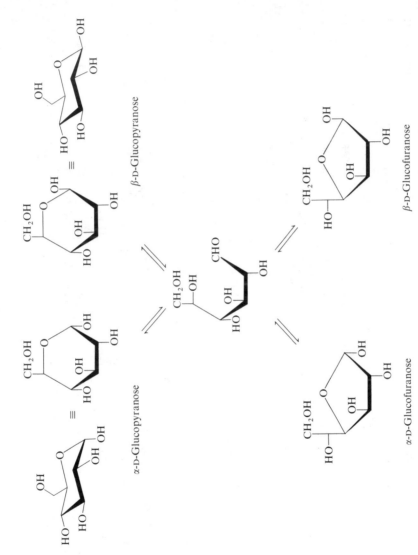

Fig. 2. Mutarotational equilibrium between the various hemiacetal ring forms of D-glucose is established through the acyclic aldehydo form. Pyranose ring forms are shown both in Haworth perspective formulas and in preferred chain forms to emphasize configurational and conformational aspects of structure.

β-D-Glucopyranose

α-D-Glucopyranose

β-D-Glucofuranose

α-D-Glucofuranose

only very small amounts of furanose forms exist in aqueous solution. The systematic nomenclature of cyclic sugars designates both ring size and configuration at the newly formed chiral center at C-1 (or C-2 for ketoses). In general, pyranose ring forms are more stable than furanose structures in aqueous solution, although not necessarily in all solvents. Reducing sugars are therefore frequently shown in this ring size, and the presence of an equilibrium mixture of α- and β-anomers may be indicated by the configurationally undefined hemiacetal hydroxyl group ($\sim$OH). Figure 3 shows most of the commonly occurring sugars that are constituents of polysaccharides of plant and animal origin, as well as many of microbial origin. However, polysaccharides from microorganisms contain many rare and unusual sugar constituents, and these are listed in Volume II, Chapter 4 on bacterial polysaccharides.

Of the sugars shown in Fig. 3 no attempt has been made to include as separate sugars those that have undergone minor modification, e.g., esterification or etherification. In most cases in which sugars of this type occur as polysaccharide constituents, the modifications have been introduced biosynthetically after incorporation of the "basic" sugar into the polysaccharide chain. In some cases, e.g., esters and acetals, the substituents are removed during hydrolysis, but acid-stable substituents such as ethers are retained on hydrolysis and give rise to "new" sugars. Table I gives examples of common functional group modifications of naturally occurring sugar residues in polysaccharides. It should be noted that the same types of functional groups are available for laboratory manipulation, although as discussed in Chapter 3 selectivity in the introduction of such modifications is not easy to achieve.

In contrast to the reducing hemiacetals (or hemiketals), which undergo rapid interconversion in solution, glycosides, the products of reaction with alcohols, (as aglycones) are nonreducing acetals. Compounds of this class, whether synthesized as isomeric mixtures in the laboratory or of natural occurrence, are stable in neutral or basic solution and are of defined ring size and anomeric configuration (Fig. 4). These structural features must therefore be included in any systematic nomenclature. The glycosidic linkage as the intersugar linkage in disaccharides and higher saccharides involves different sugar units as glycosyl substituents and aglycones (Fig. 5). Since such aglycones are polyhydroxy compounds, any one hydroxyl group of which may be involved in the glycosidic linkage, oligosaccharide nomencalture must designate linkage type, e.g., 1→2, as well as ring size and configuration for each sugar residue. Maltose is the trivial name for the disaccharide **6**, which is unambiguously called 4-*O*-α-D-glucopyranosyl-D-glucopyranose or *O*-α-D-glucopyranosyl-(1 → 4)-D-glucopyranose (shown here in both Haworth perspective and conformational formulas) and is one

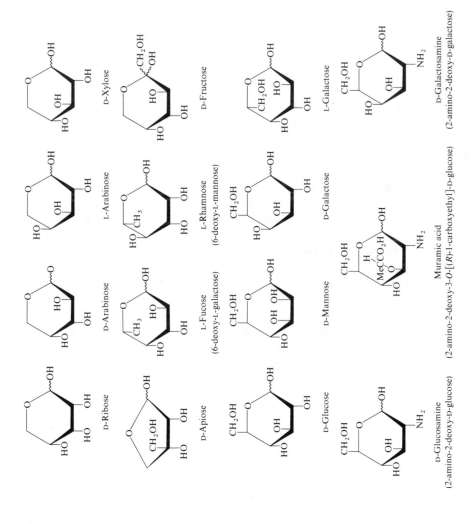

D-Xylose

D-Fructose

L-Galactose

D-Galactosamine
(2-amino-2-deoxy-D-galactose)

L-Arabinose

L-Rhamnose
(6-deoxy-L-mannose)

D-Galactose

Muramic acid
(2-amino-2-deoxy-3-O-[(R)-1-carboxyethyl]-D-glucose)

D-Arabinose

L-Fucose
(6-deoxy-L-galactose)

D-Mannose

D-Ribose

D-Apiose

D-Glucose

D-Glucosamine
(2-amino-2-deoxy-D-glucose)

10

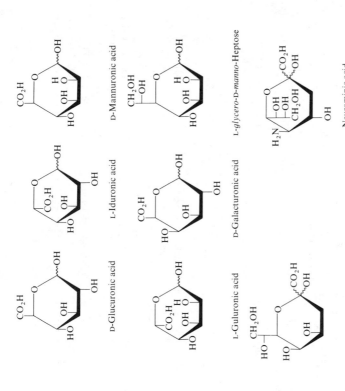

Fig. 3. Sugars of widespread occurrence as constituents of plant, animal, and some microbial polysaccharides. Sugars are all shown as hemiacetals (or hemiketals) in the pyranose form (other then for D-apiose) regardless of the ring form when glycosidically linked in polysaccharides. With the exception of muramic acid (3-*O*-lactyl-D-glucosamine), ethers, e.g., methyl ethers, are not listed separately.

D-Glucuronic acid

L-Iduronic acid

D-Mannuronic acid

L-Guluronic acid

D-Galacturonic acid

L-*glycero*-D-*manno*-Heptose

Neuraminic acid

3-Deoxy-D-*manno*-octulosonic acid (KDO)   (5-amino-3,5-dideoxy-D-*glycero*-D-*galacto*-nonulosonic acid)

11

TABLE I

Functional Group Modifications of Sugar Residues in Polysaccharides

Functional group	Derivatives			Occurrence
—OH	Ethers	O-Methyl	—OCH$_3$	Widespread
		(R)- or (S)-O-lactyl	—OCH(CH$_3$)CO$_2$H	Constituent of some sugars in microbial polysaccharides; most common example is N-acetylmuramic acid, 2-acetamido-2-deoxy-3-O-[(R)-1-carboxyethyl]-D-glucose
	Esters	O-Acetyl	—OCOCH$_3$	Widespread; other O-acyl groups infrequent
		Sulfate hemiesters	—OSO$_3^-$	Widespread, especially in algal glycans and glycosaminoglycans; D- and L-galactose 6-sulfates are precursors of the respective 3,6-anhydro-D- and -L-galactose residues in polysaccharides
		Phosphoromonoesters	—OPO$_3^{2-}$	Relatively infrequent
		Phosphorodiesters	—O[PO$_2^-$]O—	Relatively common linking oligosaccharide units
	Acetals/ketals	Pyruvyl	$\begin{matrix} O \diagdown \\ \ \ \ \ C \\ O \diagup \end{matrix} \begin{matrix} CH_3 \\ \diagup \\ \diagdown \\ CO_2H \end{matrix}$	Attached to various sugar residues, most commonly to terminal units
—NH$_2$	Amides	N-Acetyl	—NHCOCH$_3$	Widespread; other N-acyl groups are constituents of sialic acids
		N-Sulfate (sulfoamino)	—NHSO$_3^-$	Constituent of D-glucosamine in heparin
—CO$_2$H	Esters		—CO$_2$CH$_3$	Methyl D-galacturonate residues in pectins

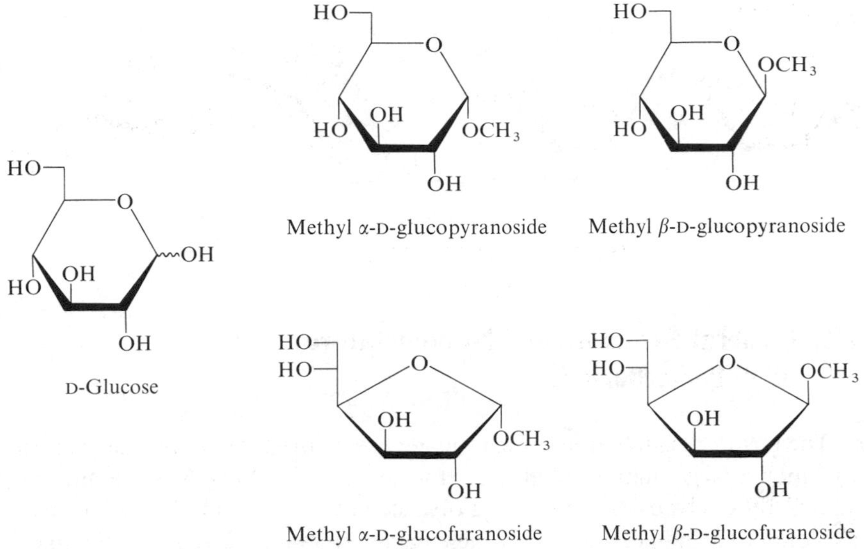

Fig. 4. Simple glycosides of individually defined ring size and anomeric configuration illustrated by the four methyl glycosides formed from D-glucose.

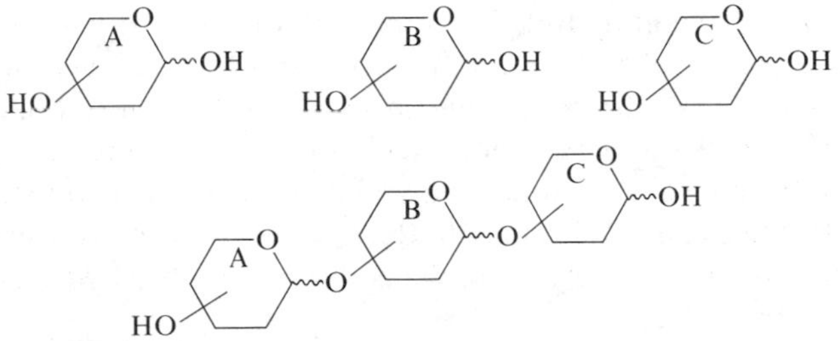

Fig. 5. Generalized formation of glycosidic linkages in a trisaccharide illustrating the roles of sugar A as glycosyl substituent (or glycosyl donor) to sugar B, which is both aglycone (or glycosyl acceptor) to sugar A and glycosyl substituent to sugar C, which in turn acts solely as aglycone (to sugar B). Once formed, each glycosidic linkage involves a glycosyl substituent of defined ring size (arbitrarily shown as pyranosides) and anomeric configuration and an aglycone of specified linkage type (e.g., 1→4).

of 20 possible reducing disaccharides containing two D-glucose units. In addition, there are another 10 possible nonreducing disaccharides based on this one sugar. Since systematic names for disaccharides and higher oligosaccharides are long and cumbersome, shorthand forms are convenient, and these are described later in connection with polysaccharide designations.

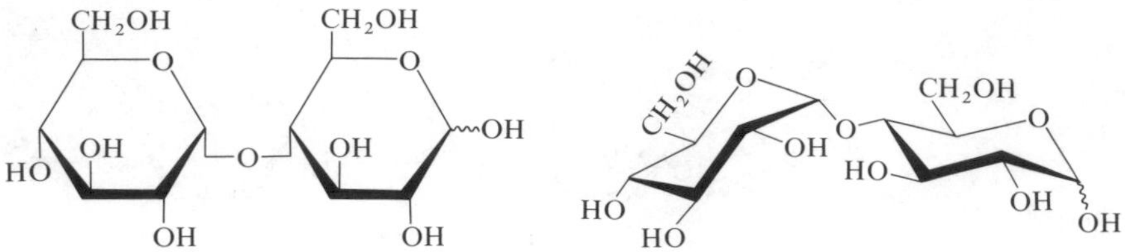

Maltose (**9**)
(4-*O*-α-D-glucopyranosyl-D-glucopyranose)

III. General Structure and Nomenclature of Polysaccharides

The term "glycan" is the systematic generic name given to polysaccharides in which a large number of glycose (monosaccharide) residues are mutually joined by *O*-glycosidic linkages. Polysaccharides (Fig. 6) may be regarded as condensation polymers in which each intersugar glycosidic linkage is formed from the glycosyl moiety of a hemiacetal or hemiketal and a hydroxyl group of another sugar unit acting as an acceptor molecule or aglycone. Since a given sugar residue may form only one glycosidic linkage with another sugar residue as aglycone but a sugar residue acting in the latter capacity carries several hydroxyl groups, one or more of which may be an acceptor of glycosyl substituents, polysaccharides may be linear or branched. With the exception of the cyclic oligosaccharides, known as cycloamyloses, there is a defined chain sense in polysaccharides from nonreducing terminus (or termini) to reducing terminus. Since the latter unit acts only as an agly-cone, a reducing sugar residue may undergo natural or inadvertent chemical modification or, alternatively, be replaced by some other hydroxylated unit,

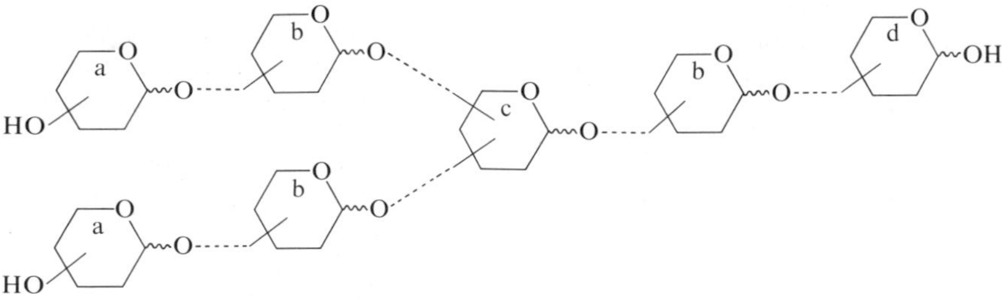

Fig. 6. Segments of a polysaccharide chain showing defined chain sense from nonreducing termini (a), via chain units (b), possibly through branch points (c), to "reducing" terminus (d), which may be a reducing hemiacetal or some other hydroxylated compound acting as glycosyl acceptor (or aglycone). Branching sugar residues serve as "polyvalent" aglycones.

e.g., D-mannitol or D-glyceric acid. A consequence of the "monovalent" character of glycosyl units but the "polyvalent" nature of aglycone units is that, whereas branching is encountered, cross-linking between adjacent chains by covalent bonds is impossible through glycosidic linkages. Inter-chain cross-linking through other types of bonds is possible in principle but has been clearly established only in bacterial cell wall peptidoglycans through polypeptide bridges in which the chain sense is reversed through ε-amino groups in lysine or diaminopimelic acid units.

Polysaccharides containing only one kind of monosaccharide unit are homopolysaccharides or homoglycans, e.g., glucan, arabinan, and galac-turonan. Although many sugars found as polysaccharide constituents occur in only one enantiomeric form, it is usual to insert enantiomeric prefixes (D or L), and, if known, anomeric prefixes (α or β) are added. At least for homopolysaccharides of uniform linkage type, the systematic name may also designate ring size (furanose or pyranose) and linkage type. Thus, cellulose is a $(1\rightarrow4)$-β-D-glucopyranan, and inulin is a $(2\rightarrow1)$-β-D-fructofuranan.

Polysaccharides containing two or more kinds of monosaccharide unit are heteropolysaccharides or heteroglycans, e.g., arabinoxylans and galac-tomannans. Enantiomeric prefixes are frequently added, but anomeric pre-fixes, although correct, are rarely appended. It may be noted that the names do not indicate the relative proportions or the distributions of the two or more sugar constituents. If the principal chain of the polysaccharide is com-posed of one kind of sugar unit only, its configurational prefix should be written last and those of other sugar residues used adjectivally in alphabetical order, e.g., D-galacto-D-xylo-D-glucan. For other heteropolysaccharides configurational prefixes are listed alphabetically, e.g., D-galacto-D-gluco-D-mannoglycan.

It is common practice to represent average or regularly repetitive segments of polysaccharide structures in shorthand forms in which, if known, linkage types, ring sizes, and anomeric configurations are added to the enantiomer-ically defined sugars. When space permits, a fully punctuated form is used with the following types of information. (a) Common sugars, e.g., pentoses and hexoses, and certain 6-deoxyhexoses are designated by the first three letters, e.g., Xyl = xylose, Gal = galactose, Rha = rhamnose, *but* Glc = glucose (Glu = glutamic acid); (b) enantiomeric prefixes (D or L) and anomeric prefixes (α or β, which should be used only in conjunction with enantiomeric prefixes) are then added; (c) ring sizes are designated by *p* (pyranose) or *f* (furanose); (d) structurally modified, but configurationally related sugars have the appropriate modification added as a suffix after the ring size, e.g., D-Gal*p*A = D-galactopyranuronic acid, α-D-Gal*p*N = 2-amino-2-deoxy-α-D-galactopyranose, β-D-Glc*p*NAc = 2-acetamido-2-deoxy-β-D-glucopyranose,

β-D-Man$p$NAcA = 2-acetamido-2-deoxy-β-D-mannopyranopyranosyluronic acid (N-acetyl-β-D-mannosaminuronic acid) (note that numerical prefixes for amino substituents are usually omitted for the commonly occurring 2-amino-2-deoxy sugars; (e) linkage types are indicated by arrows from the glycosyl group to the aglycone.

In common but not required usage, sugar sequences are read from left to right from nonreducing terminus to "reducing" terminus. The following are examples of repetitive sequences in polysaccharides:

$$\rightarrow 4)\text{-}\beta\text{-D-Glc}p\text{A-}(1\rightarrow 3)\text{-}\beta\text{-D-Glc}p\text{NAc-}(1\text{-}$$

Hyaluronic acid

(a D-glucosamino-D-glucuronoglycan; strictly speaking,
a 2-acetamido-2-deoxy-β-D-gluco-β-D-glucopyranuronoglycan)

$$\rightarrow 4)\text{-}\beta\text{-D-Xyl}p\text{-}(1\rightarrow 4)\text{-}\beta\text{-D-Xyl}p\text{-}(1\rightarrow 4)\text{-}\beta\text{-D-Xyl}p\text{-}(1\text{-}$$

2

↑

1

(4-Me)-α-D-Glc$p$A

(4-O-Methyl)-D-glucurono-D-xylan

[strictly speaking, a (4-O-methyl)-α-D-glucopyranurono-β-D-xylopyranan]

$$\rightarrow 4)\text{-}\beta\text{-D-Man}p\text{-}(1\rightarrow 4)\text{-}\beta\text{-D-Man}p\text{-}(1\rightarrow 4)\text{-}\beta\text{-D-Man}p\text{-}(1\text{-}$$

6

↑

1

α-D-Gal$p$

D-Galacto-D-mannan

(strictly speaking, an α-D-galactopyrano-β-D-mannopyranan)

When structural units are known but complete sequences have not yet been established, dashed lines designate uncertain sequences. For complex arrays of sugar residues, and if necessary for brevity, abbreviated shorthand forms may be used with omission of punctuation or of information that is either taken for granted, e.g., (1→4) linkages in starch-related polysaccharides, or simply not known. In these volumes these abbreviations are used at the discretion of the authors (and the editor), provided that in the context the information is conveyed without ambiguity.

IV. The Literature of Polysaccharide Chemistry

Polysaccharide chemistry is a broad topic that stands at the intersection of organic chemistry, biochemistry, and polymer chemistry. Structure determination uses the procedures of organic chemistry but also a variety of physical methods, especially spectroscopy. The reactions of polysaccharides are of interest to organic chemists and biochemists alike. The physical properties of polysaccharides in the solid state and in solution are of interest not only to polymer chemists seeking to understand and exploit those properties, but also to biochemists and biologists wishing to learn more about the functions of polysaccharides in their natural environment. It is not surprising, then, that different aspects of polysaccharide chemistry and biochemistry are scattered widely throughout the scientific literature. This point is emphasized by the array of papers on polysaccharides in *Chemical Abstracts*, which can be found in the following sections: Pharmacodynamics (1), General Biochemistry (6), Enzymes (7), Biochemical Methods (9), Microbial Biochemistry (10), Plant Biochemistry (11), Mammalian Biochemistry (13), Immunochemistry (15), Fermentations (16), Foods (17), Carbohydrates (33), Cellulose, Lignin, Paper, and Other Wood Products (43), Industrial Carbohydrates (44), and probably others.

Serial publications dealing exclusively with carbohydrates contain sections devoted to polysaccharides. "Advances in Carbohydrate Chemistry and Biochemistry" (Academic Press, New York) is noteworthy for its longevity (uninterrupted publication since 1946 with the appearance of two volumes each year since 1974), as well as for a consistently high standard of authoritative reviews. "Specialist Periodical Reports on Carbohydrate Chemistry" of the Royal Society of Chemistry (formerly The Chemical Society, London) provides a different orientation, with the primary emphasis on complete and up-to-date (not always realized in practice) citation of all relevant papers in the literature. Access to both of these serial publications is essential for all seeking to be specialists in the field. From the standpoint of experimental procedures the series "Methods in Carbohydrate Chemistry" (Academic Press, New York) provides detailed instructions on established methods, sometimes with helpful evaluations of alternative methods. Analytical and structural methods for polysaccharides and glycoconjugates, as well as for specifically enzymatic procedures, are covered in selected volumes (especially Volumes 28 and 50) of "Methods in Enzymology" (Academic Press, New York). Specialist monographs and occasional review articles on particular topics are too numerous to list, but references to such articles will be found in these volumes, as well as in *Chemical Abstracts* and in the above-mentioned serial publications.

Original papers on polysaccharide chemistry and biochemistry, as stated previously, appear in a wide range of primary journals. Particular mention must be made of *Carbohydrate Research*, a commercially sponsored journal covering all facets of the subject which provides a convenient single location for many significant publications. At the same time it cannot be emphasized too strongly that the chemistry of carbohydrates, including that of its macromolecules, both benefits from and contributes to many areas of chemical science. The following is a selective and by no means exhaustive list of primary journals covering both chemical (a) and biochemical (b) aspects: (a) *Journal of the Chemical Society, Perkin I, Canadian Journal of Chemistry, Acta Chemical Scandinavica*; (b) *Journal of Biological Chemistry, Biochemistry, Biochimica Biophysica Acta, Biochemical Journal, European Journal of Biochemistry, Biopolymers, Chemical and Pharmaceutical Bulletin* (Japan), *Agricultural and Biological Chemistry, Phytochemistry*.

References

1. A. Gottschalk, ed., "Glycoproteins. Their Structure and Function," 2nd ed., Parts A and B. Elsevier, Amsterdam, 1972.
2. R. C. Hughes, "Membrane Glycoproteins." Butterworth, London, 1976.
3. M. I. Horowitz and W. Pigman, eds., "The Glycoconjugates," Vols. I and II. Academic Press, New York, 1977, 1978.
4. E. G. Brunngraber, ed., "Neurochemistry of Aminosugars: Neurochemistry and Neuropathology of the Complex Carbohydrates." Thomas, Springfield, Illinois, 1979.
5. R. U. Margolis and R. K. Margolis, eds., "Complex Carbohydrates of Nervous Tissue." Plenum, New York, 1979.
6. W. J. Lennarz, ed., "The Biochemistry of Glycoproteins and Proteoglycans." Plenum, New York, 1980.
7. R. D. Marshall, *Int. Rev. Biochem.* **25**, 1–53 (1979).
8. J. Montreuil, *Adv. Carbohydr. Chem. Biochem.* **37**, 157–223 (1980).
9. L. Rodén and M. I. Horowitz, in "The Glycoconjugates" (M. I. Horowitz and W. Pigman, eds.), Vol. II, pp. 3–71. Academic Press, New York, 1978.
10. L. Rodén, in "Neurochemistry of Aminosugars" (E. G. Brunngraber, ed.), pp. 267–371. Thomas, Springfield, Illinois, 1979.
11. H. Muir and T. E. Hardingham, *MTP Int. Rev. Sci.: Biochem.*, Ser. One **5**, 153–222 (1975).
12. U. Lindahl, *Int. Rev. Sci.: Org. Chem.*, Ser. Two **7**, 283–312 (1976).
13. U. Lindahl and M. Höök, *Annu. Rev. Biochem.* **47**, 385–417 (1978).
14. J. F. Kennedy, "Proteoglycans—Biological and Chemical Aspects in Human Life." Elsevier, Amsterdam, 1979.
15. For IUPAC-IUB Tentative Rules for Carbohydrate Nomenclature, see *Eur. J. Biochem.* **21**, 455–477 (1971) or *Biochemistry* **10**, 3983–4004 (1971).
16. R. D. Guthrie, "Introduction to Carbohydrate Chemistry." Oxford Univ. Press, London and New York, 1974.
17. R. J. Ferrier and P. M. Collins, "Monosaccharide Chemistry." Penguin, London, 1972.
18. L. Hough and A. C. Richardson, in "Rodd's Chemistry of Carbon Compounds" (S. Coffey, ed.), Vol. 1F, pp. 1–595. Elsevier, Amsterdam, 1967.
19. W. Pigman and D. Horton, eds., "The Carbohydrates," 2nd ed., Vols. 1A, 1B, 2A, 2B. Academic Press, New York, 1970–1981.
20. J. F. Stoddart, "Stereochemistry of Carbohydrates." Wiley (Interscience), New York, 1971.

2

Isolation and Fractionation of Polysaccharides

GERALD O. ASPINALL

I. Introduction 19

II. Concepts of Homogeneity and Heterogeneity 20

III. Isolation and Fractionation of Polysaccharides 26

IV. Criteria for Homogeneity 31

References 32

I. Introduction

The overall objectives in isolating a polysaccharide, as with any other type of macromolecule, are to obtain that material in as high a yield and in as chemically pure and homogeneous a form as possible. Sometimes these two objectives may work in opposition to each other. The initial problems are those of isolation without degradation, purification so that the polysaccharide is obtained free of contaminating materials, e.g., lipids, proteins, and nucleic acids, and fractionation to give substances that are chemically and physically homogeneous to within acceptable limits. Before homogeneity and heterogeneity are considered, the problem of degradation will be mentioned briefly. Degradations of polysaccharides by acid and, less frequently, by base are deliberately used in the determination of structure (Chapter 3),

19

and it is obvious that inadvertent degradation by these procedures should be kept to a minimum. In some cases, however, the very possibility of degradation may not be recognized until a vulnerable chemical linkage has been identified. It may be noted in this context that preconceptions as to purity or homogeneity may render the investigator blind to the possibility that structurally disparate regions in a natural macromolecule are indeed covalently linked. An excellent example is provided by early studies on glycosaminoglycans such as heparin and chondroitin sulfate (see Chapter 5, Volume III) which occur naturally as proteoglycans in which the glycan is covalently linked to protein through a base-sensitive O-glycosidic bond. The isolation of "pure" polysaccharide involving removal of protein by enzymatic digestion followed by extraction with dilute alkali then becomes a self-fulfilling prophecy. It may be no simple matter to distinguish between macromolecules in which segments are joined by chemically sensitive covalent bonds and those in which there are strong noncovalent interactions, and thus the severity of conditions required for dissociation may cause chemical modification so that reassociation does not occur readily. Such degradation during isolation may result in heterogeneity where none existed formerly.

II. Concepts of Homogeneity and Heterogeneity (1)

Many proteins, which are synthesized under direct genetic control, are *monodisperse*; i.e., all molecules, isotopic variations apart, are identical in structure and molecular weight. However, few polysaccharides, if any, are synthesized in this manner, and, even for those that are chemically and physically homogeneous, variations occur from molecule to molecule. If these variations are continuous in respect of all parameters, such as molecular size, proportions of sugar constituents, and particular linkage types, separation into discrete molecular species is impossible and the material is said to be *polydisperse*. Macromolecules that show discontinuities in molecular size but not in chemical composition are physically heterogeneous but chemically homogeneous and may be termed *polymolecular*. Chemically heterogeneous macromolecules show discontinuities in one or more of proportions of sugar constituents, linkage types, or degrees of branching. It is possible to envisage a situation in which there are mixtures of two or more monodisperse materials, but such a *paucidisperse* preparation is unlikely to be encountered other than with proteins. Gibbons (2) has also coined the term *heterodisperse* to denote the presence of two or more polydisperse populations of molecules. The first problems are, of course, to separate mixtures of entirely different structural types, but for certain mixtures, especially of neutral polysaccharides, there may be few distinguishing features on which to base a fractionation procedure. In addition, mixtures of polysaccharides are frequently encountered in which relatively small differences in structure may result in substantial differences in properties. For example, amylose and amylopectin, the linear and branched components of starch, respectively both consist of at least 95% of 4-linked α-D-glucopyranose residues. It is necessary, therefore, to examine ways in which polysaccharide molecules may vary in size and structure.

For homopolysaccharides of uniform linkage type, the only possible variations are of molecular size such that different molecules are *polymer homologs*, among which there may be continuous variations (polydisperse or polymolecular) or discontinuous variations with consequent physical heterogeneity. When one or more additional structural features are present in a polysaccharide, variations in structure may result from different relative proportions and/or different distributions of those features. The following examples illustrate some additional types of structural feature:

1. Linear homopolysaccharides with a second linkage type are exemplified by pullulan (**1**), an α-D-glucan, in which 4- and 6-linked residues are present in the ratio of 2:1 in a highly regular sequence (*3*). A somewhat less regular arrangement is encountered in cereal α-D-glucans (**2**), in which 3-linked residues are usually separated by two or three 4-linked residues and are only rarely in adjacent positions (*4*).

2. Linear heteropolysaccharides with two sugar constituents are of several types. Hyaluronic acid (**3**) contains a strictly alternating sequence of 4-linked β-D-glucuronic acid and 3-linked 2-acetamido-2-deoxy-β-D-glucopyranose residues. It may be noted that other polysaccharides of this general type, including other glycosaminoglycans (see Chapter 5, Volume III), are based on similar alternating sequences in which partial modification of residues of one or both types has masked the repeating unit. The 4-linked β-D-hexopyranose residues of both types in plant glucomannans (**4**) (see Chapter 3, Volume II) are arranged without apparent regularity. Alginic acid (**5**)

$$-[\rightarrow 6)\text{-}\alpha\text{-D-Glc}p\text{-}(1\rightarrow 4)\text{-}\alpha\text{-D-Glc}p\text{-}(1\rightarrow 4)\text{-}\alpha\text{-D-Glc}p\text{-}(1\text{-}1_n\text{-}$$

1

$$\rightarrow 3)\text{-}\beta\text{-D-Glc}p\text{-}(1\text{-}[\rightarrow 4)\text{-}\beta\text{-D-Glc}p\text{-}(1\text{-}]_2\rightarrow 3)\text{-}\beta\text{-D-Glc}p\text{-}(1\text{-}[\rightarrow 4)\text{-}\beta\text{-D-Glc}p\text{-}(1\text{-}]_3\text{-}$$

2

$$\rightarrow 4)\text{-}\beta\text{-D-Glc}p\text{-}(1\text{-}[\rightarrow 4)\text{-}\beta\text{-D-Man}p\text{-}(1\text{-}1_n\text{-}$$

3, where $n = 0$ (infrequently), 1–6

$$-[\rightarrow 4)\text{-}\beta\text{-D-Glc}pA\text{-}(1\rightarrow 3)\text{-}\beta\text{-D-Glc}pNAc\text{-}(1\text{-}]_n\text{-}$$

4

$$-]\rightarrow 4)\text{-}\beta\text{-D-Man}pA\text{-}(1\text{-}]_x\text{-}[\rightarrow 4)\text{-}\text{L-Gul}pA\text{-}(1\text{-}]_y\rightarrow 4)\text{-}\beta\text{-D-Man}pA\text{-}(1\rightarrow 4)\text{-}\alpha\text{-L-Gul}pA\text{-}(1\text{-}$$

5

represents another type of linear heteropolysaccharide in which residues of 4-linked β-D-mannuronic acid and its C-5 epimer, α-L-guluronic acid, occur in blocks (M blocks and G blocks), as well as in less regularly distributed mixed regions (see Chapter 4, Volume II). Although alginic acid as isolated is a heteropolysaccharide, the biosynthesis to form α-L-guluronic acid residues involves epimerization of some of the β-D-mannuronic acid residues in a preassembled D-mannuronan chain.

3. Branched homopolysaccharides with frequent branching and correspondingly high proportions of a second linkage type may be illustrated by scleroglucan (**6**) (5), which is structurally regular, although not all polysaccharides in this category necessarily show the same regularity. Branched homopolysaccharides with a lower frequency of branching and necessarily carrying more extended side chains, such as amylopectin and glycogen, are discussed below.

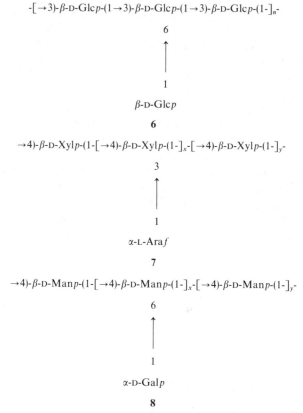

$-[\rightarrow3)\text{-}\beta\text{-}D\text{-}Glc p\text{-}(1\rightarrow3)\text{-}\beta\text{-}D\text{-}Glc p\text{-}(1\rightarrow3)\text{-}\beta\text{-}D\text{-}Glc p\text{-}(1\text{-}]_n\text{-}$

6

$\uparrow$

1

β-D-Glc$p$

6

$\rightarrow4)\text{-}\beta\text{-}D\text{-}Xyl p\text{-}(1\text{-}[\rightarrow4)\text{-}\beta\text{-}D\text{-}Xyl p\text{-}(1\text{-}]_x\text{-}[\rightarrow4)\text{-}\beta\text{-}D\text{-}Xyl p\text{-}(1\text{-}]_y\text{-}$

3

$\uparrow$

1

α-L-Ara$f$

7

$\rightarrow4)\text{-}\beta\text{-}D\text{-}Man p\text{-}(1\text{-}[\rightarrow4)\text{-}\beta\text{-}D\text{-}Man p\text{-}(1\text{-}]_x\text{-}[\rightarrow4)\text{-}\beta\text{-}D\text{-}Man p\text{-}(1\text{-}]_y\text{-}$

6

$\uparrow$

1

α-D-Gal$p$

8

4. Branched heteropolysaccharides in which homopolysaccharide chains of uniform linkage type carry second sugar constituents as single-unit side

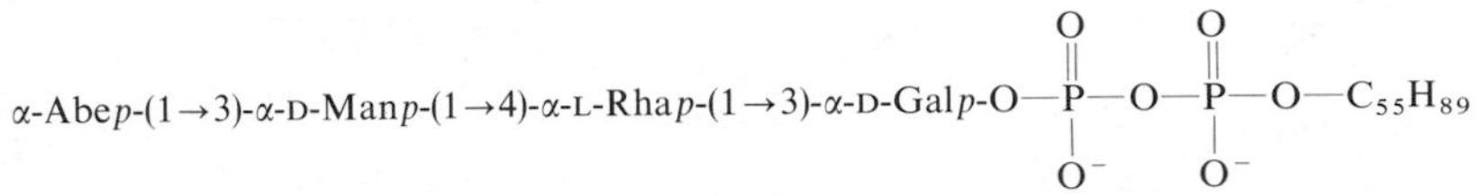

$$\alpha\text{-Abe}p\text{-}(1\rightarrow3)\text{-}\alpha\text{-D-Man}p\text{-}(1\rightarrow4)\text{-}\alpha\text{-L-Rha}p\text{-}(1\rightarrow3)\text{-}\alpha\text{-D-Gal}p\text{-O}\underset{\underset{O^-}{|}}{\overset{\overset{O}{\|}}{P}}\text{-O-}\underset{\underset{O^-}{|}}{\overset{\overset{O}{\|}}{P}}\text{-O-}C_{55}H_{89}$$

9

$$\left[\begin{array}{c} \rightarrow2)\text{-}\alpha\text{-D-Man}p\text{-}(1\rightarrow4)\text{-}\alpha\text{-L-Rha}p\text{-}(1\rightarrow3)\text{-}\alpha\text{-D-Gal}p\text{-}(1\text{-} \\ 3 \\ \uparrow \\ 1 \\ \alpha\text{-Abe}p \end{array} \right]_n$$

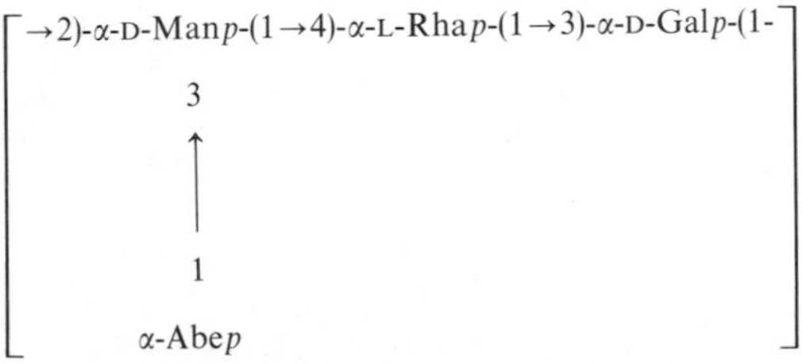

10

Fig. 1. A linear tetrasaccharide unit attached through a phosphorodiester to a carrier lipid (**9**) acts as the glycosyl donor in the biosynthesis of regular repeating units in the branched chains of O-antigen chains (**10**) in the lipopolysaccharide from *Salmonella* species [Abe, abequose (3,6-dideoxy-D-*xylo*-hexose)].

chains are exemplified by the plant arabinoxylans (**7**) and galactomannans (**8**) (Chapter 3, Volume II). For such polysaccharides of relatively simple structural type, the proportions of the different units are easily determined, but satisfactory general methods for assessing distributions of side chains have yet to be established. Most of the available evidence suggests that there is little obvious regularity of branching in these and other similar plant polysaccharides. Where strict regularity of branching is encountered, as in many bacterial polysaccharides containing a minimum of three different sugar residues (Chapter 5, Volume II), the biosynthesis involves the polymerization of preassembled oligosaccharide units (6) (Fig. 1) attached to a carrier lipid (Chapter 5).

5. In branched homo- or heteropolysaccharides with side chains containing two or more sugar units, uniformity or nonuniformity of chain length is another structural variable. When the side chains contain two or more different sugar residues, it is not uncommon to find members of a polymer-homologous series of the type A, B-A, C-B-A, etc., although direct structural proof that these structural units occupy equivalent positions in the polysaccharide structure is not easily obtained. Examples among plant polysaccharides are the side chains (**11–13**) in galactoarabinoxylans from some cereals and grasses and those (**14–16**) in xyloglucans (Chapter 3, Volume II). Questions of distribution aside, these different oligosaccharide units may represent the products of either incomplete biosynthesis or random degradation during isolation.

α-L-Ara$f$-(1-

11

β-D-Xyl$p$-(1→2)-L-Ara$f$-(1-

12

β-D-Gal$p$-(1→4)-β-D-Xyl$p$-(1→2)-L-Ara$f$-(1-

13

α-D-Xyl$p$-(1-

14

β-D-Gal$p$-(1→2)-α-D-Xyl$p$-(1-

15

α-L-Fuc$p$-(1→2)-β-D-Gal$p$-(1→2)-α-D-Xyl$p$-(1-

16

6. Branched polysaccharides with extended side chains, such as amylopectin and glycogen, have the potential for different arrangements of branches (Fig. 2) in comblike (**17**) or multiply branched treelike (**18**) structures (Chapters 3 and 4, Volume III).

These examples illustrate only a few of the possible types of structural variation that may be encountered in the same population of molecules. When one or more types of structural variation exist among molecules, a

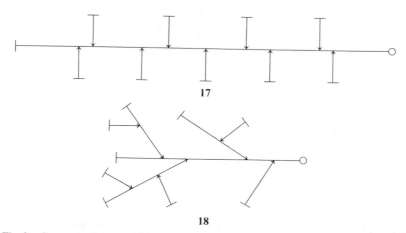

Fig. 2. Segments of two possible arrangements of chains in comb-like (or herringbone) (**17**) or multiply branched tree-like (**18**) structures in amylopectin or glycogen (⊢—— denotes non-reducing termini of 4-linked α-D-glucopyranose chains; ——→ denotes 6-linked junctions at branch points to 4-linked chains; ——○ denotes reducing terminus).

measure of heterogeneity is implied, and in principle separation into discrete fractions is possible. On the other hand, distribution of different structural features within different parts of the same molecule is an aspect of fine structure. Since separation of different, but generally similar molecules cannot always be realized in practice and since structural methods deal with statistical populations of molecules, it is often not possible to distinguish variations among from variations within molecules. In considering both structural variations among molecules and distributions of structural features within molecules, one should avoid the opposite pitfalls of (a) imposing preconceived regularities and (b) failing to perceive a hidden regularity where some units of a given type have undergone postpolymerization modification without the creation of new or the rupture of existing glycosidic bonds. When discrete, i.e., discontinuous, variations in structural features are encountered, an assessment of the extent of those variations may be an important part of the description of that population of molecules in relation to the physical properties or biological function of the material. Examples of these modifications are (a) epimerization at C-5 of uronic acid residues, e.g., β-D-mannuronic acid (**19**) to α-L-guluronic acid (**20**) in alginic acid, or β-D-glucuronic acid (**21**) to α-L-iduronic acid (**22**) in dermatan sulfate or heparin; (b) the conversion of D- or L-galactose 6-sulfate (**23**) to 3,6-anhydro-D- or L-galactose

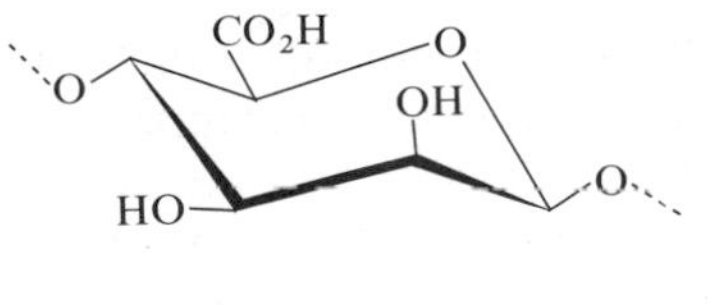

19

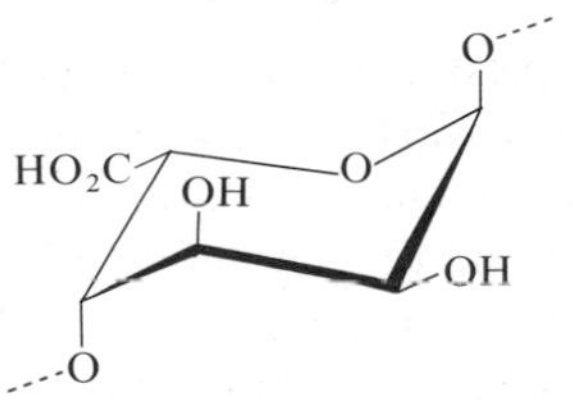

20

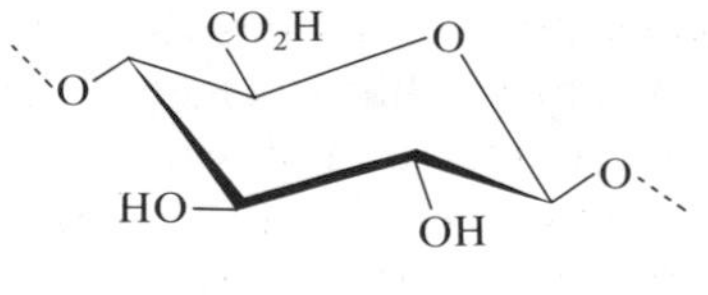

21

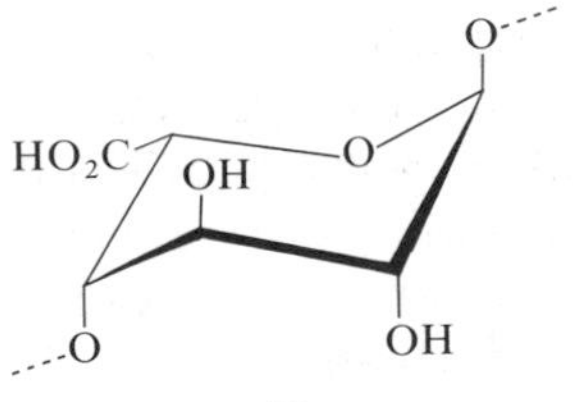

22

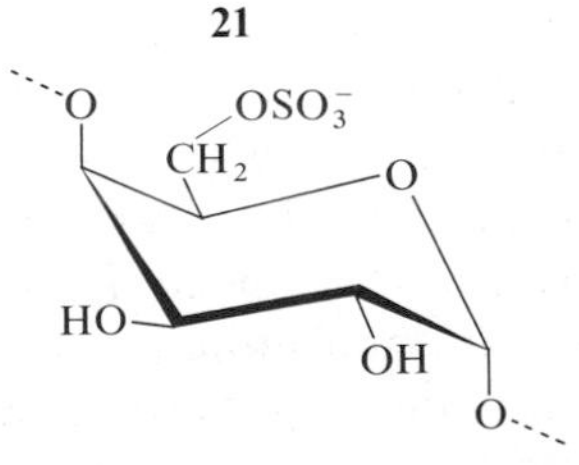

23

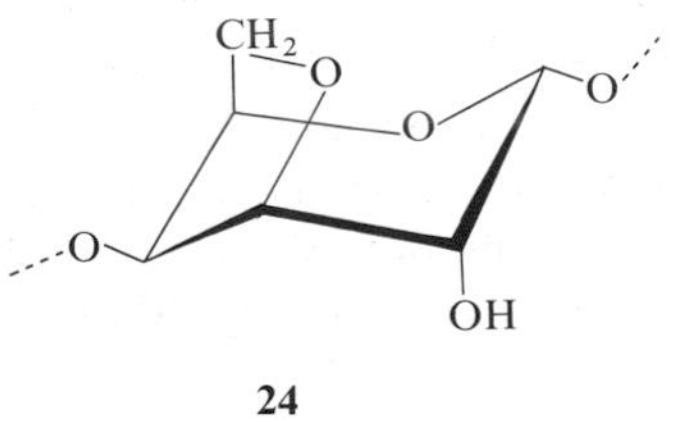

24

(24) residues in carrageenan or agarose-related polysaccharides; and (c) the introduction of substituents, especially stable nonremovable ethers.

III. Isolation and Fractionation of Polysaccharides

Methods for the isolation and fractionation of polysaccharides are too numerous for a comprehensive survey to be given here. This section, therefore, considers the general approaches taken to obtain polysaccharides in soluble form and the types of fractionation methods that are then used. In some instances, e.g., the capsular polysaccharides from bacteria and exudate gums, polysaccharides are readily solubilized and may often be simply isolated in remarkably homogeneous form. In other cases, especially when they occur as cell wall components in plants or microorganisms, whether as fibrous structural materials or in the associated matrix, polysaccharides are present in close association with other macromolecules, which are often other polysaccharides. With care being taken to minimize inadvertent cleavage of interpolymeric covalent bonds, graded extraction is then performed in attempts to render macromolecules selectively soluble, at least in less complex mixtures from which individual components may be obtained on fractionation.

The simplest extraction methods for polysaccharides are those using water alone at various temperatures. Some polysaccharides may be brought into solution by the use of polar nonaqueous solvents. For example, dimethyl sulfoxide, alone or containing 10% of water, is used to extract starch and glycogen from various sources (7). This solvent is used to isolate O-acetyl-4-O-methylglucuronoxylans from hardwoods and O-acetylgalactoglucomannans from some softwoods with retention of the labile substituents (8). N-Methylmorpholine N-oxide is a remarkable solvent for certain polysaccharides (9) including cellulose (10). Acidic polysaccharides can often be isolated with reagents that effect cation exchange. This is achieved directly with salt solutions, such that monovalent sodium or potassium replaces bivalent metal ions. Alternatively, some polysaccharides, e.g., pectins, are solubilized by complexing tightly coordinated bivalent metal ions with chelating solvents such as disodium ethylenedinitrilotetraacetate, ammonium oxalate, or sodium hexametaphosphate. In the specific instance of pectin isolation, the last-named solvent is preferred since it chelates most effectively at pH 4, and conditions likely to cause inadvertent base-catalyzed β-elimination from esterified galacturonic acid residue are avoided (11). Chaotropic agents, such as urea (8 M), guanidinium salts, and lithium isothiocyanate, which denature, dissolve, and dissociate proteins by perturbing water structure and interfering with hydrophobic bonding, have received little attention for the isolation of polysaccharides.

Extraction of polysaccharides under acidic conditions is obviously to be avoided if possible so that inadvertent cleavage of glycosidic bonds does not occur. Dilute alkali has been used extensively for polysaccharide extractions, but there is now a more general awareness of the possible structural modifications or base-catalyzed degradations that may occur (Chapter 3). As mentioned previously, the O-glycosidic linkage of glycosaminoglycans to protein is split under alkaline conditions. When, as a last resort, alkaline reagents are used, prior treatment of the material under examination with sodium borohydride or the addition of this reagent to the alkaline solution at least minimizes degradations that proceed from reducing groups.

The main function of some of the reagents employed during the extraction of plant materials or other tissues is the solubilization of noncarbohydrate material, often with accompanying degradation. Chlorous acid is one such reagent, which is used both to render lignin from plant tissue soluble in water (12) and to effect oxidation of certain amino acid units in proteins (13,14).

Some polysaccharides, such as cellulose, chitin, and yeast cell wall β-D-glucan, are isolated as insoluble residues after all other materials have been dissolved. The severity of conditions required to solubilize other components raises questions as to the extent to which the residual polysaccharides have undergone degradation. An important recent discovery is the remarkable solvent N-methylmorpholine N-oxide, which, as the hydrate, can dissolve cellulose nondegradatively without derivative formation (10). This solvent can also be used to dissolve plant cell walls completely (9).

When polysaccharides have been made soluble, fractionations are based either on selective precipitations of the polysaccharides themselves or of their salts or complexes or on chromatographic procedures. In some fractionations, e.g., those dependent on noncovalent binding of polysaccharides with lectins, the same principle is used for both precipitation and chromatographic separations, but the technique is mentioned in only one place. The complete fractionation of mixtures by precipitation methods is rarely achieved, although one component may often be isolated in reasonably homogeneous form after reprecipitation. It should be noted, however, that interactions between polysaccharides are such that coprecipitation may occur in which the less soluble component is accompanied by some of the more soluble component. In other cases, partial precipitation of only one component in the mixture will result in the supernatant solution containing a reduced but significant amount of that material. In the discussion that follows emphasis is placed, wherever possible, on the separation of mixtures of polysaccharides, but some examples refer only to purifications in which the polysaccharide is separated from noncarbohydrate material.

Typically, polysaccharides can be isolated from aqueous solution by the addition of a water-miscible solvent such as acetone or a lower alcohol. Although these precipitates, after washing to remove contaminating material, may be dried by successive washing with anhydrous organic solvents, they are obtained in a much more satisfactory physical condition for subsequent manipulations by freeze-drying from aqueous solution. Carefully controlled fractional precipitation by the addition of nonsolvent to aqueous solutions may occasionally give separations of mixtures of polysaccharides, but the method is not highly selective. Likewise, salting-out procedures, as with proteins, rarely effect a separation of neutral polysaccharides, but the preferential precipitation of the mixed-linkage β-D-glucan component of cereal gums with ammonium sulfate provides a notable exception (15).

Fractional precipitation of polysaccharides by selective salt or complex formation is generally most effective for the isolation of the component that forms a precipitate from which the parent polysaccharide can be regenerated. There are, however, rather few examples in which more than one polysaccharide has been isolated in homogeneous form by fractional precipitation alone. One such example is the fractionation of the chondroitin 4- and 6-sulfate components of proteoglycan by ethanol precipitation of the calcium salts (16). Nevertheless, in several cases a second polysaccharide is recovered from the mother liquor after precipitation of the major component.

The following examples indicate the types of salt or complex formation of which advantage may be taken in fractional precipitations. Salt formation from acidic polysaccharides (glycuronans and glycan sulfates) is illustrated by precipitations of κ-carrageenan with potassium chloride (13), of acidic polysaccharides (17,18) including pectic acid (often partially esterified) (19) with neutral cupric acetate, and of the above-mentioned chondroitin sulfates as calcium or barium salts (16). Selective precipitation with cationic detergents, such as cetyltrimethylammonium bromide (Cetavlon) and cetylpyridinium bromide (20), is exemplified in the preferential precipitation of the more highly acidic polysaccharide components of gum tragacanth (21) and *Anogeissus leiocarpus* gum (22). Cetavlon is also used in the isolation of teichoic acid-like capsular polysaccharides such as the capsular antigen from *Hemophilus influenzae* type c (23), which contains phosphorodiester bonds.

Variations of these methods can be extended to the separation of neutral polysaccharides, usually through interactions with complexes whose formation requires the presence of suitably disposed hydroxyl groups. The most widely used method is the use of alkaline copper reagents, e.g., Fehling solution (18), as in the precipitation of various mannans (24), presumably through complexation of the *cis*-2- and 3-hydroxyl groups. The formation of insoluble copper complexes with xylans (25) is less obviously explained. The formation of an insoluble complex of wood glucomannans with barium hydroxide (26) provides the basis for their separation from associated xylans. Borate complexes allow selective salt formation with cetyltrimethylammonium hydroxide to be used in the fractionation of neutral arabinogalactans (27).

Advantage is taken of an entirely different complex formation in the fractionation of starch in which the addition of hydrophobic solvents such as 1-butanol to starch dispersions results in the formation of an insoluble inclusion complex (28) with the conformationally stabilized V-amylose helix. Amylopectin, which is incapable of forming a comparably stabilized complex, remains in the supernatant solution.

A number of polysaccharides form insoluble complexes with dyes, and this type of interaction is particularly useful for the selective precipitation of neutral polysaccharides. Among the small number of examples so far demonstrated, the interactions of β-D-glucans containing 4-linked units with Calcofluor White M2R and Congo Red (29), and of arabino-3,6-galactans (30,31) with artificial carbohydrate antigens prepared by coupling diazotized 4-aminophenyl β-D-glucopyranoside with phloroglucinol (32), are notable.

Although there is a considerable region of overlap in scale of operation, precipitation methods for polysaccharide fractionation are generally suitable for separations of rather larger quantities (1 g and upward) of material, whereas chromatographic separations can be carried out on milligram amounts. Some of the chromatographic procedures are also used analytically as criteria for homogeneity. Separations may be based on differences in molecular size [gel filtration or molecular sieve chromatography (33)] or composition, most commonly in the proportions of acidic groups, in which case ion-exchange chromatography is widely used. Electrophoresis also distinguishes molecules on the basis of ionizable groups but is less readily adapted for preparative purposes.

For preparative separations acidic polysaccharides containing carboxylic acid, sulfate hemiester, or phosphorodiester groups are best fractionated by ion-exchange chromatography on polysaccharide fibers (cellulose) or cross-linked gels (dextran in Sephadex or agarose in Sepharose) carrying the required functional groups, e.g., O-(2-diethylaminoethyl) (34). A few examples chosen from a large number include the separations and purifications of *Streptococcus pneumoniae* type 1 capsular polysaccharide from DEAE–Sephadex A-50 (35) (acetate form) or DEAE–Sepharose CL 6B (36) by elution with a sodium chloride gradient in acetate buffer; lemon peel pectin from DEAE–Sephadex A-50 (formate form) with formic acid (37); the sulfated algal polysaccharide from *Ulva lactuca* from DEAE–cellulose with potassium chloride (38); and the phosphorodiester-linked capsular antigen

from *Hemophilus influenzae* type c (after regeneration from the cetyltrimethylammonium salt) from DEAE–Sepharose with a sodium chloride gradient
(*23*). The cross-linked gels are the supports of choice because of their higher
capacities, but they suffer from the disadvantage of large changes in volume
with alterations in ionic strength of the eluting solvent. Polystyrene resins, of
generally lower capacities, are less frequently used [see, however, a recom-
mended procedure for the preparation of beef lung heparan sulfate (*39*)].
Ion-exchange chromatography can be extended to the separation of neutral
polysaccharides capable of forming charged complexes, e.g., with borate
ions (*40*). Polysaccharides forming borate complexes are retained on columns
in the borate form and then are eluted with borate of increasing strength,
whereas noncomplexing polysaccharides are eluted directly with water.
Mustard seed arabinan was first purified in this way (*41*), although for
preparative purposes precipitation with cetyltrimethylammonium hydroxide
is the preferred method.

The reversible noncovalent binding of carbohydrate polymers, most
commonly to proteins, provides the basis for affinity chromatography, which
takes advantage of the high specificity found in many biological interactions.
Carbohydrate-binding proteins or lectins (*42*), most of which are specific for
individual sugars of defined glycosidic configuration, were first recognized
by their capacity to agglutinate particular cells. Concanavalin A from the
jack bean has multiple specificity, especially toward α-D-mannopyranans and
less toward α-D-glucopyranans, and precipitates such α-D-glycans from
solution. The specific capsular polysaccharide from *Streptococcus pneumoniae* type 12 was purified by precipitation with this protein (*43*). Affinity
chromatography using concanavalin A covalently bound to an inert support
such as Sepharose has been used extensively for the fractionation of *N*-
glycosidic glycopeptides containing α-D-mannopyranose residues (*44–46*)
and for the purification of glycoproteins such as ovotransferrin (*47*). Other
examples of the use of immobilized lectins in studies on glycoconjugates
include the purification of human urinary erythropoietin (*48*) with the
Phaseolus vulgaris β-D-galactopyranoside-binding lectin and the fractionation of blood group A- and B-active polyglycosyl peptides formed from
human erythrocyte membranes using the α-D-galactopyranoside- and 2-
acetamido-2-deoxy-α-D-galactopyranoside-binding lectin from *Bandeiraea
simplicifolia* (*49*). Although potentially of great value, lectin-based fractionations of polysaccharides have been seldom used to date. A recent example is
provided by the fractionation of α-D-mannopyranose-terminated heteroglycans from the fruit bodies of *Fomitopsis pinicola* using concanavalin A–
Sepharose (*50*). Another example is the isolation of arabinogalactans and
arabinogalactan–protein conjugates (*31,51*) using the β-D-galactopyranoside-

binding lectin from the small giant clam *Tridacna maxima* coupled to Sepharose 4B (*52*). A galactan-binding IgA myeloma protein, similarly coupled to Sepharose, may likewise be used for the purification of arabino-galactans (*53*).

The binding of heparin to antithrombin III is a key step in impeding the coagulation of blood by accelerating the inactivation of the proteins of the so-called coagulation cascade (*54*). This noncovalent interaction provides the basis for the fractionation of high- and low- affinity fractions of heparin by affinity chromatography on matrix-bound antithrombin (*55*).

In principle any type of noncovalent interaction between unlike macro-molecules may form the basis for the fractionation of mixtures of polysac-charides. An example of a polysaccharide–polysaccharide interaction of this type is provided by the selective adsorption of xyloglucan on a cellulose column, presumably by virtue of the affinity of 4-linked β-D-glucan chains in the soluble polysaccharide for similar chains in the insoluble support. Gel-forming interactions between unlike polysaccharides such as galacto-mannans and agarose (*56*) might form the basis for separating mixtures of interacting and noninteracting polysaccharides.

IV. Criteria for Homogeneity

The direct genetic control of protein biosynthesis results in the formation of identical molecules, provided that there is complete fidelity in the trans-lation of the genetic code. That assumption is borne out in all but extremely rare instances (*57*). In contrast, polysaccharide biosynthesis is only indirectly gene-controlled in that the enzymes with individual specificities responsible for the transfer of sugar residues from particular glycosyl donors, usually glycosyl esters of nucleotides, are synthesized under genetic control. Vari-abilities of structure (of the types discussed in Section II) from one polysac-charide molecule to another may result from (a) departure from absolute specificity of the transferases, (b) incomplete formation of segments of the structure, e.g., in the extension of side chains, or (c) only partial modification of units undergoing postpolymerization changes. Depending on whether variation from one molecule to another can be directly shown, the presence or absence of heterogeneity can be established. It must be stressed that no unambiguous proof of homogeneity is possible. All that may be achieved is the demonstration of the absence of heterogeneity by as many independent criteria as possible.

The practical problem is to establish the purity, i.e., absence of extraneous contaminants, of the substance under examination and to show that no separation is possible into fractions in which there are discontinuities in molecular size or structure. The essential evidence for the absence of heterogeneity for a polysaccharide, as for any chemical compound, is the demonstration of constancy in chemical composition and physical properties when the material is reisolated from an attempted separation or purification involving either precipitation from solution followed by regeneration or recovery from chromatographic or electrophoretic separation. Constancy of chemical composition can be shown by one or more of the following: sugar composition based on hydrolysis; functional group analysis or analysis for particular classes of sugars, e.g., hexuronic acids; spectroscopic examination, especially by nuclear magnetic resonance spectroscopy; and physical properties, e.g., optical rotation, and solution properties such as viscosity.

Many of the separation techniques discussed in the previous section in relation to preparative-scale fractionations may also be used for purely analytical purposes, often with the gain in resolution that accompanies reduction in scale. Chromatographic rather than precipitation methods are most useful for this purpose, especially those based on ion exchange, gel filtration (molecular sieve chromatography), and bioaffinity, for the first two of which high-performance chromatographic columns are now available. Sedimentation analysis in the ultracentrifuge is a widely used criterion for the absence of heterogeneity (58), although the detection of two or more components does not distinguish between chemical and physical heterogeneity. Electrophoresis may provide evidence for charge heterogeneity arising from ionizable groups in the parent polysaccharide or from suitably disposed hydroxyl groups capable of complex formation, e.g., with borate ions. Free-boundary electrophoresis (59), a purely analytical technique, is no longer widely used, but zone electrophoresis on inert supports finds applications. Zone electrophoresis of neutral polysaccharides in borate buffer may be carried out on glass fiber paper, and the corresponding preparative separations can be achieved on columns of powdered glass (60). Electrophoresis of acidic glycosaminoglycans on cellulose acetate strips is a useful analytical tool (61). Gel electrophoresis, e.g., on polyacrylamide, although widely used for the separation of glycoproteins but based mainly on the protein component (62), has not been widely used for the separation of polysaccharides.

References

1. For an excellent discussion of analogous questions for glycoproteins, see R. A. Gibbons, *in* "Glycoproteins. Their Composition, Structure and Function" (A. Gottschalk, ed.), 2nd ed., pp. 31–140. Elsevier, Amsterdam, 1972.
2. R. A. Gibbons, *Nature (London)* **200**, 665–666 (1963).

3. P. A. J. Gorin and J. F. T. Spencer, *Adv. Carbohydr. Chem.* **23**, 367–417 (1968).
4. K. C. B. Wilkie, *Adv. Carbohydr. Chem. Biochem.* **36**, 215–264 (1979).
5. J. Johnson, Jr., S. Kirkwood, A. Misaki, T. E. Nelson, J. V. Scaletti, and F. Smith, *Chem. Ind.* (*London*) pp. 820–822 (1963).
6. M. J. Osborn, *Annu. Rev. Biochem.* **38**, 501–538 (1969).
7. H. W. Leach and T. J. Schoch, *Cereal Chem.* **39**, 318–327 (1962).
8. H. O. Bouveng and B. Lindberg, *Methods Carbohydr. Chem.* **5**, 147–150 (1965).
9. J.-P. Joseleau, G. Chambat, and B. Chumpitazi-Hermoza, *Carbohydr. Res.* **90**, 339–344 (1981).
10. H. Chanzy, M. Dube, and R. H. Marchessault, *J. Polym. Sci., Polym. Lett. Ed.* **18**, 13–25 (1979).
11. R. W. Stoddart, A. J. Barrett, and D. H. Northcote, *Biochem. J.* **102**, 194–204 (1967).
12. J. W. Green, *Methods Carbohydr. Chem.* **3**, 9–21 (1963); R. L. Whistler and J. N. BeMiller, *ibid.* pp. 21–24.
13. T. J. Painter, *Methods Carbohydr. Chem.* **5**, 98–100 (1965).
14. R. R. Selvendran, A. M. C. Davies, and E. Tidder, *Phytochemistry* **14**, 2169–2174 (1975).
15. I. A. Preece and K. G. Mackenzie, *J. Inst. Brew.* **58**, 353–362 (1952).
16. K. Meyer, E. Davidson, A. Linker, and P. Hoffman, *Biochim. Biophys. Acta* **21**, 506–518 (1956); see also R. W. Jeanloz, *Methods Carbohydr. Chem.* **5**, 110–114 (1965).
17. A. J. Erskine and J. K. N. Jones, *Can. J. Chem.* **34**, 821–826 (1956).
18. J. K. N. Jones and R. J. Stoodley, *Methods Carbohydr. Chem.* **5**, 36–38 (1965).
19. G. O. Aspinall, J. A. Molloy, and J. W. T. Craig, *Can. J. Biochem.* **47**, 1063–1070 (1969).
20. J. E. Scott, *Methods Carbohydr. Chem.* **5**, 38–44 (1965).
21. G. O. Aspinall and J. Baillie, *J. Chem. Soc.* pp. 1702–1714 (1963).
22. G. O. Aspinall, J. J. Carlyle, J. M. McNab, and A. Rudowski, *J. Chem. Soc.* **C**, 840–845 (1969).
23. P. Branefors-Helander, B. Classon, L. Kenne, and B. Lindberg, *Carbohydr. Res.* **76**, 197–202 (1979).
24. T. E. Edwards, *Methods Carbohydr. Chem.* **5**, 176–179 (1965).
25. G. A. Adams, *Methods Carbohydr. Chem.* **5**, 170–175 (1965).
26. H. Meier, *Methods Carbohydr. Chem.* **5**, 45–46 (1965).
27. G. A. Adams, *Methods Carbohydr. Chem.* **5**, 75–78 (1965); see also H. O. Bouveng and B. Lindberg, *Acta Chem. Scand.* **12**, 1977–1984 (1958).
28. W. Banks and C. T. Greenwood, "Starch and Its Components." Edinburgh Univ. Press, Edinburgh, 1975.
29. P. J. Wood and R. G. Fulcher, *Cereal Chem.* **55**, 952–966 (1978); P. J. Wood, *Carbohydr. Res.* **85**, 271–287 (1980).
30. M. A. Jermyn, Y. M. Yeow, and E. F. Woods, *Aust. J. Plant Physiol.* **2**, 501–531 (1975).
31. P. A. Gleeson and A. E. Clarke, *Phytochemistry* **19**, 1777–1782 (1980).
32. J. Yariv, M. M. Rapport, and L. Graf, *Biochem. J.* **85**, 383–388 (1962); J. Yativ, H. Lis, and E. Katchalski, *ibid.* **105**, 1C–2C (1967).
33. S. C. Churms, *Adv. Carbohydr. Chem. Biochem.* **25**, 13–51 (1970); *in* "Chromatography" (E. Heftmann, ed.), 3rd. ed., pp. 637 *et seq.* Van Nostrand-Reinhold, Princeton, New Jersey, 1975.
34. H. Neukom, H. Deuel, W. J. Heri, and W. Kundig, *Helv. Chim. Acta* **43**, 64–71 (1960); H. Neukom and W. Kundig, *Methods Carbohydr. Chem.* **5**, 14–17 (1965).
35. R. C. E. Guy, M. J. How, M. Stacey, and M. Heidelberger, *J. Biol. Chem.* **242**, 5106–5111 (1967).
36. B. Lindberg, B. Lindqvist, J. Lönngren, and D. A. Powell, *Carbohydr. Res.* **78**, 111–117 (1980).
37. G. O. Aspinall, J. W. T. Craig, and J. L. Whyte, *Carbohydr. Res.* **7**, 442–452 (1968).

38. E. Percival and J. K. Wold, *J. Chem. Soc.* pp. 5459–5468 (1963).
39. A. Linker, *Methods Carbohydr. Chem.* **7**, 89–93 (1976).
40. W. Heri, H. Neukom, and H. Deuel, *Helv. Chim. Acta* **44**, 1939–1945 (1961).
41. E. L. Hirst, D. A. Rees, and N. G. Richardson, *Biochem. J.* **95**, 453–458 (1965).
42. I. J. Goldstein and C. E. Hayes, *Adv. Carbohydr. Chem. Biochem.* **35**, 127–340 (1978).
43. J. A. Cifonelli, P. Rebers, M. B. Perry, and J. K. N. Jones, *Biochemistry* **5**, 3066–3072 (1966).
44. S.-I. Ogata, T. Muramatsu, and A. Kobata, *J. Biochem.* (*Toyko*) **78**, 687–696 (1975).
45. T. Krusius, J. Finne, and H. Rauvala, *FEBS Lett.* **71**, 117–120 (1976).
46. S. Narasimhan, J. R. Wilson, E. Martin, and H. Schachter, *Can. J. Biochem.* **57**, 83–96 (1979).
47. H. Iwase and K. Hotta, *J. Biol. Chem.* **252**, 5437–5443 (1977).
48. F. Sieber, *Biochim. Biophys. Acta* **496**, 146–154 (1977).
49. J. Finne, T. Krusius, H. Rauvala, R. Kekomaki, and G. Myllyla, *FEBS Lett.* **89**, 111–115 (1978).
50. T. Usui, S. Hosokawa, T. Mizuno, T. Suzuki, and H. Meguro, *J. Biochem.* (*Tokyo*) **89**, 1029–1037 (1981).
51. P. A. Gleeson, M. A. Jermyn, and A. E. Clarke, *Anal. Biochem.* **92**, 41–45 (1979).
52. P. A. Gleeson and A. E. Clarke, *Biochem. J.* **181**, 607–621 (1979).
53. I. G. Andrew and B. A. Stone, *Proc. Aust. Biochem. Soc.* **10**, 28 (1977).
54. R. D. Rosenberg, *Fed. Proc., Fed. Am. Soc. Exp. Biol.* **36**, 10–18 (1977).
55. M. Höök, I. Björk, J. Hopwood, and U. Lindahl, *FEBS Lett.* **66**, 90–93 (1976).
56. I. C. M. Dea and A. Morrison, *Adv. Carbohydr. Chem. Biochem.* **31**, 241–312 (1975).
57. M. Yarns, *Prog. Nucleic Acid Res. Mol. Biol.* **23**, 195–225 (1979).
58. W. Banks and C. T. Greenwood, *Adv. Carbohydr. Chem.* **18**, 357–398 (1963).
59. R. L. Whistler and C. S. Campbell, *Methods Carbohydr. Chem.* **5**, 201–203 (1965).
60. D. H. Northcote, *Methods Carbohydr. Chem.* **5**, 49–53 (1965).
61. M. Breen, H. G. Weinstein, L. J. Black, M. S. Borcherding, and R. A. Sittig, *Methods Carbohydr. Chem.* **7**, 101–115 (1976).
62. R. C. Hughes, "Membrane Glycoproteins," pp. 50–52. Butterworth, London, 1976.

Chemical Characterization and Structure Determination of Polysaccharides

GERALD O. ASPINALL

	I. Introduction	36
	II. Composition and Molecular Size	37
	A. Analysis of Sugar Components and Removable Substituents .	37
	B. Determination of Molecular Size	43
	III. Problems of Structure Determination	44
	IV. Methylation	45
	A. Methylation Procedures	49
	B. Characterization and Analysis of Methylated Sugars	51
	V. Partial Depolymerization by Hydrolysis and Related Reactions .	56
	A. Fractionation and Characterization of Oligosaccharides and Degraded Polysaccharides	57
	B. Partial Acid Hydrolysis	62
	C. Partial Depolymerization under Nonaqueous Conditions	64
	D. Oxidative Hydrolysis	68
	E. Enzyme-Catalyzed Hydrolysis	69
	F. Partial Depolymerization of Permethylated Polysaccharides .	71
	G. Applications of Mass Spectrometry to the Characterization of Oligosaccharide Derivatives	73
	VI. Oxidations with Periodate and Lead Tetraacetate	81
	A. Analytical Aspects	81
	B. Degradative Aspects	86

THE POLYSACCHARIDES, VOL. 1

VII. Structural Modifications of Polysaccharides   89
    A. Reduction of and Oxidation to Uronic Acids   90
    B. Selective Substitution Followed by Structural Modification .   94
    C. Oxalane and Oxirane Ring Formation   95
    D. Location of Removable Substituents   97

VIII. Other Methods for Partial Depolymerization   100
    A. Alkaline Degradations and Other Base-Catalyzed
       Fragmentations   100
    B. Selective Cleavage of Glycosiduronic Acid Linkages   112
    C. Selective Cleavage of Aminoglycosidic Linkages
       by Deamination   118

    References   124

I. Introduction

The determination of the complete covalent or primary structure of a polysaccharide or the carbohydrate portion of a glycoconjugate is often a complex task in which answers must be given to questions of composition, constitution, and configurational aspects of stereochemistry. The structural information that can be obtained by chemical methods is the main theme of this chapter. However, in relevant sections the chapter includes accounts of applications of mass spectrometry to the characterization of carbohydrate derivatives. Other spectroscopic methods, especially infrared (IR) and ^{1}H- and ^{13}C-nuclear magnetic resonance (NMR) spectroscopy, which give much information on composition and structure, are discussed in detail in Chapter 4. Conformational aspects of stereochemistry are also considered in the chapter on spectroscopic methods, but a full account of the importance of macromolecular conformations in relation to the physical properties and biological functions of polysaccharides is given in Chapter 5. Detailed discussions of the value of enzymatic methods in compositional and structural analysis are included in Volume III.

For present purposes it is assumed that polysaccharide preparations are of sufficient purity, i.e., absence of noncovalently associated noncarbohydrate contaminants, and chemical and/or physical homogeneity to merit detailed compositional and structural investigation. Nevertheless, the ultimate counsel of perfection may sometimes be misplaced since decisions on homogeneity (or at least the absence of obvious heterogeneity) may be possible only in the light of internally consistent structural evidence. On the one hand, it is necessary to determine sugar composition and to perform linkage analysis by methylation on polysaccharides of demonstrated purity and homogeneity. On the other hand, these analyses may now be performed so rapidly and on such small quantities of material that it is a legitimate strategy to use them to monitor polysaccharide fractionations.

Previous reviews of methods for polysaccharide analysis and structure determination should be consulted for detailed references to earlier work (*1–5*). Attention is directed to the series "Methods in Carbohydrate Chemistry" (Academic Press) and to certain volumes (4, 8, 28, and 50) in the series "Methods in Enzymology" (Academic Press) for detailed experimental procedures.

II. Composition and Molecular Size

As with any other class of substances, the determination of chemical composition and molecular size of polysaccharides is part of their overall characterization. In the previous chapter it was stressed that most, if not all, polysaccharides are polydisperse, so that information on both molecular size and composition can be obtained only as averages. Insofar as a polysaccharide contains repetitive features, whether of a regular type as in most polysaccharides of microbial origin or only in a statistical sense as in many plant polysaccharides, compositional analysis gives essential information on the relative proportions of sugar constituents. The determination of molecular weights is discussed in the context of structure elucidation.

A. Analysis of Sugar Components and Removable Substituents

The determination of polysaccharide composition requires the identification and, as far as possible, quantitative estimation of the sugar constituents. In the analysis of sugars liberated on hydrolysis, those with stable substituents, e.g., ethers, should be regarded as separate sugar constituents, even if substituted and nonsubstituted sugars have a common biosynthetic origin with etherification occurring at a postpolymerization stage. In addition, analysis is necessary for removable substituents, e.g., *O*-acyl (most commonly *O*-acetyl) substituents, *N*-acetyl, sulfate, and phosphate groups, and ketals (as of pyruvic acid). Certain of these substituents can be analyzed nondestructively by IR and NMR spectroscopy (Chapter 4). The importance of both ^{1}H- and ^{13}C-NMR spectroscopy cannot be overemphasized since the early recognition of unusual features will indicate the need for suitable chemically based analytical procedures. References to many microbial polysaccharides for which NMR spectroscopy has played a major role in compositional and structural analysis are given in Chapter 5, Volume II.

The results of compositional analysis should be interpreted in a wider context that incorporates information on detailed structure and mode of biosynthesis. Indeed, such information may dictate the most meaningful procedures for analysis of composition. The following are some general points to be borne in mind in devising analytical methods appropriate for particular polysaccharides. The concept of the "repeating unit" may have validity when, as with many microbial polysaccharides, a major part of the polysaccharide structure is assembled from oligosaccharide units attached to a carrier lipid (see Chapter 5, Volume II and Chapter 2, Volume III). In such cases compositional analysis is required only with an accuracy sufficient to distinguish correct from incorrect integral ratios of sugar constituents. On the other hand, for many plant polysaccharides, the concept of a repeating unit may have only statistical significance. The significance of minor constituents must be assessed in each case so that a decision can be made as to the accuracy of analysis required. Are the minor components merely impurities that have escaped removal during isolation and fractionation of the polysaccharide? Do these minor structural features represent modifications at a postpolymerizational level, and, if so, are they evenly distributed in all molecules or do they arise only from a subset of molecules? Examples of polysaccharides in which regular repeating units are masked by postpolymerization modifications are discussed elsewhere, but they include alginic acid, the sulfated D-galactans (carrageenans) and the D-galacto-L-galacto-glycans (agarose and related polysaccharides) from seaweeds, and glycosaminoglycans such as heparin and dermatan sulfate. Do minor constituents represent a biosynthetically distinct subunit of the complete polysaccharide with provides a connecting link between regions of the entire macromolecule? Examples of such minor but integral subunits exist in the "linkage region" of proteoglycans wherein a short segment separates the structurally regular glycosaminoglycan and the protein core, and the sugar residues of the core region of lipopolysaccharides from gram-negative bacteria which separate the repeating O antigen and the lipid A of the backbone.

It is important to stress that satisfactory methods of compositional analysis for individual polysaccharides can be developed only as the need is recognized. Compositional analysis, detailed structure determination, and elaboration of biosynthetic pathways are mutually interacting facets of polysaccharide chemistry, no one of which should take precedence over the others.

Analyses of certain broad classes of sugars can be performed spectrophotometrically (6,7). Many of these methods involve treatment with acid, so that prior hydrolysis of polysaccharide or glycoconjugate samples is not necessary. Hydrolysis is accompanied by decomposition to give chromogens (usually furan or pyrrole derivatives), which condense with specific reagents to give colored products. These methods are valuable for preliminary characterizations, and especially for monitoring polysaccharide fractionations, but the methods are empirical, require calibration with known compounds, and do not often differentiate among members of the same class of compound. Furthermore, the specificities of the methods must be carefully checked for interference by other classes of sugars. The following are among the commonly used colorimetric methods: (a) anthrone–sulfuric acid (8) and phenol–sulfuric acid (9) for neutral sugars; (b) orcinol–sulfuric acid (10), a broad-spectrum reagent, which is frequently used for liquid chromatography analysis; (c) carbazole–sulfuric acid (11) and m-hydroxydiphenyl–sulfuric acid (12) for uronic acids; (d) cysteine–sulfuric acid for 6-deoxyhexoses (13); and (e) p-dimethylaminobenzaldehyde–hydrochloric acid (14) for sialic acids.

The least drastic conditions required to achieve total depolymerization of polysaccharides vary markedly from one material to another, and ideally the process goes to completion without loss of hydrolysis products. In practice, however, some loss of sugars may occur and/or the reaction may not proceed to completion because of the resistance to hydrolysis of certain linkages. A balance must therefore be struck between maximum depolymerization and minimum destruction of sugars, and in some cases it may be necessary to use different conditions for the hydrolysis and analysis of different sugar components in the same polysaccharide.

Ketose-containing polysaccharides are examples of polysaccharides that undergo complete hydrolysis under extremely mild conditions, for example, fructans with 0.1 M oxalic acid at 70°C for 1 h (15) and sialic acid polymers with 0.05 M sulfuric acid at 80°C for 1 h (16). In general, aldose-containing polysaccharides, such as glucans, arabinoxylans, and galactomannans, can be hydrolyzed completely with minimum loss of sugar with 0.5 or 1.0 M sulfuric acid at 100°C for 4–6 h (17) or with 1 M trifluoroacetic acid at 120°C for 1 h (18). However, other sugars, such as 2-deoxyaldoses, ketoses including sialic acids, and 3,6-anhydro-D- and -L-galactose, are largely destroyed under these conditions, and entire components may be lost. The decomposition products are rarely fully characterized, although furan (and occasionally pyrrole) derivatives are formed. These compounds with selected reagents act as chromogens in reactions of uncertain stoichiometry which provide the basis for the previously mentioned spectrophotometric determinations of particular classes of sugars. With some of these sugars the problem of decomposition may be overcome by protection as soon as the glycosidic linkage is cleaved. Thus, sialic acids may be stabilized during methanolysis, giving methyl glycoside methyl esters, which may be estimated by gas–liquid chromatography by (GLC) of O-trimethylsilyl (TMS) derivatives (19). 3,6-Anhydrohexoses, although often giving mixtures of types of compound, can be protected and isolated as derivatives by methanolysis to give mixtures of methyl glycosides (furanosides are formed when possible) and acyclic dimethylacetals or by mercaptolysis to give dialkyldithioacetals (see Section V,C,3).

The apparent loss of substantial quantities of certain sugars results from the pseudoequilibrium that is established when sugars are heated in dilute acid. The major products of such reactions are 1,6-anhydrohexopyranoses, (20) but anhydride formation is only of quantitative significance for those sugars, such as L-idose and L-gulose, that readily adopt the $^1C_4$ conformation. These particular sugars have not yet been found as constituents of poly-saccharides, although they are formed on reduction of the corresponding hexuronic acid constituents of dermatan sulfate (21) and alginic acid (22), respectively.

Incomplete hydrolysis is encountered with glycosiduronic acid and 2-amino-2-deoxyglycosidic linkages, and, indeed, advantage may be taken of such resistance to hydrolysis to isolate oligosaccharides as partial hydrolysis products. In the case of uronic acids as polysaccharide constituents, complete liberation is rarely possible without accompanying decomposition. When hydrochloric acid (12% or higher concentration) is used for hydrolysis, destructive decarboxylation occurs and can be used for quantitative estimation (23,24). Alternatively, various procedures may be used to effect reduction to the corresponding hexose, e.g., D-glucuronic acid to D-glucose, although this approach is of limited value if the hexose is also a constituent of the acidic polysaccharide. The reduction of hexuronic acid to hexose residues may be achieved before hydrolysis by successive treatments of the poly-saccharide with a water-soluble carbodiimide and sodium borohydride (25), or, after hydrolysis to acidic oligosaccharides, by treatment of TMS deriv-atives with lithium aluminum hydride (or deuteride) followed by simul-taneous deprotection and further hydrolysis (26). The sugars thus formed or given on hydrolysis of the carboxyl-reduced polysaccharide can be estimated in one of the ways outlined below.

Amino sugars found as polysaccharide constituents are most commonly 2-amino-2-deoxyhexoses, which are present to a large extent as N-acetyl derivatives. Since acid hydrolysis of 2-acetamido-2-deoxyhexoses results in N-deacetylation, the resulting 2-amino-2-deoxyglycosides are resistant to hydrolysis, and treatment of the polysaccharide or glycoconjugate with 4 M hydrochloric acid at 100°C for 6 h is required for complete depolymerization (27). An alternative approach is to perform basic N-deacetylation before hydrolysis or to allow N-deacetylation to take place during hydrolysis of all but the 2-amino-2-deoxyglycosidic linkages and then to submit the amino sugars to nitrous acid deamination. This process (see Section VIII,C) leads to the formation of 2,5-anhydrohexoses with concomitant cleavage of all re-maining 2-amino-2-deoxyglycosidic linkages (28). The anhydrohexoses so formed may be estimated, although it should be borne in mind that their formation is not quantitative, and analytical methods based on this reaction should be calibrated with authentic materials.

Methods for the analysis of individual components of mixtures of sugars fall into two broad categories, namely, those based on liquid chromatographic separations, usually of reducing sugars themselves, and those using GLC of suitably volatile derivatives. The separation of common neutral sugars can be achieved on columns of anion-exchange resins in the borate form, from which the effluent can be monitored by an appropriate colorimetric analysis, e.g., with orcinol–sulfuric acid. This procedure forms the basis of the commercially available autoanalyzers (29,30). The common amino sugars can be analyzed by separation from neutral sugars on cation-exchange resins and then estimated by the use of an amino acid analyzer (27). Increasingly, high performance liquid chromatograph (HPLC) columns are being developed for the separation and quantitative analysis of sugars (31). Separations may be based on cation-exchange resins (in various cationic forms) (32) using water as eluant or on partition chromatography using aqueous organic solvents such as acetonitrile–water.

Gas–liquid chromatographic methods for carbohydrate analysis involve the formation of derivatives of sufficient volatility and adequate thermal stability (33). Trimethylsilyl ethers and acetate or trifluoroacetate esters are the most common derivatives. Unless depolymerization requires the generation of sugars protected at the reducing group, e.g., the formation of methyl glycosides on methanolysis, reducing sugars are converted to acyclic derivatives before hydroxyl groups are substituted. The two most widely used derivatives are alditol acetates, from reduction with sodium borohydride (or sodium borodeuteride to label the former reducing group) and acetylation (34), and acetylated aldonitriles, which are formed by reaction with hydroxylamine in pyridine with later addition of acetic anhydride (35). This approach eliminates the problem of multiple derivative formation when different ring forms and/or different anomeric forms are generated from reducing sugars or from equilibrium mixtures of methyl glycosides. The acyclic derivatives are characterized by retention times and, if necessary for sugars with unusual structural features, from mass spectra. It should be mentioned, however, that ketoses (glyculoses) give rise to epimeric mixtures of alditols on reduction.

Although many polysaccharides are composed only of sugars of common occurrence, unusual sugars, e.g., deoxygenated sugars, those of uncommon configurational type, and those with stable ether substituents are not infrequently encountered, especially as constituents of microbial polysaccharides (see Chapter 5, Volume II). Such sugars may be detected first as "unknown peaks" in the chromatographic separation, and the mass spectrum from combined GLC–mass spectrometry may give a clue to the structural features. Although the majority of analyses are performed on milligram quantities (or less), a suspected unknown or rarely encountered sugar should be isolated on a preparative scale for more complete structural characterization by spectroscopic (particularly NMR) and, if necessary, degradative methods. Confirmatory synthesis may be desirable to ensure correct configurational assignments.

In general, chromatographic separation methods and spectroscopic analyses do not distinguish between enantiomers. Many sugars, for example, glucose, mannose, and xylose, have been found naturally in only the one enantiomeric D form, whereas others, such as galactose, arabinose, rhamnose, and fucose, are known in both enantiomeric forms, although not necessarily as constituents of polysaccharides and rarely as constituents of the same polysaccharide. Enantiomeric differentiation of sugars of known configurational type can be achieved on milligram quantities by circular dichroism of alditol acetates (36). For even smaller amounts of material, and in favorable cases for mixtures of sugars, enantiomers can be distinguished by conversion to equilibrium mixtures of glycosides of chiral alcohols, e.g., (+)- or (−)-2-butanol or (+)- or (−)-2-octanol, followed by GLC using capillary columns of volatile derivatives such as acetates or TMS ethers (37,38). In this procedure, enantiomeric sugars furnish, from a given chiral alcohol, diastereomeric mixtures of derivatives whose chromatographic separations provide a characteristic fingerprint.

For a few sugars quantitative estimation together with enantiomeric characterization can be coupled in enzymatic reaction. Thus, the enzyme D-glucose oxidase can be used for quantitative assay of this sugar in a mixture with others (39). L-Fucose dehydrogenase likewise distinguishes this sugar from its enantiomer (40).

Methods for the analysis of removable substituents are numerous [for a review, see Aminoff et al. (7)], and only a few specific references are cited here. More importantly, some aspects of their ease of removal and identification are stressed. Direct NMR spectroscopic analysis of organic substituents in intact polysaccharides, although limited in sensitivity but frequently adequate for quantitation, has the great advantage of uniquely identifying ether, ester, and acetal substituents. Chemical analysis of substituents is dependent on their complete removal and on the specificity of the analytical procedure. Thus, pyruvic acid, the only common ketalically linked substituent, is readily liberated on acid hydrolysis and can be specifically determined using pyruvate–lactate dehydrogenase (41). O-Acyl groups are generally labile to base, and the liberated acid may be determined titrimetrically or by GLC. Although O-acetyl groups are most commonly encountered, the identity of the O-acyl substituent should not be assumed. Similarly, the identities of N-acyl substituents, which are less readily liberated, should be individually ascertained. O-Sulfate (42) and O-phosphate (43) groups may be analyzed after hydrolysis. O-Alkyl groups may be determined by the Zeisel procedure, although titrimetric methods do not differentiate the common O-methyl from other O-alkyl groups. Methyl ester groups of galacturonic acid units in pectins, the only common substituents of this type, are determined by saponification (44) or more specifically through the liberated methanol (45). Methyl glycosides of the "reducing" terminal sugar residue have been reported only for the methylmannose polysaccharide from *Mycobacterium smegmatis* (46), where methanol is liberated on acid hydrolysis.

B. Determination of Molecular Size

Insofar as the polydispersity of polysaccharides represents a continuous variation in molecular size, experimentally determined values for molecular weight are obtained as averages, which are dependent on the method used. The most useful molecular weight determinations give either number-average values M_n or weight-average values M_w. The determination of both types of average molecular weight gives an indication of the molecular weight distribution since the greater the differences between M_n and M_w, the greater is the polydispersity of the sample.

Molecular weight determinations are clearly essential in relation to the physical properties of many industrially important polysaccharide derivatives. Reasonably accurate values for molecular weights may also be of significance in relation to specific biological activities, if a minimum size is necessary for that activity, but for many polysaccharides this information has not yet been regarded as essential and is often not available. Although information on molecular size may be of limited direct value in the determination of composition and structure, this information is critical in distinguishing between linear structures and those with a low degree of branching. Chemical evidence for branching is readily obtained from methylation analysis data (Section IV) when, for large molecules with a high degree of branching, the proportions of nonreducing end groups and branch points are identical. However, for polysaccharides with a low degree of branching, it may not be possible to obtain unambiguous evidence for branching through chemical procedures that are extended to the limits of their sensitivity. Thus, the detection of apparent branches in methylation analysis may be due to difficulties in obtaining complete methylation or to limited demethylation during hydrolysis. Again, a small proportion of "branch points" may be implied through incomplete periodate oxidation of otherwise vulnerable residues. In such instances comparisons of number-average proportions of nonreducing end groups from chemical analysis with physically determined number-average molecular weights will provide an estimate of the number of nonreducing end groups per molecule and hence provide evidence for or against branching.

Methods are available, e.g., reaction with ^{14}C-labeled cyanide (47) or reduction with sodium borohydride followed by periodate oxidation to furnish formaldehyde (1 or 2 moles depending on linkage type) (48), for the determination of the degree of polymerization (and hence molecular weight) by chemical analysis of reducing end groups. These results may be of value provided (a) that the limits of experimental accuracy are recognized and (b) that there is reasonable evidence, rather than a mere tacit assumption, that such reducing groups are present in unmodified form. Some polysaccharides, e.g., laminaran (49) and the partially methylated polysaccharides from *Mycobacterium* (46,50), contain chains some or all of which carry nonreducing units at the "reducing" terminus.

Critical accounts of the problems encountered in the determination of molecular weights and in the interpretation of measurements are given by Greenwood (51) for polysaccharide derivatives and by Gibbons (52) for glycoproteins and other glycoconjugates. Summaries of experimental procedures suitable for polysaccharide derivatives are given in the articles cited from "Methods in Carbohydrate Chemistry." Number-average molecular weights can be obtained by membrane osmometry ($M_n > 20,000$), vapor pressure osmometry (53), and isothermal distillation (54) ($M_n < 20,000$), (55,56) and weight-average molecular weights can be obtained from light scattering (56). Ultracentrifugation gives various molecular weight averages depending on the techniques used and the methods for calculating results. Sedimentation measurements, taken in conjuction with the determination of diffusion constants, rarely give simple molecular weight averages. Sedimentation equilibrium measurements or, better and more rapidly, the approach to equilibrium by the technique of pseudoequilibrium give values for M_w (56). Whereas these methods for molecular weight determination give theoretically based average values, the widely used procedure of gel permeation chromatography (molecular sieve chromatography (57–59) is a secondary technique that requires calibration with samples of generally similar chemical structure and independently determined molecular weights.

III. Problems of Structure Determination

Whereas covalent or primary structure elucidation is largely solved for proteins and nucleic acids when the sequences of amino acid and nucleotide residues, respectively, have been determined, determination of the structure of polysaccharides (or the carbohydrate units of other glycoconjugates) also

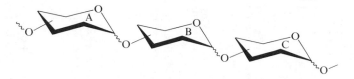

1. Nature and ring size, i.e., pyranose (p) (as shown) or furanose (f) for each sugar residue
2. Linkage type for each intersugar glycosidic linkage, i.e., 1→2, 1→3, etc.
3. Anomeric configuration (α or β) of glycosidic linkage for each sugar residue
4. Sequence of sugar residues, e.g., in the above system (if a regular repeat), -A-B-C- or -B-A-C-

Fig. 1. Information required for structure determination of polysaccharides.

requires for each sugar unit that the ring size, the configuration of the glycosidic linkage, and the linkage type (or site of substitution) be established (Fig. 1) and that information on sequence also take account of branching and the arrangement of branches in multibranched structures. All polysaccharides contain repetitive features, but whereas some contain a regular periodic repeating unit, notably most microbial polysaccharides whose biosynthesis involves polymerization of preassembled oligosaccharide units, regularity of structural features is less apparent or even unproved for other polysaccharides. Thus, for a number of plant polysaccharides there is no clear demonstration of regularity in the following structural features: second linkage types inserted in linear homopolysaccharide chains, attachment of side chains, and attachment of substituents such as *O*-acetyl groups.

It will be obvious that no one structural method can give answers to all these questions, but Table I summarizes the information gained from the methods described in later sections of this chapter together with some of the limits of each structural method. It should be stressed that each approach has limits, so whenever information on particular aspects of structure can be obtained by more than one method, it is desirable to seek confirmation by alternative approaches.

IV. Methylation

Methylation analysis is the oldest, but still by far the most widely used procedure for linkage analysis in carbohydrate polymers (*60–62*). The method is based on the complete etherification of free sugar hydroxyl groups, i.e., those not involved in ring formation, intersugar glycosidic linkages, or carrying substituents stable to conditions used for methylation of the polysaccharide and for subsequent hydrolysis of the methylated derivative. With important exceptions (see Section VIII,A) glycosidic linkages are

TABLE I

Structural Information Derived from Chemical Methods in Polysaccharide Chemistry

Procedure	Information	Theoretical limits	Experimental limits
Methylation	Ring size, substitution pattern, quantitation of individual sugar residues	Does not differentiate between 4-linked pyranosides and 5-linked furanosides	Completeness of etherification; deetherification during hydrolysis without decomposition; loss of more volatile cleavage products leading to underestimate of nonreducing end groups
Periodate oxidation			
Analytical	Quantitation of certain types of sugar units	Without supplementary data results are rarely unambiguous	Non-Malapradian overoxidation; underoxidation from protection of vulnerable units by interresidue hemiacetal formation or hydrogen bonding
Degradative (Smith degradation)	Distribution of periodate-resistant sugar residues; preparative-scale isolation of products may give evidence for individual anomeric configurations		Lack of complete selectivity in hydrolysis of acyclic acetals; acid-catalyzed transacetalation leading to formation of cyclic acetals
Partial depolymerizations			
Partial hydrolysis with mineral acid, acetolysis, etc., action of endo glycan hydrolases	Sequences of sugar residues; preparative-scale isolation of products may also give evidence for individual anomeric configurations	Products isolated are dependent on relative rates of cleavage without decomposition of glycosidic linkages	Chromatographic technique for separation of oligosaccharides; acid-catalyzed reversion (partial hydrolysis); anomerization (acetolysis)

Stepwise degradations Smith degradation	See under Periodate oxidation		Selectivity and completeness of reaction at each step
Svensson degradation	Sequences of sugar residues		Exposure of hydroxyl groups
Sequential enzymatic hydrolysis	Sequences of sugar residues		Variations in specificity of exo-glycosidases toward different aglycones
Functionally specific non-hydrolytic selective degradations			
Base-catalyzed β-eliminations from hexuronate residues	Sites of attachment of hexuronic acid residues; variable information on remaining units	Further degradation from exposed reducing groups	For methylated derivatives, complete alkylation without accompanying premature degradation
Base-catalyzed sulfone degradation	Location of primary hydroxyl groups not involved in glycosidic linkages		Selectivity of substitution of primary hydroxyl groups
Nitrous acid deaminations	Sugar sequences in the vicinity of amino sugar residues	Reaction pathways dependent on location and orientation of amino groups	Generation of amino groups from N-acetyl derivatives
Nonhydrolytic cleavage of glycosiduronic acid linkages by Hofmann and related degradations and by oxidative decarboxylation	Sugar sequences in the vicinity of hexuronic acid residues including characterization of glycosyl substituents		Completeness of original cleavage reactions
Chromium trioxide oxidation	Configurations of glycosidic linkages based on selective oxidation of equatorially oriented acetylated glycopyranosides	Conformational stability of glycosides	Generality not yet established

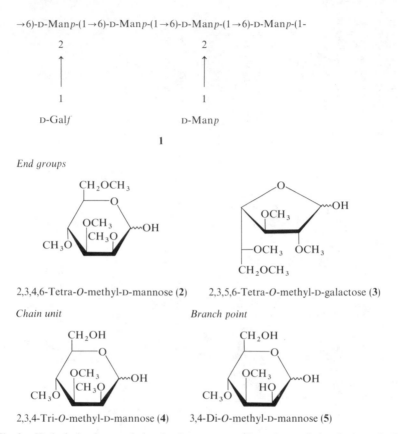

→6)-D-Man*p*-(1→6)-D-Man*p*-(1→6)-D-Man*p*-(1→6)-D-Man*p*-(1-

2   2

1   1

D-Gal*f*   D-Man*p*

1

End groups

CH₂OCH₃

2,3,4-Tetra-*O*-methyl-D-mannose (**2**)   2,3,5,6-Tetra-*O*-methyl-D-galactose (**3**)

Chain unit   *Branch point*

2,3,4-Tri-*O*-methyl-D-mannose (**4**)   3,4-Di-*O*-methyl-D-mannose (**5**)

Fig. 2. Hydrolysis of permethylated galactomannan (**1**) to give methylated sugars (**2**–**5**).

stable to the strongly basic conditions often required for alkylation. Since methyl ethers are stable to hot dilute acid, characterization of the hydrolysis products will then identify those hydroxyl groups formerly involved in inter-sugar linkages. Thus, from the partial structure (**1**) in Fig. 2, hydrolysis of the permethylated derivative would give the methylated sugars (**2**–**5**) and thereby establish the nature of those sugar residues occurring as nonreducing end groups, chain units, and branch points. For all these sugars the presence of a methoxyl group at C-4 or C-5 defines the ring size as pyranose or fura-nose, respectively. However, certain partially methylated sugars carry methoxyl groups at neither of these positions, and here methylation analysis does not provide information on ring size or linkage type, so that evidence on these points must be sought independently. An example is provided by 2,3,6-tri-*O*-methyl-D-galactose (**6**), which could (and does) arise from both

CH$_2$OCH$_3$

HO

OCH$_3$

OH

OCH$_3$

2,3,6-Tri-O-methyl-D-galactose (6)

4-O-substituted D-galactopyranose [→4)-D-Gal$p$-(1←] and 5-O-substituted D-galactofuranose [→5)-D-Gal$f$-(1←] residues. A recent method of general application for solving this problem is discussed in Section V,F.

A. Methylation Procedures

Etherification of polysaccharides is dependent on a sufficient degree of ionization of hydroxyl groups to achieve alkoxide formation with enhanced nucleophilicity toward the alkylating agent, usually methyl iodide or dimethyl sulfate. Effective reaction is also dependent on the polysaccharide being soluble in a convenient polar solvent. For most methylations the method of choice is that of Hakomori, in which the polysaccharide dispersed in dimethyl sulfoxide (DMSO) is first treated with sodium methylsulfinyl-methanide (sodium dimsyl) and then allowed to react with methyl iodide (60). The strong base, which is the conjugate base of the solvent (pK_a 35), ensures complete alkoxide formation, and efficient etherification is achieved in one step. Recently, potassium dimsyl has been shown to be more effective for some methylations (63,64). Alternative methylation methods (65), which presumably involve less complete alkoxide formation, include various procedures developed by Kuhn and collaborators (66) in which N,N-dimethyl-formamide or DMSO is used as dipolar aprotic solvent with methyl iodide or dimethyl sulfate as alkylating agent and silver oxide or barium oxide (sometimes with added barium hydroxide) as base. The once widely used procedure of W. N. Haworth, which employs dimethyl sulfate and aqueous 30% sodium hydroxide, now finds application only in special circumstances. The original Purdie methylation method, in which methyl iodide is both solvent and alkylating agent with silver oxide as base, is generally less effective, and its use is restricted to materials already partially methylated and therefore soluble in methyl iodide.

The Hakomori methylation method also leads to esterification of uronic acid residues, but a note of warning must be added here since uronic esters are very susceptible to base-catalyzed β-elimination (for an account of degradations based on this property see Section VIII,A). In practice, however, alkoxides are so much more strongly nucleophilic toward the alkylating agent than carboxylate ions that the net base concentration may be markedly decreased before esterification occurs and substantial degradation during methylation averted. Repetition of the operation to achieve full alkylation must be avoided since extensive degradation would take place on the addition of fresh base. Thus, for acidic polysaccharides it is important that complete methylation be achieved in a single operation.

Hakomori methylation results in the removal of O-acyl substituents, but acetals and ketals (e.g., of pyruvic acid) are stable to the basic conditions. The presence of dimsyl ion ensures that acetamido groups are ionized and hence that N-methylation occurs. Some N-methylation occurs with the various Kuhn alkylation procedures. The much less efficient Haworth method results in significantly less base-catalyzed β-elimination from uronic acid residues, provided, of course, that ester groups, e.g., in pectins, are first saponified. Subsequent esterification may be effected by treatment of the etherified polysaccharide with diazomethane or by the Purdie procedure. Studies on oligosaccharides containing 2-acetamido-2-deoxyhexose residues indicate that etherification can be achieved using the Haworth method without significant accompanying N-methylation (67).

The completeness of methylation of polysaccharides can be ascertained (a) by methoxyl determination if a large enough quantity of the methylated derivative is available for microanalysis or (b) more simply by the absence of O—H stretching vibrations in the IR spectrum. Analysis of the methylated sugars formed on hydrolysis of a fully methylated polysaccharide should, by its internal consistency, give confirmation of full etherification since there should be equimolar proportions of nonreducing end groups and branch points. The significance of minor products formed on hydrolysis of a methylated polysaccharide should be assessed with caution in the light of structural evidence obtained by independent means. In particular, care should be taken not to assign branch points to products of incomplete methylation, which may arise either from undermethylation of the polysaccharide or from limited demethylation during hydrolysis.

The above discussion assumes that the polysaccharide or glycoconjugate is soluble in DMSO. Some methylations have been achieved with partially or completely insoluble substrates, e.g., with mixtures of polysaccharides in plant cell wall preparations. However, for individual polysaccharides, alternative methods to achieve solubility should be adopted if necessary. For Haworth-type methylations polysaccharides may be acetylated in formamide solution (68), and the resulting acetylated polysaccharide may be simultaneously de-O-acetylated and methylated in an organic solvent such as tetrahydrofuran. Recently, N-methylmorpholine N-oxide (MMNO) has been shown to dissolve nondegradatively polysaccharides such as cellulose that are insoluble in DMSO alone, so that methylation can then be effected in MMNO–DMSO mixtures (69). A recent alternative approach to methylation, which has been successfully applied to lipopolysaccharides, uses the highly reactive methyl trifluoromethanesulfonate (triflate) as alkylating reagent in the presence of 2,6-di-*tert*-butylpyridine and trimethyl phosphate as solvent (70).

B. Characterization and Analysis of Methylated Sugars

The complete characterization of a permethylated polysaccharide requires the identification and quantitative analysis of all the sugar derivatives formed on depolymerization. The problems posed by the resistance to hydrolysis of glycosiduronic acids and 2-amino-2-deoxyhexosides are the same as in the case of the parent polysaccharides. In the case of uronic acid-containing polysaccharides complete depolymerization is usually possible only after reduction to the corresponding hexose residues. The reduction may be performed with lithium aluminum hydride (or deuteride) in tetrahydrofuran or similar solvent either on the permethylated polysaccharide (71) or on the products of partial depolymerization, with protection of reducing sugars if necessary. Reduction with the labeled reagent provides a convenient means of distinguishing those hexose derivatives formed from uronic acids with incorporation of two deuterium atoms ($-CO_2H \rightarrow -CD_2OH$) when mass spectrometry is used for identification. It may also be pointed out that derivatives of the types commonly used for the characterization of neutral methylated sugars, namely, partially methylated alditol acetates and partially methylated acetylated aldononitriles, are insufficiently volatile to be suitable for uronic acid analysis. The problem of incomplete depolymerization of aminodeoxyglycosides can be largely averted by using acetolysis and thus minimizing N-deacetylation (72).

Before the analysis of methylated sugar derivatives by GLC–mass spectrometry became widely adopted, mixtures of methylated sugars formed in preparative quantities were separated (a) by partition chromatography on filter sheets or columns of cellulose (73) or (b) by adsorption chromatography on columns of charcoal or charcoal–Celite mixtures (74). The individual sugars can then be identified by the formation of crystalline derivatives with characteristic physical constants, e.g., melting points, optical rotations, and X-ray powder diagrams (75). Characterization may be aided by identification of the parent sugar formed on demethylation with boron trichloride (76) and by examination of the products of periodate oxidation (77). With the resolving power of GLC columns coupled with the use of mass spectrometry to characterize substitution patterns in sugar derivatives, preparative-scale isolation of methylated sugars is now rarely necessary. Methyl ethers of all commonly occurring sugars as derivatives may usually be separated by using a variety of different liquid phases or by choosing alternative derivatives (78).

When sugars of unusual type or uncommon stereochemical configuration are encountered as polysaccharide constituents, the various methyl ethers can be synthesized by random partial methylation of glycosides of defined ring size (79) to serve as reference compound derivatives of the fully and partially methylated sugars whose methylation patterns may be assigned on the basis of their mass spectra.

In general, it is not necessary to define enantiomeric configurations for methylated sugars when only one enantiomer of a given sugar is present as a constituent of the polysaccharide under examination. Although examples are rare, a few polysaccharides are known in which both enantiomers are constituents of the same polysaccharide, e.g., D- and L-galactose in some algal polysaccharides from red seaweeds (80) and in the galactan from the snail *Helix pomatia* (81), and D- and L-fucosamine (2-amino-2,6-dideoxygalactose) in the lipopolysaccharide from *Pseudomonas aeruginosa* (82). Special procedures may be necessary in such cases to differentiate between derivatives of the two enantiomers.

The depolymerization of methylated polysaccharides is best effected by hydrolysis with the subsequent formation of acyclic derivatives from the reducing sugars, thus avoiding the formation of multiple derivatives from the same sugar. Reducing sugars, on acetylation or on conversion to TMS ethers, form α- and β-anomers and may, if unsubstituted on O-4 and O-5, give rise to both furanose and pyranose ring forms. For this reason, methanolysis of methylated polysaccharides, with formation of equilibrium mixtures of methyl glycosides, is now less frequently performed. However, exceptional situations may be encountered in which it is necessary to protect sugar derivatives, which would otherwise undergo decomposition. Examples are methylated polysaccharides containing residues of sialic acids (83) or 3,6-anhydrohexoses (84).

Many methylated polysaccharides are much less soluble in hot than in cold aqueous solvents. Accordingly, it is usually convenient to carry out partial hydrolysis in an organic solvent such as formic acid and then to complete the hydrolysis in dilute aqueous acid. Partially methylated alditol acetates, formed on reduction with sodium borohydride followed by acetylation, are the most widely used derivatives for the characterization of methylated sugars. The mass spectra of these compounds are normally simple to interpret, with fragmentation patterns characteristic of constitution, and especially of substitution pattern, but rather insensitive to stereochemical differences (85). Molecular ions are not seen in electron impact spectra taken at 70 eV, but molecular weights can usually be obtained by extrapolation from fragment ions coupled with an intelligent use of GLC retention time

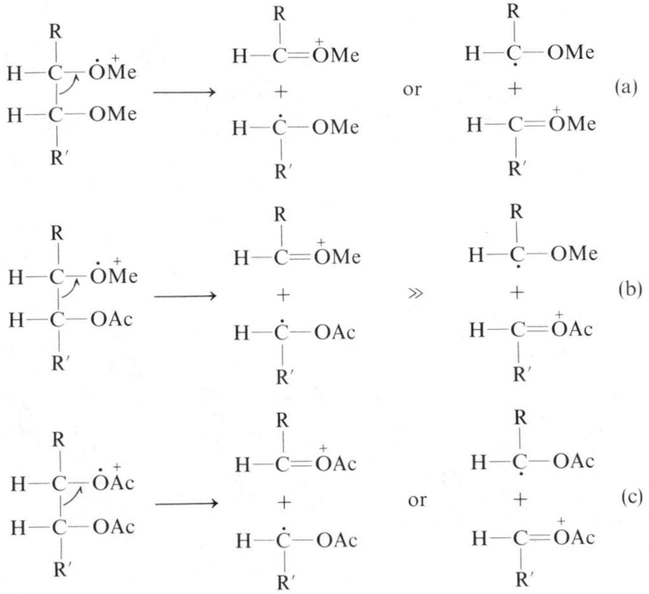

$$
\begin{array}{ccc}
\begin{array}{c}
R \\
| \\
H-C \overset{\cdot\,+}{\underset{\nearrow}{-}} \ddot{O}Me \\
| \\
H-C-OMe \\
| \\
R'
\end{array}
&
\longrightarrow
&
\begin{array}{ccc}
\begin{array}{c}
R \\
| \\
H-C=\overset{+}{O}Me \\
+ \\
H-\dot{C}-OMe \\
| \\
R'
\end{array}
& \text{or} &
\begin{array}{c}
R \\
| \\
H-\underset{\cdot}{C}-OMe \\
+ \\
H-C=\overset{+}{O}Me \\
| \\
R'
\end{array}
\end{array}
& \text{(a)}
\end{array}
$$

$$
\begin{array}{ccc}
\begin{array}{c}
R \\
| \\
H-C \overset{\cdot\,+}{\underset{\nearrow}{-}} \ddot{O}Me \\
| \\
H-C-OAc \\
| \\
R'
\end{array}
&
\longrightarrow
&
\begin{array}{ccc}
\begin{array}{c}
R \\
| \\
H-C=\overset{+}{O}Me \\
+ \\
H-\dot{C}-OAc \\
| \\
R'
\end{array}
& \gg &
\begin{array}{c}
R \\
| \\
H-\underset{\cdot}{C}-OMe \\
+ \\
H-C=\overset{+}{O}Ac \\
| \\
R'
\end{array}
\end{array}
& \text{(b)}
\end{array}
$$

$$
\begin{array}{ccc}
\begin{array}{c}
R \\
| \\
H-C \overset{\cdot\,+}{\underset{\nearrow}{-}} \ddot{O}Ac \\
| \\
H-C-OAc \\
| \\
R'
\end{array}
&
\longrightarrow
&
\begin{array}{ccc}
\begin{array}{c}
R \\
| \\
H-C=\overset{+}{O}Ac \\
+ \\
H-\dot{C}-OAc \\
| \\
R'
\end{array}
& \text{or} &
\begin{array}{c}
R \\
| \\
H-\underset{\cdot}{C}-OAc \\
+ \\
H-C=\overset{+}{O}Ac \\
| \\
R'
\end{array}
\end{array}
& \text{(c)}
\end{array}
$$

Fig. 3. Primary fragment ions arising by α-cleavage in the mass spectra of partially methylated alditol acetates.

data. Direct observation of molecular ions is frequently possible in chemical ionization mass spectra using isobutane as reagent gas (86).

Primary fragment ions from partially methylated alditol acetates (Fig. 3) arise by α-cleavage with, in general, preferred formation of (a) ions of lower molecular weight, (b) ions from cleavage between two methoxyl-bearing carbon atoms, with no marked preferences between the two methoxyl-bearing cations, and (c) ions from cleavage between a methoxyl-bearing and an acetoxyl-bearing carbon atom with marked preference for the methoxyl-bearing species to carry the positive charge, but (d) ions formed by scission between two acetoxyl-bearing carbon atoms are generally of low abundance. There is little tendency for chain scission to take place adjacent to a deoxygenated carbon atom, but the presence of such a unit is usually apparent from the increase in m/e values by 14 amu. Some typical fragment ions are shown in Fig. 4 for 1,5-di-O-acetyl-2,3,4,6-tetra-O-methyl-D-mannitol-1-d (7). Primary fragment ions undergo a series of subsequent eliminations to give secondary fragments, including losses by β-elimination of acetic acid (m/e 60) or methanol (m/e 32), losses by α-elimination of acetic acid (m/e 60) but not of methanol, and losses via cyclic transition states of formaldehyde, methoxymethyl acetate, or acetoxymethyl acetate. A very valuable collection of mass spectra together with retention time data has been published for

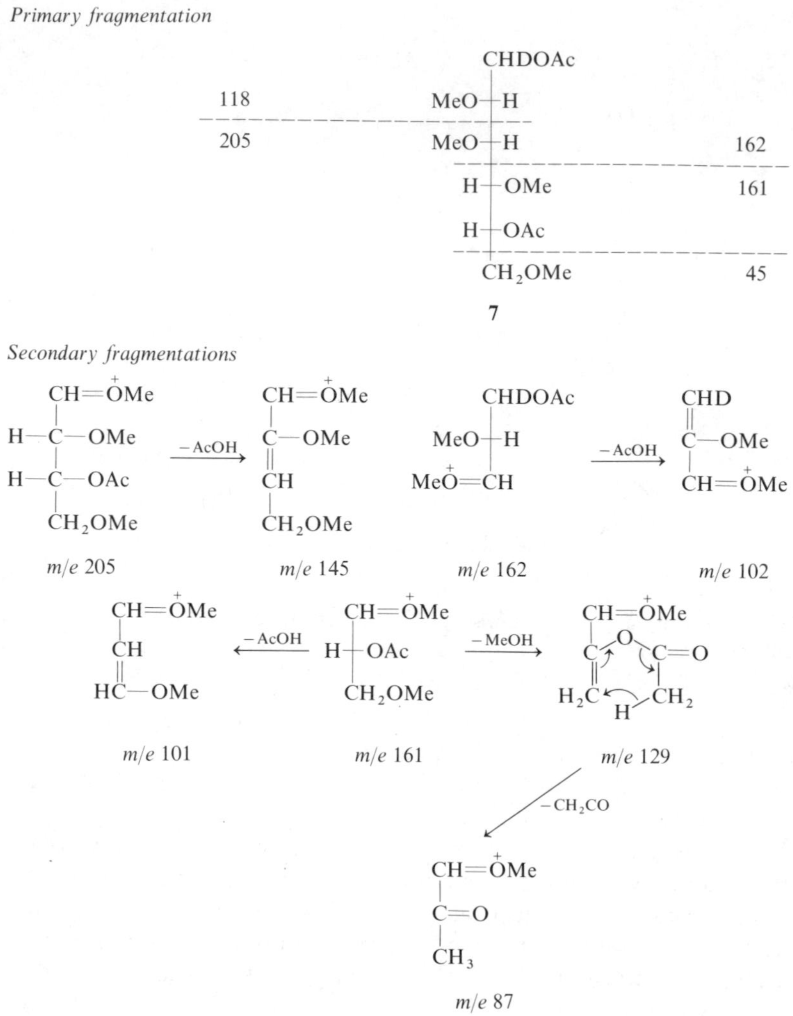

Primary fragmentation

$$
\begin{array}{c}
\text{CHDOAc} \\
118 \quad\quad \text{MeO}-\text{H} \\
\text{-----------} \\
205 \quad\quad \text{MeO}-\text{H} \quad\quad 162 \\
\text{H}-\text{OMe} \quad\quad 161 \\
\text{H}-\text{OAc} \\
\text{-----------} \\
\text{CH}_2\text{OMe} \quad\quad 45 \\
\mathbf{7}
\end{array}
$$

Secondary fragmentations

$$
\begin{array}{ccc}
\overset{+}{\text{CH}}=\text{OMe} & \overset{+}{\text{CH}}=\text{OMe} & \text{CHDOAc} & \text{CHD} \\
\text{H}-\text{C}-\text{OMe} & \text{C}-\text{OMe} & \text{MeO}-\text{H} & \text{C}-\text{OMe} \\
\text{H}-\text{C}-\text{OAc} \xrightarrow{-\text{AcOH}} & \text{CH} & \text{MeO}=\text{CH} \xrightarrow{-\text{AcOH}} & \text{CH}=\overset{+}{\text{OMe}} \\
\text{CH}_2\text{OMe} & \text{CH}_2\text{OMe} & & \\
\end{array}
$$

m/e 205   *m/e* 145   *m/e* 162   *m/e* 102

$$
\begin{array}{ccc}
\overset{+}{\text{CH}}=\text{OMe} & \overset{+}{\text{CH}}=\text{OMe} & \overset{+}{\text{CH}}=\text{OMe} \\
\text{CH} \xleftarrow{-\text{AcOH}} & \text{H}-\text{OAc} \xrightarrow{-\text{MeOH}} & \text{C}-\text{O}-\text{C}=\text{O} \\
\text{HC}-\text{OMe} & \text{CH}_2\text{OMe} & \text{H}_2\text{C}-\text{CH}_2 \\
\end{array}
$$

m/e 101   *m/e* 161   *m/e* 129

$$-\text{CH}_2\text{CO}$$

$$
\begin{array}{c}
\overset{+}{\text{CH}}=\text{OMe} \\
\text{C}=\text{O} \\
\text{CH}_3 \\
\end{array}
$$

m/e 87

Fig. 4. Fragment ions from 1,5-di-*O*-acetyl-2,3,4,6-tetra-*O*-methyl-D-mannitol-1-*d* (**7**).

partially methylated alditol acetates from methylated hexoses, pentoses, 6-deoxyhexoses, and some heptoses and dideoxyhexoses (*61*). Mass spectra of these and a variety of other carbohydrate derivatives have been reviewed (*85*). Some salient features are outlined below. Recently, an account has been published of GLC data and mass spectra of derivatives of aminohexoses and aminohexitols (*87*).

The main limitation of the use of partially methylated alditol acetates for the characterization of methylated sugars lies in the structural symmetry

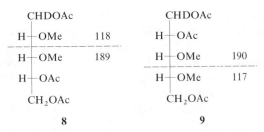

Fig. 5. Primary mass spectral fragment ions from 2,3-di-*O*-methylpentitol triacetate-1-*d* (8) and 3,4-di-*O*-methylpentitol triacetate-1-*d* (9).

that may exist when the primary hydroxyl group (O-5 in pentoses and O-6 in hexoses) is not etherified. This difficulty, however, can be overcome by introducing deuterium at C-1 by reduction of the sugar with sodium borodeuteride. Thus, in the example shown in Fig. 5, the partially methylated alditol acetates from 2,3- (8) and 3,4-di-*O*-methylpentitols (9) can be differentiated by observation of the relevant isotopic shifts in primary fragment ions.

Partially methylated acetylated aldononitriles are acyclic derivatives readily formed from reducing sugars by reaction with hydroxylamine in pyridine, followed by the addition of acetic anhydride to effect elimination of acetic acid from oxime acetates and acetylation of unsubstituted hydroxyl groups. These derivatives, although less extensively used, appear to give good GLC separations, and their mass spectra can be readily interpreted without the problem of structural symmetry (*88*). Two examples (10 and 11) are shown in Fig. 6.

Partially ethylated alditol acetates may be used as alternative sugar derivatives when separation difficulties are encountered with particular combinations of the corresponding methylated compounds (*89,90*). More

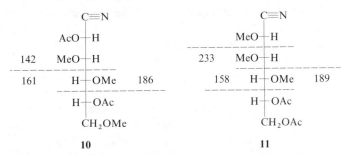

Fig. 6. Primary mass spectral fragment ions from 2,5-di-*O*-acetyl-3,4,6-tri-*O*-methyl-D-mannononitrile (10) and 5,6-di-*O*-acetyl-2,3,4-tri-*O*-methyl-D-mannononitrile (11). (Nitrile-containing ions are generally of low abundance.)

importantly, ethylation or trideuteriomethylation is used to label limited numbers of hydroxyl groups liberated in the partial fragmentation of permethylated polysaccharide by controlled partial hydrolysis or functionally specific degradations (62). These controlled depolymerizations of methylated polysaccharides have a dual advantage over the corresponding reactions of the parent polysaccharides in that (a) specific sites of cleavage can be recognized by derivative formation at both reducing groups (or their equivalent) and exposed aglyconic hydroxyl groups and (b) reactions can be performed in nonpolar organic solvents. Several examples of this approach are given in Sections V,F and VIII.

V. Partial Depolymerization by Hydrolysis and Related Reactions

The principle of arresting the hydrolysis of polysaccharides at some convenient stage before complete depolymerization has occurred and then characterizing the products is frequently used. The isolated degraded polysaccharides and oligosaccharides on characterization then give more detailed, and often unambiguous, information on sequences and anomeric configurations, in addition to providing confirmation of linkage types. In the case of linear polysaccharides of uniform linkage type, and assuming that all glycosidic bonds are equally susceptible to purely random cleavage, partial hydrolysis leads to the formation of polymer-homologous series of oligosaccharides. Similarly, for multilinkage homo- or heteropolysaccharides, if the susceptibilities to hydrolysis of different linkages are approximately equal, a representative selection of all possible oligosaccharides is liberated. These compounds can then be individually characterized provided that sufficiently selective separation procedures are available. In practice, however, rates of hydrolysis of different glycosidic linkages are often sufficiently different that not all possible oligosaccharide sequences can be isolated by a single type of depolymerization. Thus, the failure to isolate a particular disaccharide or trisaccharide need not be taken as firm evidence for the absence of that linkage type unless supported by independent observations.

Random partial hydrolysis of a polysaccharide of any complexity requires a relatively large quantity of material. Since fragments formed during the process may be degraded further as reaction proceeds, the yields of oligosaccharides, which are isolated, are often low. For enzymatic hydrolysis higher yields of oligosaccharides can be obtained by carrying out the reaction in a dialysis membrane so that dialyzable fragments of low molecular weight are removed as soon as they are formed (91). Likewise, for partial acid hydrolysis, an equivalent procedure involves the use of a water-soluble polystyrene sulfonic acid as catalyst (**92**). However, only very acid-labile polysaccharides, such as fructans, can be degraded in this way, since the cellophane membrane would itself be hydrolyzed under more drastic conditions. Alternatively, hydrolysis can be carried out in a stepwise manner with dialysis at the end of each reaction period to remove products of low molecular weight (**93**).

Substantial differences in rates of hydrolysis of glycosidic linkages are frequently encountered, and advantage may be taken of these differences to achieve relatively selective fragmentations. Alternative types of acid-catalyzed depolymerization, e.g., acetolysis, may be used to effect different degrees of selectivity in bond cleavage. Enzymatic hydrolysis provides an extreme case of selectivity in hydrolysis, and this topic is discussed in full in Chapter 1, Volume III. In this chapter the topic receives only brief mention by way of comparison with partial hydrolysis procedures.

A. Fractionation and Characterization of Oligosaccharides and Degraded Polysaccharides

Depending on the extent of hydrolysis (or related reaction) and on the distributions and relative susceptibilities to cleavage of the various glycosidic linkages, partial depolymerization of polysaccharides gives rise to mixtures of sugars of relatively low molecular weight together with, in some cases, degraded polysaccharides. In only a few instances have preparative-scale separations of individual oligosaccharides been achieved beyond the hexasaccharide level. In those cases in which the more acid-sensitive linkages are located mainly in the outer chains, the removal of these units may be carried out selectively with only limited cleavage of internal linkages. A few of the latter linkages are split, but the residual degraded polysaccharide, although of diminished molecular size, can be isolated as a polymeric material, e.g., by gel filtration or by precipitation from aqueous solution with a water-miscible nonsolvent such as ethanol or acetone. Examples of the selective removal of outer chain units include (a) the controlled hydrolysis, often "autohydrolysis" for acidic polysaccharides, of L-arabinofuranosyl residues in plant polysaccharides (see Fig. 7) (*94*), (b) the liberation of D-galactofuranosyl from fungal galactomannans (*95*), (c) the selective hydrolysis of terminal 3,6-dideoxyhexose units in certain lipopolysaccharides (*96*), and (d) the removal of sialic acid residues from glycoproteins and glycolipids (*15*).

Degraded polysaccharides isolated from partial acid hydrolysis are, in general, more polydisperse than the parent polysaccharides. Apart from using such materials for further partial hydrolysis to generate oligosaccharides from interior chains, the most valuable structural information is

α-D-Gal*p*-(1→3)-α-L-Ara*f*-(1→3)-L-Ara

13

α-D-Gal*p*-(1→3)-L-Ara

12

-α-L-Ara*f*-(1 → 6)-α-D-Gal*p*-(1 → 3)-α-L-Ara*f*-(1 → 3)-α-L-Ara*f*-(1 → 3)-L-Ara*f*-(1-

17

α-L-Ara*f*-(1→6)-α-D-Gal*p*-(1→3)-L-Ara

14

α-L-Ara*f*-(1→3)-L-Ara

15

α-L-Ara*f*-(1→6)-α-D-Gal*p*-(1→3)-α-L-Ara*f*-(1→3)-L-Ara

16

β-L-Ara*f*-(1→2)-β-L-Ara*f*-(1→2)-β-L-Ara*f*-(1→2)-β-L-Ara*f*-(1→2)-L-Ara

20

β-L-Ara*f*-(1→2)-β-L-Ara*f*-(1→2)-L-Ara

19

β-L-Ara*f*-(1→2)-L-Ara

18

α-L-Ara*f*-(1→4)-L-Ara*p*

21

-β-L-Ara*f*-(1→2)-β-L-Ara*f*-(1→2)-β-L-Ara*f*-(1→2)-β-L-Ara*f*-(1→2)-α-L-Ara*f*-(1→4)-β-L-Ara*p*-(1→3)-L-Ara*f*-(1-

26

β-L-Ara*p*-(1→3)-L-Ara

22

α-L-Ara*f*-(1→4)-β-L-Ara*p*-(1→3)-L-Ara

23

β-L-Ara*f*-(1 → 2)-α-L-Ara*f*-(1 → 4)-β-L-Ara*p*-(1 → 3)-L-Ara

24

Fig. 7. Oligosaccharides (**12–16** and **18–25**) formed on "autohydrolysis" from segments (**17** and **26**) of outer chains and acidic oligosaccharides (**27–31**) liberated under more drastic conditions from mesquite gum.

β-L-Araƒ-(1→2)-β-L-Araƒ-(1→2)-α-L-Araƒ-(1→4)-β-L-Ara$p$-(1→3)-L-Ara

25

(4-Me)-β-D-Glc$p$A-(1→6)-D-Gal

27

β-D-Glc$p$A-(1→6)-D-Gal

28

(4-Me)-β-D-Glc$p$A-(1→6)-β-D-Gal$p$-(1→6)-D-Gal

29

(4-Me)-α-D-Glc$p$A-(1→4)-D-Gal$p$

30

(4-Me)-α-D-GlpA-(1 → 4)-β-D-Gal$p$-(1 → 3)-D-Gal

31

Fig. 7. (*continued*)

derived from comparisons of overall structural features of the parent and degraded polysaccharides, e.g., by methylation analysis or by periodate oxidation studies. Thus, with a knowledge of the nature of the low molecular weight fragments liberated, comparative methylation analyses will indicate the positions to which the removed sugar residues were formerly attached. However, when more than one type of unit is removed from more than one site of attachment, it is not possible to assign linkage sites unambiguously to particular acid-sensitive substituents. This situation is encountered among many exudate gums and structurally related plant polysaccharides (*94*). Figure 7 summarizes studies on mesquite gum in which "autohydrolysis" liberates a complex mixture of oligosaccharides **16** and **18** from cleavage of arabinofuranosyl linkages, which may be considered to arise from peripheral chains **17** and **26**, respectively, and gives rise to a degraded polysaccharide essentially devoid of arabinose residues, which is suitable for direct examination and for the generation of further partial hydrolysis products (**27–31**) (*97*).

Depending on the complexity of the mixture of sugars, the preparative-scale separation of oligosaccharides for individual characterization can be achieved by one or a combination of filter sheet chromatography and various types of column chromatography. For mixtures of neutral oligosaccharides, column separations can be performed by chromatography on suitably cross-linked cation-exchange resins by development with water (*98*), by molecular sieve chromatography (gel filtration) on Sephadex (a cross-linked dextran), or Biogel (a polyacrylamide gel) (57–59), by partition chromatography on cellulose (73) or by adsorption chromatography on charcoal (frequently in admixture with Celite to increase flow rates) (74) by elution with ethanol–water mixtures. For mixtures of acidic oligosaccharides, ion-exchange chromatography on weak base anion exchangers in their acetate or formate form with elution with increasing concentrations of acetic or formic acids is an important separation procedure (99). For this purpose diethylaminoethylcellulose and diethylaminoethyl–Sephadex are convenient materials of high capacity (100). Mixtures of acidic oligosaccharides may also be separated by ion-exchange chromatography using sodium tetraborate as eluant (101). Analyses of oligosaccharides are carried out by various forms of HPLC (31,32) and are likely to be developed further in the near future.

Several reference works and review articles provide a comprehensive coverage of the literature on oligosaccharides and their crystalline derivatives up to about 1965 (102–104). Increasingly, however, in the past 15 years or so, oligosaccharide characterization in connection with polysaccharide structure determination has been performed on amounts as low as the milligram level or less. The reduction in scale of operation has been possible for two reasons. First, since other small-scale methods for polysaccharide analysis, especially methylation, have become available and therefore make it possible to obtain information on much smaller quantities of material and, second, because chromatographic methods have provided adequate criteria of purity for oligosaccharides without the necessity for crystalline derivative formation. The results from any individual method that can be performed with such small amounts of material, when taken in isolation, rarely provide complete proof of structure and must be interpreted in the light of independently acquired data. It must be emphasized again that, with special exceptions, chromatographic separations and spectroscopic analyses do not distinguish between enantiomers.

For the complete characterization of oligosaccharides, in addition to the overall determination of composition and linkage types, it is necessary to establish sugar sequences and to assign anomeric configurations to individual glycosidic bonds. For the higher oligosaccharides this objective is usually realized by reference to previously characterized di- and trisaccharides. There are many alternative procedures for determining aspects of oligosaccharide structures, and reference may be made to several reviews (102–105). Only those methods that lend themselves to small-scale manipulations are discussed here. A section is devoted to the mass spectrometry of oligosaccharide derivatives. Other aspects of spectroscopy, especially NMR spectroscopy, are discussed in Chapter 4.

With the exception of sucrose and higher oligosaccharides related to sucrose, as well as O-glycosyl derivatives of D-mannitol, all oligosaccharides isolated as partial hydrolysis products are reducing substances and will be considered accordingly. Nevertheless, it should be pointed out that O-glycosyl derivatives of glycerol, tetritols, and higher alditols can be formed from polysaccharides by other reaction sequences such as the Smith degradation (Section VII,B) and after reduction of pentodialdoses formed from the selective cleavage of glycosiduronic acid derivatives (Section VIII,B). For reducing disaccharides, reducing and nonreducing residues can be distinguished by oxidation with bromine to aldonic acids (or their lactones) or by reduction with sodium borohydride (or borodeuteride), followed in either case by hydrolysis. The sole reducing sugar arises from the nonreducing unit, and the aldonic acid or alditol from the reducing unit. The disaccharide alditol, preferably with incorporation of deuterium from sodium borodeuteride to differentiate C-1 from the other terminus, is the derivative of choice for other purposes. The disaccharide alditols contain a single anomeric center for which ^{1}H- and ^{13}C-NMR spectroscopy provide evidence for configurational assignments. Configurational assignments at anomeric centers may also be made from optical rotational data (106), if adequate quantities are available, or from the results of hydrolysis with specific glycosidases (107). Linkage types may be determined by methylation analysis of the deuterium-labeled disaccharide alditols. Alternatively, mass spectrometry of the intact permethylated derivatives, from direct insertion or through GLC–mass spectrometry, gives information on linkage types but without differentiation between stereoisomeric sugar units. Structure determinations based on glycol cleavage with periodate or lead tetraacetate can be used only if adequate quantities of material are available, and unambiguous conclusions are rarely possible. Nevertheless, it should be pointed out that controlled degradation by oxidation of reducing residues with lead tetraacetate followed by reduction with sodium borohydride to O-glycosylglycerols gives valuable data in the correlation of anomeric configurations in disaccharides (108,109).

For trisaccharides and higher oligosaccharides evidence for sequences of sugar residues (or of different linkage types in homooligosaccharides) can be obtained by chromatography (on paper or on thin-layer chromatographic plates) of the simpler oligosaccharides formed from the parent oligosaccharides and their nonreducing derivatives (e.g., alditols) on partial depolymerization. Depending on the susceptibilities of differently substituted reducing groups to degradation by dilute alkali, alkaline degradation may proceed in a stepwise manner from the reducing terminus (see Section VIII,A,1). Figure 8 summarizes these approaches to chemical degradations. Mass spectral fragmentations of oligosaccharide derivatives (Section V,G) may also give sequential information if constituent units provide different molecular weight increments (85).

1. Composition: Hydrolysis to A, B, C, and D; hydrolysis of oligosaccharide alditol to A, B, C, and D* (alditol from reducing terminus)
2. Ring size and linkage types: Hydrolysis of permethylated derivative (best performed on oligosaccharide alditol-1-*d* from reduction with sodium borodeuteride)
3. Sequences of sugar residues: (a) Partial hydrolysis of oligosaccharide and derived oligosaccharide alditol (assuming similar susceptibilities of each glycosidic linkage to hydrolysis) ⟶ A-B + C-D*
   $$A\text{-}B\text{-}C + D^*$$
   $$A + B\text{-}C\text{-}D^*$$
   $$A + B^*C + D^*$$
   (b) Alkaline degradation from reducing terminus (at room temperature reaction proceeds in a stepwise manner until a 2-O-substituted reducing group is encountered) ⟶ A-B-C ⟶ A-B ⟶ A

Fig. 8. Summary of chemical procedures for oligosaccharide structure determination.

B. Partial Acid Hydrolysis

Differences in rates of hydrolysis of glycosidic linkages among sugars of the same general type and ring size are usually small, although α-D-glycopyranosidic linkages are generally more easily hydrolyzed than β-D-glycosidic linkages, and bonds of the (1 → 6) type are usually more resistant to hydrolysis than others. Furanosides, however, are hydrolyzed faster than the corresponding pyranosides by factors of 10–10^3. When, as is frequently the case for L-arabinofuranosides and D-galactofuranosides, residues of this type occupy terminal or near-terminal positions in polysaccharide structures, they can be removed selectively under mild conditions of hydrolysis with limited disruption of the residual polysaccharide structure. Deoxy sugars, especially when the deoxygenated carbon atoms are in the ring, are also readily liberated from terminal sites even when in pyranosidic linkage.

Glycosiduronic acids are much more resistant to hydrolysis at low pH than the corresponding neutral glycosides. Graded hydrolysis of glycuronic acid-containing polysaccharides conveniently leads to the isolation of acidic disaccharides (aldobiouronic acids) and higher oligosaccharides. The acid stability of aldobiouronic acids, which is not clearly understood (*110*), decreases at higher pH to the extent that the innermost glucopyranosiduronic acid linkage (**32**) in the heparin–protein "linkage region" is sufficiently easily hydrolyzed at pH 3 (Fig. 9) to permit the isolation of the oligosaccharide **35** in addition to **33** and **34** (*111*). However, at yet higher pH values (4–5) hydrolysis of glycosiduronic acids is accompanied by considerable epimerization and degradation (*112*).

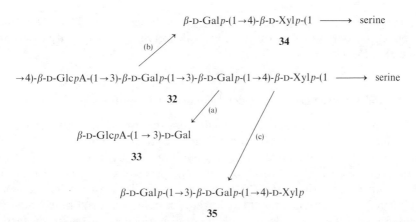

β-D-Gal$p$-(1→4)-β-D-Xyl$p$-(1 ⟶ serine

34

→4)-β-D-Glc$p$A-(1→3)-β-D-Gal$p$-(1→3)-β-D-Gal$p$-(1→4)-β-D-Xyl$p$-(1 ⟶ serine

32

(a)

β-D-Glc$p$A-(1 → 3)-D-Gal

33

(c)

β-D-Gal$p$-(1→3)-β-D-Gal$p$-(1→4)-D-Xyl$p$

35

Fig. 9. Some structurally significant oligosaccharides (**33–35**) arising from overlapping segments of the heparin–protein linkage region (**32**) from partial acid hydrolysis, (a) with 1 *N* hydrochloric acid, (b) at pH 1.55, and (c) at pH 3.

2-Amino-2-deoxyglycosides are very resistant to acid hydrolysis since the inductive effect of the already protonated 2-amino group results in reluctant further protonation of the glycosidic oxygen (*113*). In the majority of native polysaccharides amino sugars occur as their *N*-acetyl derivatives. If acid hydrolysis of the acetamido group is slow relative to glycoside hydrolysis or if the group is maintained by using acetolysis, marked resistance to hydrolysis is not observed. However, such polysaccharides can be N-deacetylated by hydrazinolysis (*114*) or by treatment with sodium hydroxide in DMSO (*115*). The resulting 2-amino-2-deoxyglycosidic linkages are then almost completely resistant to acid hydrolysis, and aminodisaccharides [*O*-(2-amino-2-deoxyglycosyl)glycoses] can be readily isolated as products of partial hydrolysis. Thus, Fig. 10 shows that, whereas partial acid hydrolysis

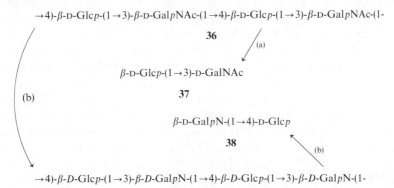

→4)-β-D-Glc$p$-(1→3)-β-D-Gal$p$NAc-(1→4)-β-D-Glc$p$-(1→3)-β-D-Gal$p$NAc-(1-

36

(a)

β-D-Glc$p$-(1→3)-D-GalNAc

37

(b)

β-D-Gal$p$N-(1→4)-D-Glc$p$

38

(b)

→4)-β-*D*-Glc$p$-(1→3)-β-*D*-Gal$p$N-(1→4)-β-*D*-Glc$p$-(1→3)-β-*D*-Gal$p$N-(1-

Fig. 10. Preferential formation of disaccharides **37** and **38** on partial acid hydrolysis of carboxyl-reduced chondroitin (**36**), (a) directly and (b) after N-deacetylation.

of carboxyl-reduced chondroitin (36) gives disaccharide 37 as the main product, partial acid hydrolysis after N-deacetylation gives only disaccharide 38 (*116,117*).

C. Partial Depolymerization under Nonaqueous Conditions

Two possible advantages may result from the use of nonaqueous conditions for partial depolymerization. Certain sugars, e.g., 3,6-anhydrohexoses, are readily destroyed under normal conditions of hydrolysis but may be "trapped" by derivative formation. In some procedures, such as acetolysis, the relative susceptibilities of different glycosidic linkages to cleavage may be altered to an extent that the changed "cracking pattern" may lead to a significantly different collection of oligosaccharides as products of partial depolymerization. Acetolysis, specifically, is a convenient procedure when the parent polysaccharide, e.g., cellulose, is insoluble in most aqueous systems but the acetylated derivative is readily soluble in acetic anhydride–acetic acid mixtures.

1. Acetolysis

Partial acetolysis involves treatment of a polysaccharide, or preferably its acetylated derivative since acetylation occurs anyway under acid-catalyzed conditions, with acetic anhydride–sulfuric acid mixtures, frequently with added acetic acid. The resulting mixtures of sugars are then de-O-acetylated with sodium or barium methoxide to give mixtures of mono- and oligosaccharides. The relative rates of cleavage of different glycosidic bonds during acetolysis may be sufficiently different and even reversed from those observed during partial acid hydrolysis that alternative selections of possible oligosaccharides can be isolated. Four examples may be cited. In contrast to partial hydrolysis in which (1 → 6) linkages between hexose residues are more resistant than other linkage types, these linkages are preferentially split during acetolysis. Accordingly, acetolysis has been extensively utilized to characterize the profile of side chains (40–42) attached to a (1 → 6)-linked core in mannans (39) from yeasts and fungi (Fig. 11) (*118*). Discrimination can be made between other interhexose linkage types less frequently, but desulfated λ-carrageenan has been shown to undergo preferential acetolysis of 3-O-substituted β-D-galactopyranosyl (1 → 4) bonds, whereas the opposite preference is shown during partial acid hydrolysis for 4-O-substituted α-D-galactopyranosyl (1 → 3) bonds (*119*). 6-Deoxyhexopyranosyl bonds, especially those of L-fucose (*120*) and L-rhamnose (*121*) in terminal positions, are rather readily split during partial acid hydrolysis but are sufficiently stable to acetolysis to be isolated as oligosaccharide constituents. Whereas the

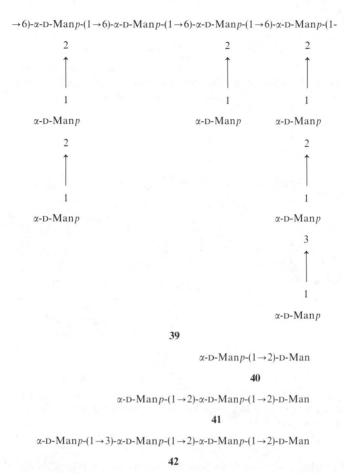

→6)-α-D-Manp-(1→6)-α-D-Manp-(1→6)-α-D-Manp-(1→6)-α-D-Manp-(1-

2   2   2

↑   ↑   ↑

1   1   1

α-D-Manp   α-D-Manp   α-D-Manp

2   2

↑   ↑

1   1

α-D-Manp   α-D-Manp

3

↑

1

α-D-Manp

39

α-D-Manp-(1→2)-D-Man

40

α-D-Manp-(1→2)-α-D-Manp-(1→2)-D-Man

41

α-D-Manp-(1→3)-α-D-Manp-(1→2)-α-D-Manp-(1→2)-D-Man

42

Fig. 11. Selective cleavage by acetolysis of (1→6) bonds in yeast mannans (**39**) with liberation of oligosaccharides **40**–**42**.

L-fucopyranosyl linkages in tragacanthic acid (**43** in Fig. 12) are most readily cleaved during partial acid hydrolysis giving a fucose-free degraded tragacanthic acid (**44**), acetolysis permits the isolation of the disaccharide **45** with an intact nonreducing fucopyranose end group (*120*). Even more surprisingly, sialic acid (neuraminic acid) linkages, which are extremely labile to aqueous acid often with decomposition of the liberated sugar, may remain intact during acetolysis and thus permit the isolation of sialic acid-containing oligosaccharides (*122*). Similar oligosaccharides have been isolated from brain ganglioside (*123*) and from glycopeptides formed from human serotransferrin (*124*).

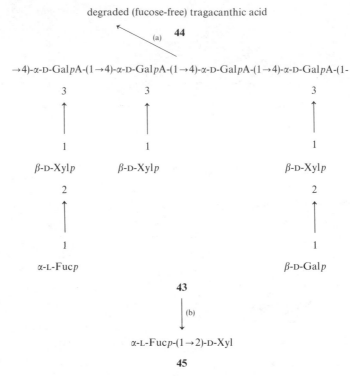

degraded (fucose-free) tragacanthic acid

→4)-α-D-GalpA-(1→4)-α-D-GalpA-(1→4)-α-D-GalpA-(1→4)-α-D-GalpA-(1-

Fig. 12.   Controlled partial acid hydrolysis (a) selectively cleaves α-L-fucopyranosyl linkages in tragacanthic acid (**43**) with formation of degraded polysaccharide **44**, but acetolysis (b) permits the isolation of disaccharide **45**.

Anomerization may occur during acetolysis and is probably most important for (1 → 6) linkages, where retardation from electronegative substituents is least felt (*125*). Since the thermodynamically more stable α-anomers are favored, the formation of oligosaccharides that are configurationally unrelated to the parent polysaccharide is more likely to be observed for polysaccharides containing a predominance of β linkages. Significant anomerization has been observed during the acetolysis of a gum containing (1 → 6)-linked β-D-hexopyranose residues (*126*).

2. Trifluoroacetolysis

The potential applications of trifluoroacetolysis using mixtures of trifluoroacetic anhydride and trifluoroacetic acid have been explored primarily for glycoconjugates and *N*-acetylhexosamine-containing oligosaccharides (*127–129*). To a much greater extent than in acetolysis, the first effect of

α-D-Manp-(1→3)-β-D-Manp-(1→4)-D-GlcpNAc

46

α-D-Manp-(1→3)-D-Man

47

Fig. 13.   Degradation of oligosaccharides with reducing 2-acetamido-2-deoxyhexose residues during trifluoroacetolysis to give the next lower oligosaccharide as the trifluoroacetylated derivative, e.g., **46** → **47**.

trifluoroacetylation is to stabilize the glycosidic linkages of O-trifluoroacetyl sugars (*130*). The following examples are of reactions showing some selectivity. Whereas O-trifluoroacetylation takes place at room temperature, at higher temperatures transamidation occurs with N-acetylated 2-amino-2-deoxyglycosides with the formation of N-trifluoroacetyl derivatives (*131*). Since N-detrifluoroacetylation can be achieved with either methanolic ammonia or sodium borohydride in aqueous ethanol, the sequence of reactions can be used to effect N-deacetylation with only limited cleavage of glycosidic linkages. However, when oligosaccharides containing 3- or 4-linked reducing 2-acetamido-2-deoxy sugar residues are treated in this way, degradation by elimination may take place with loss of the reducing residue and exposure of the next sugar unit in sequence (e.g., **46** → **47** in Fig. 13)(*129*). N-Glycosidically linked oligosaccharide chains in glycoproteins may be liberated by trifluoroacetolysis, but extended reaction will result in further degradation (as above) from the reducing N-acetyl-D-glucosamine terminus (*127*).

3. Mercaptolysis and Methanolysis

A few types of sugar residues undergo decomposition during hydrolysis but may be protected by derivative formation during acid-catalyzed depolymerization in a nonaqueous medium. The use of such an approach to the isolation of oligosaccharides containing such acid-sensitive residues is clearly feasible only if the units occupy internal positions. Algal polysaccharides such as agarose with 3,6-anhydro-L-galactose residues, and κ- and alkali-modified κ-carrageenan with 3,6-anhydro-D-galactose residues (see Chapter 4, Volume II), furnish high yields of the disaccharide repeating units with correspondingly little cleavage of the more stable glycosidic linkages when mercaptolysis or methanolysis is performed under controlled conditions. Mercaptolysis, carried out under partially aqueous conditions, affords disaccharides (e.g., **48**) as their acyclic dialkyldithioacetals (*132,133*). Methanolysis also results in the formation of acyclic derivatives, namely, dimethylacetals, as the major products together with some methyl pyranosides (e.g., **49**) (methyl 3,6-anhydrogalactofuranoside formation is sterically impossible)(*84*).

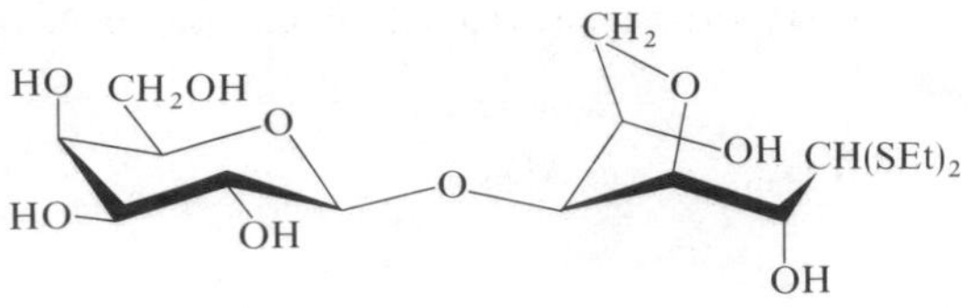

CH$_2$
HO   CH$_2$OH
O
OH   CH(SEt)$_2$
HO—
OH
OH

48

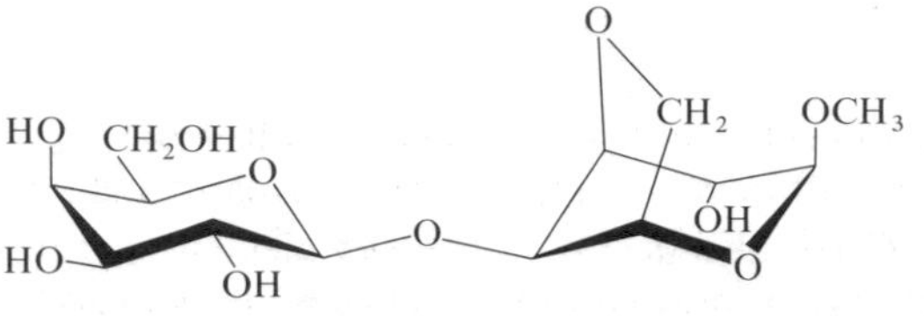

O
HO   CH$_2$OH
O
CH$_2$ OCH$_3$
OH
HO—
OH
O

49

CH$_3$
HO$_2$C
O
O
O
O
CH$_2$
CH(OCH$_3$)$_2$
OH
HO—
OH
O
H

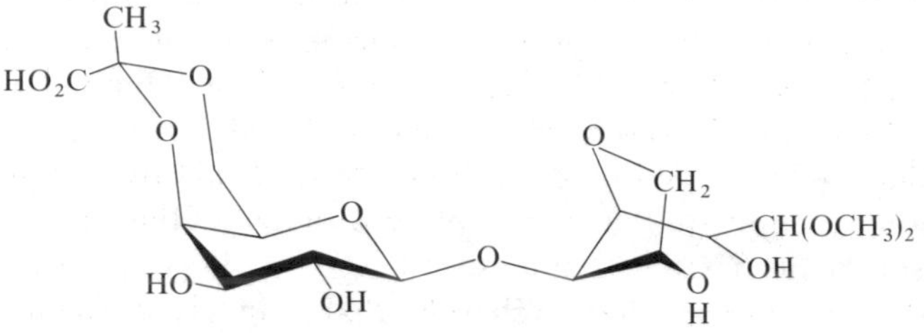

50

Methanolysis conditions, in the case of agarose, are sufficiently mild to permit the isolation of agarobiose dimethylacetal as a cyclic ketal of pyruvic acid (**50**) (*134*).

D. Oxidative Hydrolysis

Oxidative hydrolysis in the presence of bromine may be used to give stable oxidation products, e.g., aldaric acids from liberated uronic acids. A particularly valuable application of the technique is in the oxidative hydrolysis of carrageenans to give good yields of oligosaccharide sulfates, with stabilization of the labile 3,6-anhydro-D-galactose units and retention of sulfate ester substituents (*135*).

Equatorially oriented acetylated glycopyranosides undergo oxidative cleavage with chromium trioxide in acetic acid with a high degree of stereo-selectivity. The axially oriented acetylated glycosides do not react. The reaction can be used to obtain information on the anomeric configuration of glycosidic linkages in acetylated polysaccharides by carrying out sugar analyses before and after treatment with chromium trioxide. The validity of this procedure is dependent on the conformational stability of the commonly occurring aldopyranosides and may be unreliable for conformationally flexible sugars, e.g., glycosides of 3,6-dideoxy-D-*arabino*-hexose (*136*).

E. Enzyme-Catalyzed Hydrolysis

A detailed account of enzymatic methods in polysaccharide chemistry is given in Chapter 1, Volume III. This section therefore includes only a few comments on the use of enzymes in the context of partial depolymerization. All glycan hydrolases (or glycanases) and glycosidases are specific for the sugar unit undergoing hydrolysis and for its anomeric configuration. Exoglycosidases, whether used individually or generated in a stepwise manner by sequential induction, act on polysaccharide (or other glycoconjugate) substrates by the removal of nonreducing terminal units, usually without regard to the type of linkage to the next innermost residues. Nevertheless, it is important for one to ensure that complete removal of terminal residues of a given type has occurred before drawing conclusions or proceeding to further enzyme degradation (*170*). Exoglycanases likewise act from nonreducing termini but are highly specific as to linkage type and may give rise to products of greater complexity than simple monosaccharides. Examples are provided by phosphorylase and β-amylase, which act on amylose and the outer chains of amylopectin and glycogen with the liberation of α-D-glucopyranosyl phosphate and maltose, respectively. The specificities of these enzymes require, furthermore, that the substrates contain chains of monotonously linked units such that the enzymes can approach only to within a limited distance of other structural features, e.g., a different linkage type and/or a branch point. In contrast, endo enzymes are not limited by action pattern and thus can cleave unbranched regions of both internal and external chains of the requisite type, but they are similarly restricted as to proximity of approach to branch points or other different structural features.

Two examples of such endoglycanases may be cited. The cellulase from *Streptomyces* sp. QM B814 cleaves the glycosidic bond between two unbranched 4-linked β-D-glucopyranose residues. Its action on the linear mixed-linkage β-D-glucans (**51**) from oats and barley (*137*) is illustrated in Fig. 14. The xylanase from a similar source has an analogous specificity with respect to unbranched 4-linked β-D-xylopyranose residues, as shown (Fig. 15) by its action on wheat flour arabinoxylan (**52**) (*138*).

The last example illustrates one of the greatest values of enzymatic hydrolysis in the determination of polysaccharide structures, namely, the capacity to cleave specifically more acid-resistant glycosidic linkages in polysaccharides containing otherwise acid-sensitive linkages, e.g., L-arabinofuranosyl linkages in the example cited. A similar example is encountered in the action

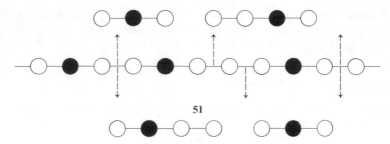

Fig. 14. Oligosaccharides formed by the action of cellulase from *Streptomyces* sp. OM B814 on mixed-linkage cereal β-D-glucan (**51**). Key: ○, 4-linked β-D-glucopyranose residue; ●, 3-linked β-D-glucopyranose residue.

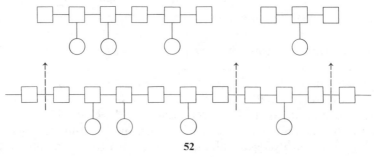

Fig. 15. Action of xylanase on wheat flour arabinoxylan (**52**). Key: □, 4-linked β-D-xylo-pyranose residue; ○, α-L-arabinofuranose residue linked (1 → 3), or (1 → 2), to β-D-xylopyranose residue.

of a β-D-galactosidase on agarose with the liberation of neoagarobiose, neoagarotetraose (**53**), and higher oligosaccharides containing the extremely acid-sensitive 3,6-anhydro-L-galactopyranose residues (*139*).

53

Highly specific endoglycanases are derived from bacteriophages (*140*). These enzymes act on capsular polysaccharides, giving oligosaccharides that represent the repeating unit of the polysaccharide, although not neces-

α-D-Glcp-(1→4)-β-D-GlcpA-(1→2)-α-L-Rhap-(1→3)-α-D-Glcp-(1→3)-α-L-Rhap-(1→3)-D-Gal

54

→3)-α-L-Rhap-(1→3)-β-D-Galp-(1→4)-α-D-Glcp-(1-

3

↑

1

α-D-Glcp-(1→4)-β-D-GlcpA-(1→2)-α-L-Rhap

55

Fig. 16. Linear hexasaccharide (**54**) formed from the branched *Klebsiella* type 18 capsular polysaccharide (**55**) by the action of a bacteriophage endo-β-D-galactosidase.

sarily the biosynthetic repeating unit, which is assembled from a lipid-bound intermediate (*141,142*). An example is provided in Fig. 16 by the formation of the linear oligosaccharide (**54**) from the branched capsular polysaccharide (**55**) of *Klebsiella* type 18 (*142*).

F. Partial Depolymerization of Permethylated Polysaccharides

In principle, the partial hydrolysis of permethylated polysaccharides is a very valuable approach to polysaccharide structure, but, until recently, partial hydrolyses of methylated polysaccharides were carried out in only two situations: to isolate partially methylated acid oligosaccharides (usually only aldobiouronic acids) and to remove acid-sensitive terminal residues and then identify sites of former attachment after realkylation of exposed hydroxyl groups. Now, realkylation with trideuteriomethyl iodide or ethyl iodide coupled with the use of GLC–mass spectrometry to identify sugar derivatives after hydrolysis greatly increases the scope of this approach.

A major obstacle to the more widespread use of partial hydrolysis of methylated polysaccharides has been the absence of satisfactory methods for separating mixtures of partially methylated oligosaccharides other than those containing uronic acid residues. Permethylated disaccharide alditols and some corresponding trisaccharide derivatives can be separated and analyzed by GLC–mass spectrometry, but this is probably the upper limit of molecular size for this technique. Thus, partial hydrolysis of the methylated derivative of *Klebsiella* type 38 capsular polysaccharide (**56** in Fig. 17) followed by reduction with sodium borodeuteride and further trideuterio-methylation gives the fully alkylated disaccharide alditols **57** and **58** (*143*). Recently, a general solution to the separation problem has been developed by Valent *et al.* (*144*) with fractionations by HPLC on reversed-phase

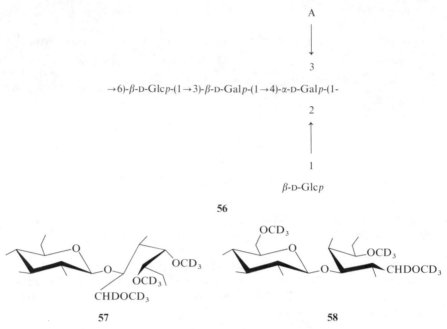

$$A$$
$$\downarrow$$
$$3$$

$\rightarrow6)$-β-D-Glc$p$-$(1\rightarrow3)$-β-D-Gal$p$-$(1\rightarrow4)$-α-D-Gal$p$-$(1$-

$$2$$
$$\uparrow$$
$$1$$

β-D-Glc$p$

56

57   **58**

Fig. 17. Methylated disaccharide alditols **57** and **58** formed by partial acid hydrolysis of the permethylated derivative of *Klebsiella* type 38 capsular polysaccharide (**56**), followed by reduction with sodium borodeuteride and trideuteriomethylation (A, 3-deoxy-L-*glycero*-pentulosonic acid). Here and elsewhere, unless otherwise stated, undesignated substituents are OCH_3.

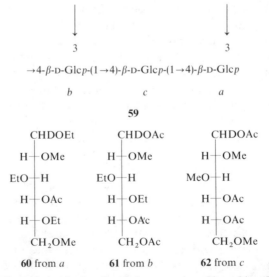

$$3 \qquad\qquad\qquad 3$$

$\rightarrow4$-β-D-Glc$p$-$(1\rightarrow4)$-β-D-Glc$p$-$(1\rightarrow4)$-β-D-Glc$p$

$$b \qquad\qquad c \qquad\qquad a$$

59

CHDOEt	CHDOAc	CHDOAc
H——OMe	H——OMe	H——OMe
EtO——H	EtO——H	MeO——H
H——OAc	H——OEt	H——OAc
H——OEt	H——OAc	H——OAc
CH_2OMe	CH_2OAc	CH_2OMe
60 from *a*	**61** from *b*	**62** from *c*

Fig. 18. Partially methylated trisaccharide from a region of branching (**59**) is formed *inter alia* on partial acid hydrolysis of methylated xanthan gum. Further treatment with (1) $NaBD_4$, (2) EtI, NaH, DMSO, (3) H_3O^+, (4) $NaBD_4$, and (5) Ac_2O/Pyr affords partially alkylated alditol acetates **60**–**62** from residues *a*, *b*, and *c*, respectively. Arrows indicate positions at which glycosyl substituents were formerly attached.

columns limited only by chromatographic resolution rather than molecular size. The following sequence of operations is employed: (a) partial hydrolysis of the methylated polysaccharide; (b) reduction with sodium borohydride; (c) ethylation; (d) HPLC separation of the mixed alkylated oligosaccharide alditols; and (e) alkylated sugar analysis for each separated derivative by GLC–mass spectrometry of the products after hydrolysis, reduction, and acetylation. An example of one such trisaccharide fragment (59) from the partial hydrolysis of methylated lichenan and of the partially alkylated alditol acetates (60–62) formed from it is shown in Fig. 18.

Implicit in the example just cited is the principle employed in a general procedure developed in the same laboratory for the unambiguous determination of ring size (145). The characterization of methylated sugars such as 2,3,6-tri-O-methyl-D-glucose does not indicate whether the parent sugar residue was a 4-linked D-glucopyranose residue or a 5-linked D-glucofuranose residue. Random partial hydrolysis of a methylated polysaccharide is followed by reduction with sodium borodeuteride, O-ethylation, total hydrolysis, and GLC-mass spectrometry analysis of the derived partially alkylated alditol acetates. Even though there may be substantial differences in rates of hydrolysis of different linkages, the ring sizes of all sugars, not already established by methylation, will be indicated by the formation of derivatives containing ethoxyl groups, *either* at C-4 or C-5 *and/or* at both C-1 and C-4 or both C-1 and C-5. Figure 19 shows the possible fates of residues of 2,3-di-O-methyl-L-arabinose in a methylated polysaccharide, partial hydrolysis of which may give oligosaccharide fragments with residues of this sugar at the reducing terminus (63 or 64) and/or at the nonreducing terminus (65 or 66). From the sequence of reactions cited above these residues will give, respectively, the partially ethylated methylated alditol acetates 67 or 68, or 60 or 70. For another approach to the determination of ring size see Ref. 145a.

G. Applications of Mass Spectrometry to the Characterization of Oligosaccharide Derivatives

Mass spectrometry is a major tool in the characterization of oligosaccharide derivatives. Several standard works deal with the technical aspects of mass spectrometry as applied to various classes of natural products (146,147), and specific applications to carbohydrate derivatives have been reviewed in several publications (85,148–150). This section deals almost exclusively with permethylated derivatives since these compounds play such a major role in various aspects of structure determination of complex carbohydrates. It should be pointed out, however, that rather similar principles are followed in the interpretation of the mass spectra of other carbohydrate derivatives, such as TMS ethers and acetate esters.

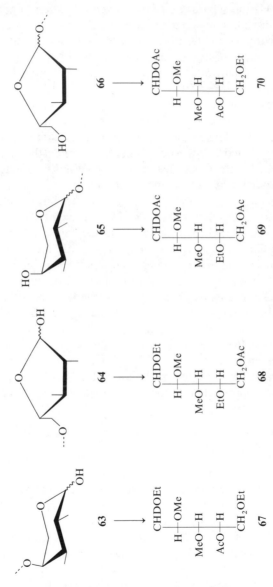

Fig. 19. Determination of ring size of 4- or 5-linked methylated sugars from their isolation as reducing (**63** or **64**) or nonreducing (**65** or **66**) residues in oligosaccharides from partial acid hydrolysis of permethylated polysaccharides. Example: 2,3-di-*O*-methyl-L-arabinose residues treated with (1) NaBD₄, (2) EtI, NaH, DMSO, (3) H₃O⁺, (4) NaBD₄, and (5) Ac₂O/Pyr.

Mass spectrometry as commonly practiced uses electron impact most frequently as the ionization mode. Carbohydrate derivatives rarely give molecular ions in electron impact spectra, although molecular weights may often be inferred from various fragment ions. Detailed structural information is best obtained from electron impact spectra. Softer methods of ionization, including chemical ionization using reagent gases such as isobutane (151,152), field ionization, and field desorption (153), may give molecular or quasi-molecular ions and in general give much more abundant ions in the high-mass range (500–2000), but their simpler spectra provide less information on structure. These methods will become of greater importance with the development of chromatographic methods for the fractionation of higher oligosaccharides. At present, oligosaccharide derivatives of molecular weights of up to 1000 are convenient for mass spectral analysis, and derivatives of molecular weights of up to 2000 have been reported for permethylated glycoconjugates, such as gangliosides (154) and oligosaccharide-containing glycopeptides from extensive proteolytic digestion of glycoproteins (155).

Recently Ballou and collaborators fractionated a polymer-homologous series of 3-O-methylmannose polysaccharides (46) from *Mycobacterium smegmatis* into discrete compounds by HPLC (156) and then examined the field desorption mass spectra of the unsubstituted compounds (157). Quasi-molecular ions were observed up to $m/e \sim 2500$ for a tetradecasaccharide. The appearance of fragment ions from the cleavage of successive glycosyl bonds provides the basis for complete sequence determination up to the decasaccharide level.

In terms of thermal stability and ease of interpretation of spectra, per-methylated compounds and TMS ethers are the derivatives of choice for mass spectrometry. At present, the separation and characterization of volatile derivatives by combined GLC–mass spectrometry are widely used for disaccharides and some trisaccharides. However, the resolution of mixtures in the trisaccharide range is not always very good. In general, higher oligo-saccharides are separated as the parent compounds, and individually pre-pared derivatives are introduced into the mass spectrometer by direct insertion. With the increasing potential of HPLC for the separation of oligosaccharide derivatives and the availability of commercial instruments for combined HPLC–mass spectrometry, mass spectral analysis of mixtures of higher molecular weight carbohydrates will be extended.

The interpretation of the mass spectra of oligosaccharide derivatives follows the general principles that have been elaborated for simple cyclic and acyclic carbohydrates (85,148). These principles may be illustrated with reference to permethylated oligosaccharide alditols since these are often the most convenient derivatives to prepare from reducing oligosaccharides isolated as partial hydrolysis products. As in the case of partially methylated

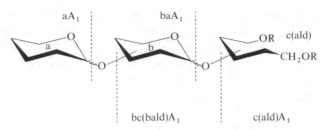

Fig. 20. Mass spectral nomenclature for fragment ions from oligosaccharide derivatives.

alditol acetates, reduction is best performed with sodium borodeuteride in order to avoid ambiguities arising from otherwise structurally symmetric terminal units. Spectra may be analyzed by considering, in turn, fragment ions derived from reducing and nonreducing terminal units and then fragment ions arising from internal units, often by more complex fragmentation pathways.

A nomenclature scheme for oligosaccharide derivatives was developed by Chishov and Kochetkov (*148*) and later modified by Kováčik *et al.* (*158,159*). In this scheme for unbranched oligosaccharides, sugar residues are designated by lowercase letters starting from the nonreducing end group (a) along to the unit derived from the reducing terminus—in the example shown in Fig. 20, the alditol residue (c). In this example fragment ions are generated by a variety of bond cleavage processes, and the various pathways are denoted by uppercase letters (A, B, etc.). Additional numerical subscripts indicate successive ions along a given pathway. Some examples are shown below and incorporate the further conventions that the first lowercase letter indicates the ring unit that has undergone cleavage and that subsequent lowercase letters denote unaltered residues that remain attached as substituents.

Methylated alditol units in oligosaccharides are cleaved in a manner similar to that described previously for partially methylated alditol acetates. On the other hand, substituted cyclic carbohydrates undergo a variety of types of fragmentation, the most important of which are summarized in Fig. 21 for simple permethylated glycosides. Uppercase letters are assigned to different fragmentation pathways in which the original or rearranged molecular ion undergoes cleavage with loss of a neutral fragment.

The most easily recognizable fragment ions in the mass spectra of permethylated oligosaccharide alditols are those from nonreducing units, i.e., the A_1 ions at m/e 219 for permethylated hexose, m/e 189 for permethylated deoxyhexose, m/e 175 for permethylated pentose, m/e 233 for permethylated hexuronic ester, and m/e 260 for permethylated *O*- and *N*-2-acetamido-2-deoxyhexose residues. Likewise, from 1-*d*-labeled alditol residues from reducing termini, the following fragment ions are readily recognized: m/e

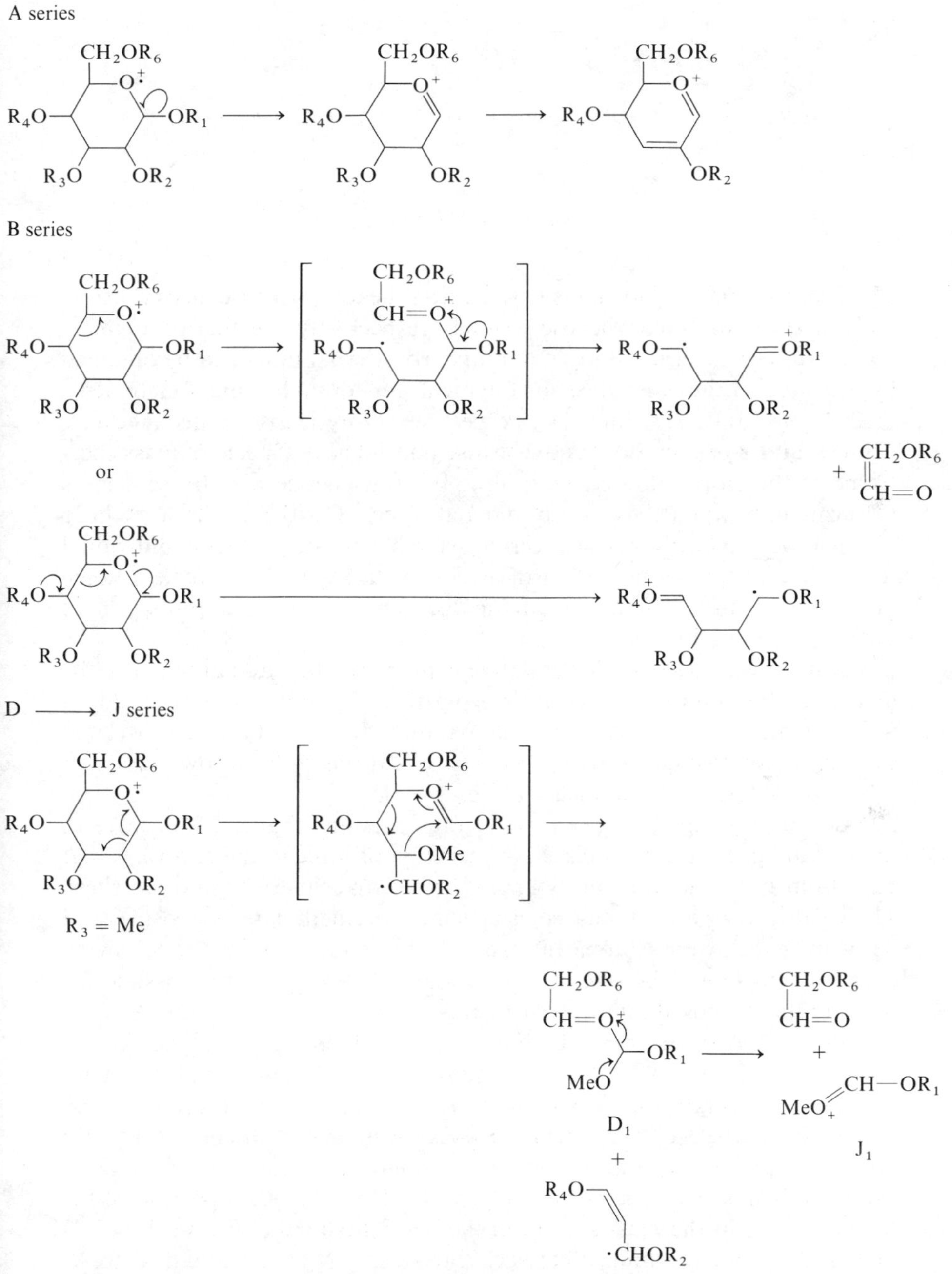

A series

B series

or

D ⟶ J series

$R_3 = Me$

D_1

J_1

+

Fig. 21. Some mass spectral fragmentation pathways for typical, e.g., permethylated, glycopyranosides.

E series

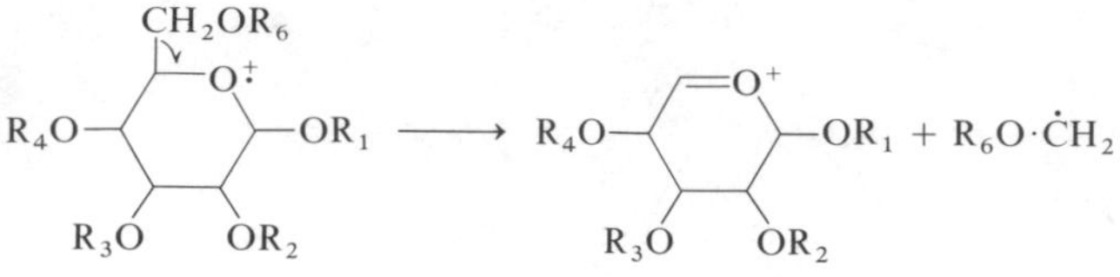

Fig. 21. (*continued*)

236, 206, 192, and 277 for units derived from hexose, pentose, deoxyhexose, and 2-acetamido-2-deoxyhexose residues, respectively. Furthermore, information on the general nature of unbranched internal units can be obtained by detecting in the m/e values for fragment ions of higher mass increments of 204, 174, 160, 218, and 245, respectively, for hexose, 6-deoxyhexose, pentose, and 2-acetamido-2-deoxyhexose units. These aspects of mass spectrometry, therefore, allow conclusions to be drawn concerning the sequences of sugar units when these are of different types. On the whole, very little information can be obtained concerning stereochemistry, either of individual residues or of the configuration of glycosidic linkages. Nevertheless, some information can be obtained on linkage types from an examination of fragmentation pathways.

When the oligosaccharide derivative is the permethylated alditol-1-d, the linkage to the alditol is determined from the fragmentation of the alditol unit, in which scission takes place between methoxylated and glycosylated carbon atoms. Primary fragmentation gives cations of both low mass, e.g., m/e 45, 46, 89, and 90, and high mass, e.g., M-45.

Assessing the significance of those fragment ions that are indicative of internal linkage types requires a consideration of some of the pathways that arise from bond cleavage in or immediately adjacent to pyranoside rings. These pathways were established using simple permethylated glycosides and derivatives that were isotopically labeled at specific sites. In the following discussion it should be noted that alternative structures are possible for many of the proposed ionic intermediate species.

Ions of the A series are oxocarbonium ions, which arise from cleavage of the glycosyl (C-1)–oxygen bond. These A_1 ions undergo a number of consecutive fragmentations, but, among the possible A_2 ions, that resulting from elimination of the 3-O substituent is usually the most prominent. The relative abundances of the A_1 and A_2 ions therefore give a strong indication of the presence or absence of a 3-linked glycosyl substituent. Comparison of the fragment ions in the mass spectra of the permethylated alditols (**71** and **72** in Fig. 22) from the milk oligosaccharides lacto-N-tetraose and lacto-N-neotetraose (*160*) shows that the relative abundances of baA_1 ions at m/e

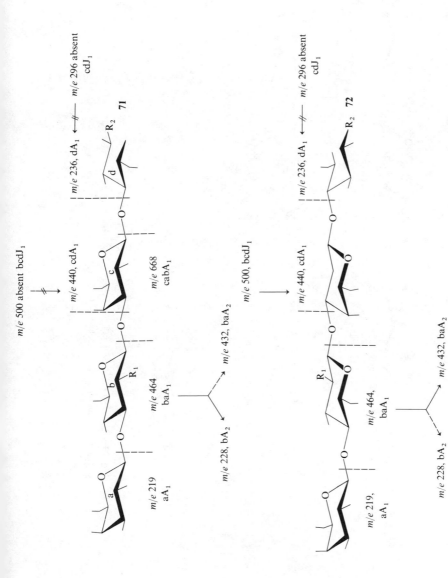

Fig. 22. Some useful features in the mass spectral fragmentations of permethylated alditols-1-d (**71** and **72**) from lacto-N-tetraose and lacto-N-neotetraose.

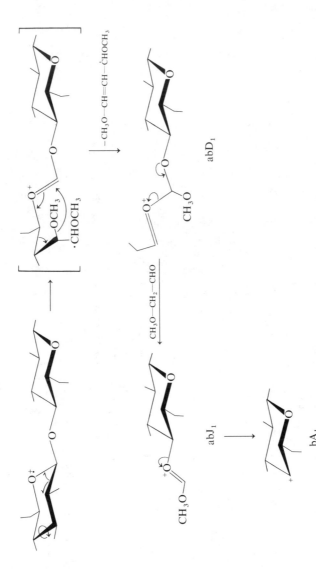

Fig. 23. Fragment ions of the D and J series in the mass spectra of permethylated oligosaccharide derivatives (shown for a permethylated disaccharide glycoside).

464 and the most prominent baA_2 ions at m/e 228 and 432, respectively, are strongly indicative of the nature of the 3-linked substituent.

Ions of the D series arise from scission of both the C-1—C-2 and C-4—C-5 bonds. The D_1 ion is of diagnostic significance in the fragmentation of 6-linked disaccharide glycosides. The D_1 ion is further degraded with formation of the J_1 ion. The formation of the J_1 ion is of particular significance in the mass spectra of oligosaccharide derivatives (Fig. 23). The formation of the D_1 ion involves the migration of a methoxyl group from C-3 to C-1 of the ruptured pyranoside ring, and the derived J_1 ion then loses the remaining stub of the ruptured ring (of 60 amu) with liberation as an ion of the aglyconic substituent. This is a major pathway for the production of the latter ion. The absence of ions of high intensity at m/e values corresponding to J_1 ions provides strong evidence for the presence of a 3-linked glycosyl substituent in the ring undergoing cleavage. Again, comparison of the mass spectra of the permethylated alditols from lacto-N-tetraose and lacto-N-neotetraose is instructive since the former does not but the latter does show an intense ion at m/e 500 ($bcdJ_1$) as a precursor of the cdA_1 ion at m/e 440.

Other structural situations in which mass spectral fragmentations are of value include the differentiation of 2- and 4-linked residues in methylated di- and trisaccharides (*149*) and the recognition of branched oligosaccharide alditols in which the alditol unit carried two glycosyl substituents (*161*).

VI. Oxidations with Periodate and Lead Tetraacetate

A. Analytical Aspects

Glycol cleavage via oxidation of water-soluble carbohydrates by sodium metaperiodate is a widely used analytical procedure. For compounds of low molecular weight the essentially quantitative nature of the reaction is of considerable value with respect to the reduction of the oxidant to iodate in the course of oxidizing 1,2-diol and 1,2,3-triol groups (*162*) and the liberation of formic acid (usually from triol cleavage in pyranoside rings) (*162*) and, less frequently, of formaldehyde from exocyclic diol (—CHOH—CH$_2$OH) groups (*163*). These analyses can be performed simply by titrimetric or spectrophotometric methods to determine oxidant reduced, by acid–base titration to measure formic acid liberated (although the possibility of other titratable acid should not be ignored), and by various colorimetric methods for formaldehyde. In addition, the presence of sugar residues that are substituted in a manner that leaves no diol groups susceptible to oxidation may be ascertained by liberation of the sugar after hydrolysis. The most satisfactory procedure here is to reduce the periodate-oxidized carbohydrate

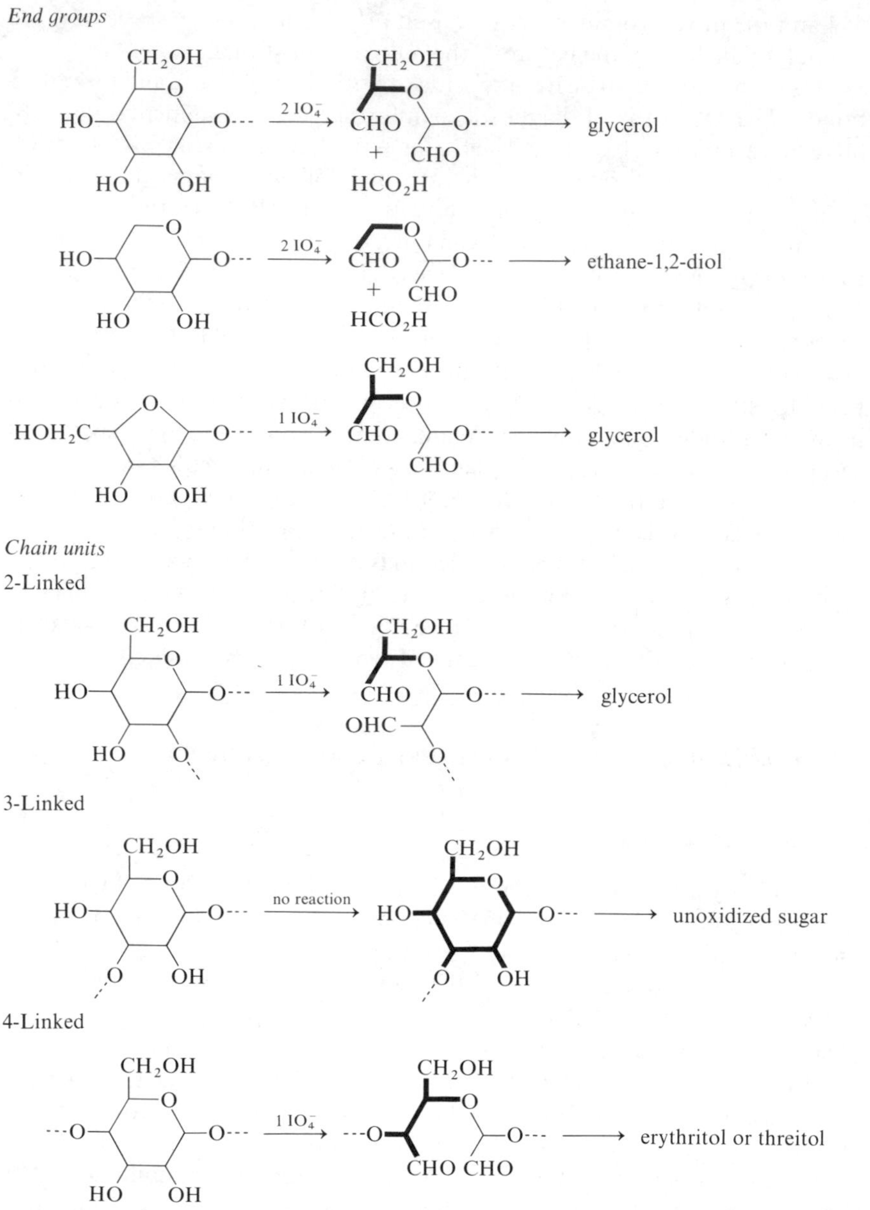

End groups

Chain units

2-Linked

3-Linked

4-Linked

Fig. 24. Reaction of variously linked sugar residues with periodate indicating, where appropriate, moles of oxidant consumed and one-carbon fragments liberated, and products formed from those portions of the original sugar units shown in heavy print on reduction with sodium borohydride followed by complete hydrolysis.

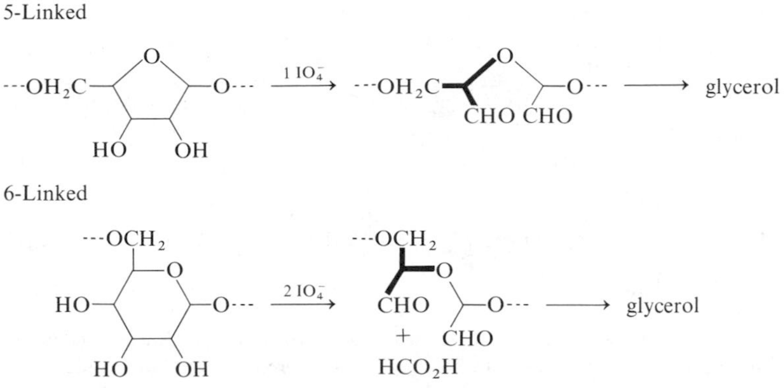

5-Linked

6-Linked

Fig. 24. (*continued*)

(as a "polyaldehyde") with sodium borohydride and then hydrolyze the resulting polyalcohol to give reducing sugar together with tetritols (erythritol or threitol), glycerol, or ethanediol from oxidized sugar rings (*164*). It will be evident from the products isolated that much stereochemical information is lost as a consequence of the oxidation, so that the origins of the latter fragments cannot be unambiguously assigned. The oxidations of some typical ring systems are shown in Fig. 24.

Formaldehyde is infrequently formed as an oxidation product from sugar residues in polysaccharides. Insofar as acyclic diols are oxidized more rapidly than cyclic diols, formaldehyde formation is often indicative of the presence of suitably substituted hexofuranose, heptopyranose, or higher-carbon sugar residues. By using a limited quantity of reagent to effect selective oxidation of such side-chain units followed by reduction with sodium borohydride, one can achieve polysaccharide modification with conversion of hexo-furanose and heptopyranose residues to the corresponding pentofuranose (*164*) and hexopyranose (*165*) residues (Fig. 25).

The extension of these simple analytical procedures to the reactions of polysaccharides gives a general indication of the nature of the linkages in otherwise unknown polysaccharides, but at best the results are rarely open to unambiguous interpretation without independent structural evidence. The purely analytical use of periodate oxidation is of greatest value in comparing quantitative aspects of polysaccharides of the same general type. However, recent studies on dextrans indicate apparent discrepancies between structural conclusions drawn from periodate and methylation linkage analyses (*166*)

Nonideal behavior of polysaccharides during periodate oxidation arises from both "overoxidation" and "underoxidation" (*109*). Overoxidation is most frequently encountered when oxidation gives rise to tartronic acid

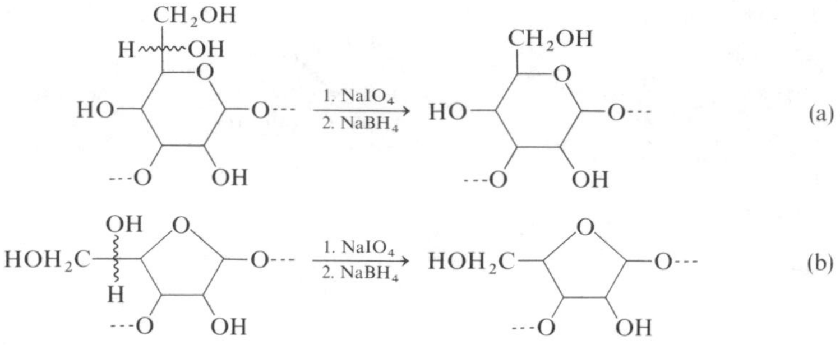

Fig. 25.   Selective degradation of side chains by periodate oxidation followed by reduction with formation of (a) hexopyranosides from heptopyranosides and (b) pentofuranosides from hexofuranosides. Even in the absence of ring substituents, e.g., at O-3 (as shown), the use of limited quantities of oxidant permits the preferential oxidation of exocyclic over endocyclic diol groups.

half-aldehyde derivatives, from hexuronic acid end groups, or to tartron-dialdehyde derivatives, for example, from hexofuranosides or heptopyranosides. Such nonspecific oxidations, which are minimized in the pH range 2.2–4, may lead through a succession of steps to the exposure of previously protected diol groups. Overoxidation may also result from the periodate oxidation of reducing groups (167) but is generally of limited importance for polysaccharides if reaction is carried out without great excess of reagent and at pH 3.6 so that the initially formed formyl ester is not hydrolyzed at a significant rate.

The incomplete oxidation of ostensibly vulnerable sugar residues has two main causes. Most commonly, hemiacetal formation between aldehyde fragments in oxidatively cleaved residues and hydroxyl groups in adjacent but not yet oxidized residues protects the latter units from oxidation. Painter and collaborators have examined the extent of incomplete oxidation in several homopolysaccharides of uniform linkage type (168–170) and have shown that the highest degree of incomplete oxidation occurs in 4-linked polysaccharides whose sugar residues do not carry primary hydroxyl groups at C-6 (171). The hemiacetal linkages are disrupted on reduction with sodium borohydride, and the protected but formerly vulnerable diol groups are reexposed for oxidation. Depending on the polysaccharide, the oxidation–reduction sequence may have to be repeated twice before full oxidation is achieved. The protection of neighboring residues by an oxidized unit in a D-mannuronic acid-rich segment of alginic acid (168) is illustrated in Fig. 26.

The second cause of protection against periodate oxidation arises from hydrogen bonding between one of a pair of hydroxyl groups, normally susceptible to oxidation, and a suitably disposed acetamido group on a

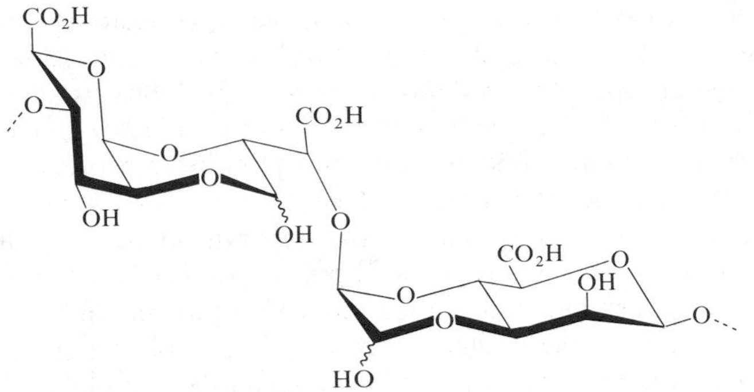

Fig. 26. Protection of potentially vulnerable diol groups from oxidation with periodate by interresidue hemiacetal formation with a neighboring oxidized residue in a D-mannuronic acid-rich segment of alginic acid.

neighboring sugar residue. This situation is encountered in certain glyco-saminoglycans (see Chapter 5, Volume III) and is dependent on the linkage types and anomeric configurations of the sugar residues to permit cooperative interresidue hydrogen bonding. At low pH and temperature the oxidation of 4-linked β-D-glucuronic acid residues is negligible when flanked by 3-linked residues of a 2-acetamido-2-deoxy-β-D-hexopyranose as in chondroitin 4-sulfate (β-D-GalpNAc) (172), for which a proposed hydrogen-bonding scheme is shown in Fig. 27, and in hyaluronic acid (β-D-GlcpNAc). In contrast to chondroitin sulfate, dermatan sulfate, in which α-L-iduronic acid residues replace those of β-D-glucuronic acid, is rapidly oxidized by periodate (173). Periodate-resistant α-D-glucuronic acid residues are encountered in regions of heparin in which these residues are flanked by those of 2-amino-2-deoxy-α-D-glucopyranose or 2-deoxy-2-sulfamino-α-D-glucopy-ranose but not (64) those of 2-acetamido-2-deoxy-α-D-glucopyranose (174). Although no detailed hydrogen-bonding scheme has been proposed to

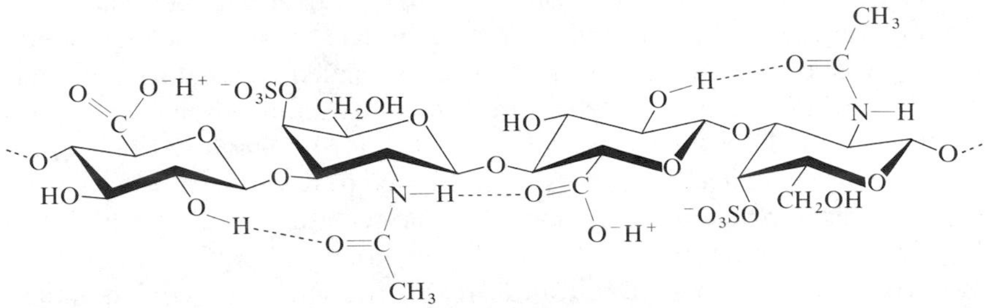

Fig. 27. Proposed scheme of interresidue hydrogen bonding with acetamido groups to account for the lack of susceptibility of 4-linked β-D-glucopyranosyluronic acid residues in chondroitin 4-sulfate to periodate oxidation at low pH.

account for this observation, the importance of cooperative interresidue hydrogen bonding is indicated since periodate-resistant glucuronic acid residues become susceptible to oxidation after scission of the periodate-oxidized polysaccharide in alkaline medium (174). Considerable advantage has been taken of these and other differences in susceptibility to periodate oxidation to effect selective cleavage of the structurally variable regions of those glycosaminoglycans such as heparin and heparan sulfate (175), and mixed chondroitin sulfate–dermatan sulfate copolymers (176), for which individual residues in the repeating disaccharide units have been modified.

Mainly because of the limited range of suitable solvents for both polysaccharide and reagent, oxidations of polysaccharides with lead tetraacetate are infrequently performed (108,109). However, oxidized polysaccharides, which are not easily isolated from aqueous solution, can be recovered after oxidation with lead tetraacetate in DMSO by precipitation with ethanol (177).

B. Degradative Aspects

Aside from the use of limited oxidative cleavage to trim exocyclic chains in sugar residues, to which reference has been made previously, the most important degradative sequence based on the periodate oxidation of polysaccharides is the Smith degradation (178). It was mentioned above that sugar residues resistant to periodate oxidation can be estimated by reduction of periodate-oxidized polysaccharides with sodium borohydride followed by hydrolysis to give the unattacked sugars together with the residual stubs of oxidized units (164). The Smith degradation refers to the controlled acid hydrolysis of reduced oxidized polysaccharides in which hydrolysis of acyclic acetals from cleaved sugar units occurs with dilute acid at room temperature without significant hydrolysis of glycosidic (even furanosidic) linkages. The reaction sequence results in the isolation from a polysaccharide of those sugar residues that resisted oxidative cleavage by periodate. Depending on the relative placing of such periodate-resistant sugar residues, the degradation may result in the formation of isolated units of low molecular weight in which the sugar residues are present as simple glycosides of fragments such as glycerol or a tetritol, or as degraded polysaccharides, which can be analyzed in detail and possibly subjected to repeated Smith degradation sequences in a stepwise manner. The former situation is exemplified in Fig. 28 by the formation of 2-O-β-D-glucopyranosyl-D-erythritol (74) from Smith degradation of oat β-D-glucan (73) (178). In contrast, the corresponding degradation (Fig. 29) of the branched β-D-glucan (75) from *Sclerotium glucanicum* results in removal of side chains with formation of an otherwise unmodified 3-linked β-D-glucan (76) (179).

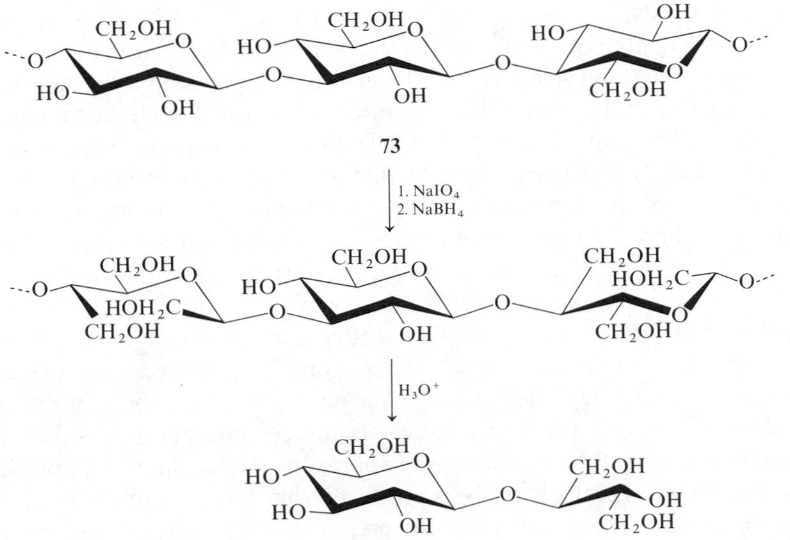

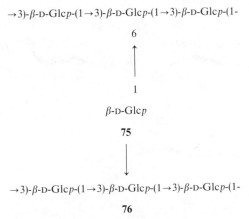

Fig. 28. Smith degradation of oat β-D-glucan (**73**) involving periodate oxidation followed by reduction and controlled acid hydrolysis of acyclic acetal linkages to give 2-*O*-β-D-glucopyranosyl-D-erythritol (**74**) from those regions in the structure in which 3-linked residues are flanked on either side by 4-linked residues.

→3)-β-D-Glc*p*-(1→3)-β-D-Glc*p*-(1→3)-β-D-Glc*p*-(1-

6

↑

1

β-D-Glc*p*

75

↓

→3)-β-D-Glc*p*-(1→3)-β-D-Glc*p*-(1→3)-β-D-Glc*p*-(1-

76

Fig. 29. Smith degradation (periodate oxidation, reduction, followed by controlled acid hydrolysis) of the branched β-D-glucan (**75**) affords the linear 3-linked main chain (**76**).

The success of the Smith degradation depends, as outlined above, on ensuring that all potentially vulnerable diol and triol groups have been oxidized but also on selectivity in the acid hydrolysis step. This step is not always easy to monitor in order to ensure that removal of cleaved fragments is complete but that no detectable hydrolysis of normal glycosidic linkages has occurred. It has been observed that the fragments resulting from the oxidative scission of hexuronic acids are appreciably more resistant to hydrolysis than the corresponding fragments from neutral sugar residues. Complete removal of cleaved fragments is particularly important if Smith degradations are to be performed in a stepwise manner on highly branched glycans with intermediate isolations of degraded polysaccharides. Dutton and Gibney (180) devised a method for monitoring the reaction based on the separation of degraded polysaccharide by precipitation from soluble products of low molecular weight. Gas–liquid chromatography is then used to analyze directly the TMS derivatives of the low molecular weight products and then to analyze the hydrolysis products from fractions of both high and low molecular weight. In addition, preparative gel permeation chromatography can be used to ensure complete hydrolysis in the generation of degraded polysaccharides (181).

Another complication in some Smith degradations arises during the acid treatment wherein the glycolaldehyde moiety (from C-1 and C-2 of oxidized residues) undergoes acid-catalyzed transacetalation to give cyclic acetals. For example, in the above-mentioned Smith degradation of oat β-D-glucan, the six-membered O-2′-hydroxyethylidene acetal (77) is formed as a by-product (178). Five-membered O-2′-hydroxyethylidene acetals can also be formed from β-D-glycans, and compounds of this type (e.g., 78) are the main by-products from α-D-glycans (182).

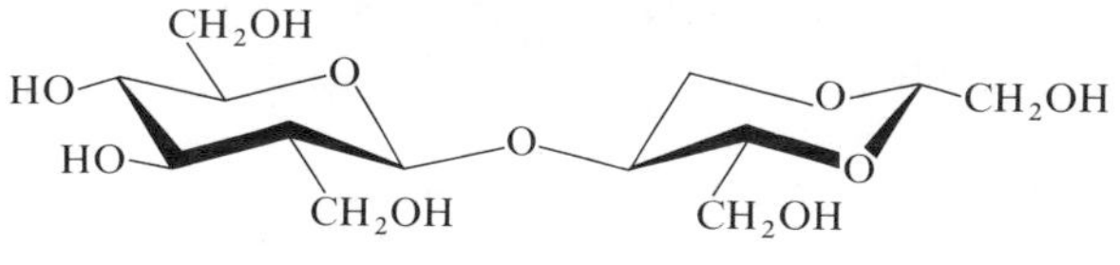

77

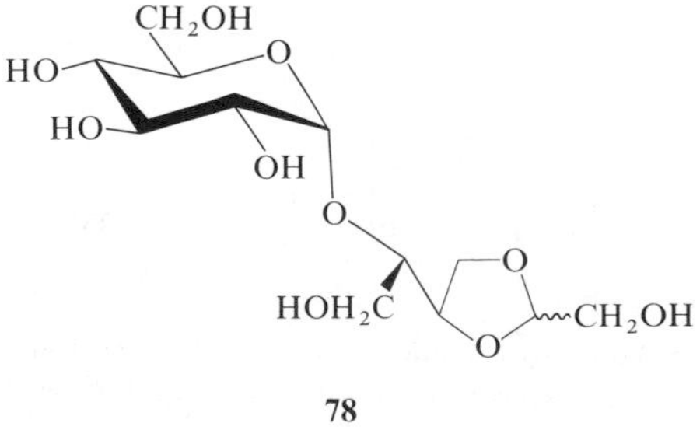

78

→2)-α-L-Rhap-(1→2)-β-D-Ribf-(1→3)-α-L-Rhap-(1→3)-α-L-Rhap-(1-

79

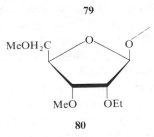

MeOH₂C

MeO   OEt

80

Fig. 30. Application of a modified Smith degradation to the lipopolysaccharide with repeating unit **79** cleaves the chain with formation of a peralkylated oligosaccharide with nonreducing end groups (**80**) whose alkylation pattern points to the former site of substitution.

A modification of the Smith degradation, introduced by Lindberg and co-workers (*183*), overcomes two of the difficulties in the reaction sequences by methylating the reduced oxidized polysaccharide ("polyalcohol") before the acid hydrolysis step. First, methylation analysis, i.e., hydrolysis, reduction, and acetylation to give partially methylated alditol acetates from nonoxidized sugar residues, will confirm that these residues are of a type containing no diol or triol groups rather than those that became resistant because of interactions with neighboring sugar residues. Second, the mild acid hydrolysis may then be performed with no hydroxyl groups available for intramolecular transacetalation, thus ensuring complete removal of glycolaldehyde fragments from oxidized residues. This variation of the Smith degradation provides additional structural information in that hydrolysis of the permethylated reduced oxidized polysaccharide liberates aglyconic hydroxyl groups, which can be identified by realkylation with trideuteriomethyl iodide or ethyl iodide, as in the formation (Fig. 30) of terminal residues of **80** from the *Klebsiella* O group lipopolysaccharide (**79**) (*183*).

An alternative procedure for isolating periodate-resistant regions in polysaccharides, known as the Barry degradation, involves treatment of the oxidized polysaccharide with phenylhydrazine (*184*) or *N,N*-dimethylhydrazine (*165*). This method has been largely superceded by the experimentally cleaner Smith degradation.

VII. Structural Modifications of Polysaccharides

Structural modifications of polysaccharides are of several types and preferably are achieved without substantial depolymerization. Changes in the physical properties of polysaccharides that take place on derivatization or on controlled treatment with acids, bases, or oxidizing agents have received much attention from the standpoint of generating new materials of potential commercial importance. These aspects of structural modification are discussed with respect to specific substances in the chapters on cellulose (Chapter 2, Volume II) and starch (Chapter 3, Volume III) and more generally in the chapters on polysaccharide conformations (Chapter 5, this Volume) and on the industrial utilization of polysaccharides. (Chapter 7, Volume II). In this section attention is directed toward modifications that are of value in connection with structure determination. These modifications are of four types: (a) those leading to alterations in the relative susceptibilities of different glycosidic linkages to hydrolysis, (b) those providing for the selective introduction of functional groups from which specific degradations can be initiated, (c) those leading to the formation of oxide (oxalane and oxirane) rings through the removal of suitably placed natural sulfate or selectively introduced sulfonate ester groups, and (d) those resulting in the removal of natural substituent ester and acetal groups so as to provide evidence for their location in the parent polysaccharides.

A. Reduction of and Oxidation to Uronic Acids

The resistance of glycosiduronic acids to complete hydrolysis without appreciable accompanying decomposition presents difficulties in compositional analysis of polysaccharides and in linkage analysis with methylated derivatives. Although advantage may be taken of the resistance of these linkages to hydrolysis to isolate acidic oligosaccharides as products of partial depolymerization, it is correspondingly difficult, if not impossible, to isolate directly oligosaccharides with different sugar sequences containing the more acid-sensitive glycosidic linkages. Although some alterations in "cracking patterns" can be achieved by using different procedures for acid-catalyzed depolymerization, a more satisfactory approach is to reduce uronic acid to the corresponding hexose residues and then carry out partial fragmentation of the carboxyl-reduced polysaccharide. This objective is best realized using the method of Taylor and Conrad in which the acidic polysaccharide in aqueous solution is treated with a water-soluble carbodiimide to give a postulated O-acylisourea, which is then reduced with sodium borohydride (185). Both stages in the reaction sequence require careful pH control, but in the experience of several workers repetition of the operation may be necessary to effect reasonably complete reduction. Otherwise, the procedure appears to be free of complications. Several alternative methods have been used, but all involve esterification of the uronic acid residues and/or protection of hydroxyl groups with removable substituents so that the reduction step can be performed in organic solvents, such as tetrahydrofuran, 2-methoxyethyl ether, or similar ethers (186). Reduction of permethylated polysaccharides, usually with lithium aluminum hydride, proceeds smoothly in tetrahydrofuran or similar solvent (71).

It is important to bear in mind that information from studies on carboxyl-reduced acidic glycans may be limited if the parent polysaccharide contains both a hexuronic acid and the configurationally related hexose as constituents, unless independent evidence is available to differentiate the original from the newly formed hexose residues.

An example of the difference in the pattern of oligosaccharides formed on partial depolymerization of a carboxyl-reduced and the parent acidic polysaccharide is shown in Fig. 31 for the interior chains (81) of leiocarpan A,

$$-[\rightarrow 4)\text{-}\beta\text{-}D\text{-}GlcpA\text{-}(1\rightarrow 2)\text{-}\alpha\text{-}D\text{-}Manp\text{-}(1\text{-}]_n\text{-} \hspace{2cm} (a)$$

81

$$\downarrow$$

$$\beta\text{-}D\text{-}GlcpA\text{-}(1\rightarrow 2)\text{-}D\text{-}Man$$

82

$$+$$

$$\beta\text{-}D\text{-}GlcpA\text{-}(1\rightarrow 2)\text{-}\alpha\text{-}D\text{-}Manp\text{-}(1\rightarrow 4)\text{-}\beta\text{-}D\text{-}GlcpA\text{-}(1\rightarrow 2)\text{-}D\text{-}Man$$

83

$$-[\rightarrow 4)\text{-}\beta\text{-}D\text{-}Glcp\text{-}(1\rightarrow 2)\text{-}\alpha\text{-}D\text{-}Manp\text{-}(1\text{-}]_n\text{-} \hspace{2cm} (b)$$

84

$$\downarrow$$

$$\beta\text{-}D\text{-}Glcp\text{-}(1\rightarrow 2)\text{-}D\text{-}Man \quad + \quad \alpha\text{-}D\text{-}Manp\text{-}(1\rightarrow 4)\text{-}D\text{-}Glcp$$

85   **86**

$$Glcp\text{-}(1\rightarrow 2)\text{-}Manp\text{-}(1\rightarrow 4)\text{-}Glcp \quad + \quad Manp\text{-}(1\rightarrow 2)\text{-}Glcp\text{-}(1\rightarrow 2)\text{-}Man$$

87   **88**

$$Glcp\text{-}(1\rightarrow 2)\text{-}Manp\text{-}(1\rightarrow 4)\text{-}Glcp\text{-}(1\rightarrow 2)\text{-}Manp \quad + \quad Manp\text{-}(1\rightarrow 2)\text{-}Glcp\text{-}(1\rightarrow 2)\text{-}Manp\text{-}(1\rightarrow 4)\text{-}Glcp$$

89   **90**

$$Glcp\text{-}(1\rightarrow 2)\text{-}Manp\text{-}(1\rightarrow 4)\text{-}Glcp\text{-}(1\rightarrow 2)\text{-}Manp\text{-}(1\rightarrow 4)\text{-}Glcp$$

91

Fig. 31. Partial acid hydrolysis of leiocarpan A with preferential cleavage of neutral glycosidic linkages in the interior chains (81) affords only acidic oligosaccharides (82 and 83). In contrast, partial acetolysis of the carboxyl-reduced polysaccharide (84) followed by de-O-acetylation furnishes oligosaccharides (85–91) from indiscriminate scission of both types of neutral glycosidic linkage.

which is composed of alternating D-glucuronic acid and D-mannose residues. Whereas the parent polysaccharide gives only oligosaccharides (**82** and **83**) with mannose reducing groups, the carboxyl-reduced polysaccharide (**84**) affords oligosaccharides (**85**–**91**) with mannose and with glucose reducing groups (*187*).

The opposite reaction, namely, the introduction of uronic acid residues into polysaccharides by selective oxidation of primary hydroxyl groups, has been used to advantage. In unsubstituted polysaccharides a very slow reaction occurs in aqueous solution with oxygen and a platinum catalyst. The reaction does not go to completion, and some depolymerization occurs,

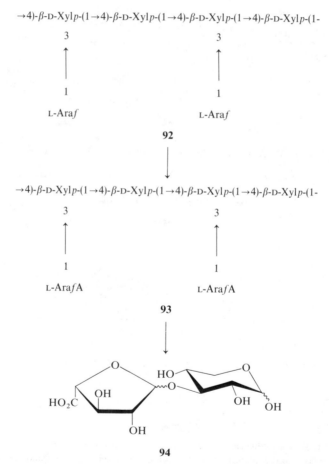

Fig. 32. Catalytic oxidation of primary hydroxyl groups in rye flour arabinoxylan (**92**) followed by partial acid hydrolysis of the oxidized polysaccharide (**93**) affords the aldobiouronic acid (**94**).

possibly due to the even slower introduction of some labile keto groups by oxidation of secondary hydroxyl groups (*188*). The first applications of this reaction showed that arabinofuranosyluronic acid linkages were stabilized relative to both arabinofuranosyl and xylopyranosyl linkages. Thus, whereas partial acid hydrolysis of rye flour arabinoxylan (**92** in Fig. 32) selectively removes arabinofuranosyl side chains, partial acid hydrolysis of the oxidized polysaccharide (**93**) under much more severe conditions affords the aldobiouronic acid 3-*O*-(L-arabinofuranosyluronic acid)-D-xylose (**94**) (*189*). The effective use of the corresponding oxidation of primary hydroxyl groups in hexopyranose residues requires a polysaccharide with only a small proportion of such units. The method has been used with bacterial dextrans, which contain a high proportion of 6-linked α-D-glucopyranose residues, to establish the length of side chains terminated by α-D-glucopyranose residues (*190–192*).

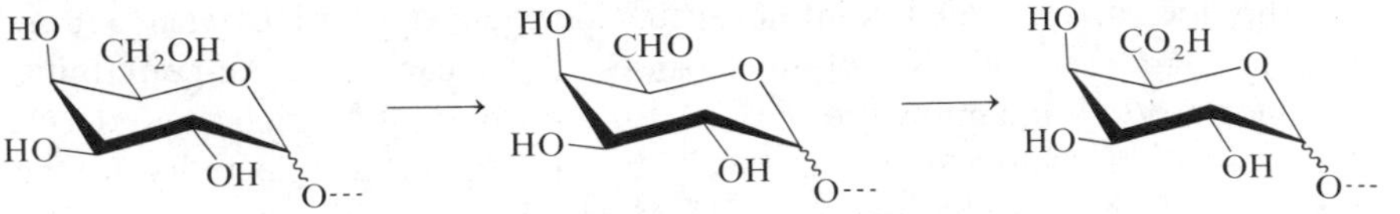

95 **96** **97**

Fig. 33. Enzymatic oxidation of nonreducing D-galactopyranose end groups (**95**) to give those of the 6-aldehydo sugar (**96**) and then, by chemical oxidation, those of D-galacturonic acid (**97**).

Terminal D-galactopyranose residues (**95** in Fig. 33) can be selectively oxidized to D-*galacto*-hexodialdo-1,5-pyranose residues (**96**) using galactose oxidase, and further treatment with sodium hypoiodite then gives residues of D-galacturonic acid (**97**) (*193*). An example of the use of this reaction sequence to obtain new structural information was provided in studies on the specific capsular polysaccharide from *Streptococcus pneumoniae* type 14 (**98**) (*194*). The permethylated oxidized polysaccharide was then subjected to base-catalyzed degradation (see Section VIII,A,2) with removal of D-galacturonic acid (from D-galactose) end groups.

$$\rightarrow4)\text{-}\beta\text{-D-Glc}p\text{-}(1\rightarrow6)\text{-}\beta\text{-D-Glc}p\text{NAc-}(1\rightarrow3)\text{-}\beta\text{-D-Gal}p\text{-}(1\text{-}$$

4

↑

1

β-D-Gal*p*

98

In otherwise protected methylated polysaccharides, primary hydroxyl groups can be oxidized with chromium trioxide in acetic acid. The arabino-galactan A from Japanese larch was successively tritylated, methylated, detritylated, and oxidized. Although subsequent experiments did not yield new structural information, partial hydrolysis of the oxidized methylated polysaccharide was shown to give a mixture of methylated aldobiouronic acids (*195*).

B. Selective Substitution Followed by Structural Modification

The introduction of functional groups at selected sites for the initiation of specific degradations commonly involves the immobilization of other positions by selective substitution. Numerous examples of the use of blocking groups, followed by alterations in stereochemistry and/or functionality at other loci, are provided in monosaccharide chemistry with extensive ram-ifications in the synthesis of nucleosides (*196*), carbohydrate-containing antibiotics (*197*), and many groups of natural products for which sugar deriv-atives serve as chiral synthons (*198–201*). Comparable selectivity in the reactions of polysaccharides is much more difficult to achieve. Effective structural modification, even simple substitution, requires that all or virtually all units of a given type are uniformly modified since it is not possible to separate modified and unmodified units in the same polysaccharide chain. Although satisfactory methods are available for the differentiation of primary and secondary hydroxyl groups, little success has attended efforts to dis-criminate in polysaccharides between secondary hydroxyl groups, for exam-ple, between equatorially and axially oriented substituents, or to block pairs of hydroxyl groups, e.g., by cyclic acetal or ketal formation. Furthermore, the selection of suitable solvents for selective transformations is by no means a trivial problem, and at present chromatographic procedures for the puri-fication of products are limited in their capacity to separate closely related derivatives.

The great majority of structural modifications consequent on selective substitution are performed on derivatives of primary hydroxyl groups. The most widely used and generally most effective procedure for selective sub-stitution of primary hydroxyl groups is O-tritylation. Other substituents are then introduced on secondary hydroxyl groups for direct use or, more commonly, for protection, so that selective de-O-tritylation may be followed by the introduction of functionally reactive substituents, e.g., sulfonate esters, at C-6. This indirect procedure, for amylose, gives a higher degree of net selective substitution at C-6 than is obtained by direct sulfonylation (*202*).

Polysaccharides carrying good leaving groups as substituents at C-6 are most useful for transformations involving nucleophilic displacements. Reactions performed with 6-O-sulfonyl derivatives include (a) intramolecular displacement in base to furnish 3,6-anhydrohexose residues and (b) displacement by iodide, which in turn may be displaced by sulfinate with the formation of 6-C-sulfonyl (or sulfone) derivatives (see Section VIII,A,3). For the latter reaction direct iodination at C-6 can be achieved with N-iodosuccinimide and triphenylphosphine without protection of secondary hydroxyl groups (203). Direct bromination using the Vilsmeier-type reagent, methanesulfonyl bromide–N,N-dimethylformamide, can also be performed without protection of secondary hydroxyl groups (204). An improved synthesis of a 6-deoxy analogue of amylose involves subsequent de-O-formylation, acetylation, and treatment with sodium borohydride (204).

For some of these transformations leading to the formation of polysaccharide derivatives suitable for subsequent controlled fragmentation, protection of secondary hydroxyl groups by methylation leads to the isolation of degradation products in which exposed hydroxyl groups indicate points of cleavage. Primary hydroxyl groups can also be introduced into permethylated glycuronans by reduction of hexuronic acid residues with lithium aluminum hydride.

Polysaccharide modifications in relation to structure determination may be performed with any of the following objectives. (a) The formation of 3,6-anhydrohexose or hex-5-enopyranose residues introduces extremely acid-sensitive glycosidic linkages from which selective cleavage will take place (see Sections VII,C and VIII,B,3). (b) Oxidation of exposed hydroxyl groups, preferably by chlorine–DMSO, affords carbonyl groups from which base-catalyzed degradations can be initiated (see Section VIII,A,4). (c) Conversion of primary hydroxyl groups, directly or indirectly via sulfonate esters, to 6-deoxy-6-iodo derivative, followed by further displacement with sodium p-toluenesulfinate, furnishes sulfones that are then available for base-catalyzed degradation (see Section VIII,A,3).

C. Oxalane and Oxirane Ring Formation

3,6-Anhydrohexose residues were first recognized as natural constituents of algal polysaccharides, derivatives of D-galactose and L-galactose occurring, respectively, in polysaccharides of the carrageenan and agarose families. As mentioned previously, glycosidic linkages of these derivatives are extremely sensitive to acid cleavage, and, unless cleavage is accompanied by simultaneous glycoside, acetal, or thioacetal formation, decomposition takes place. The 3,6-anhydrohexose units in polysaccharides are formed naturally from

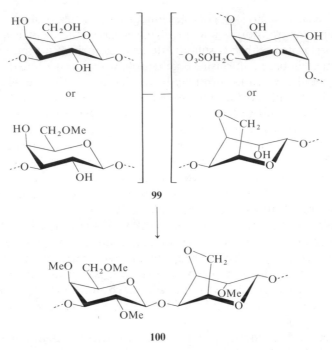

Fig. 34. Conversion of the masked repeating unit (**99**) of porphyran to that (**100**) of methylated agarose.

the corresponding hexose 6-sulfates, and enzymes that catalyze these transformations in the D-galactose and L-galactose series have been isolated (*205,206*). Similarly, in the laboratory, in polysaccharides of natural occurrence, residues of hexose 6-sulfates that carry, in a cis relationship, unsubstituted 3-hydroxyl groups undergo ready conversion the 3,6-anhydrides on treatment with base, preferably sodium hydroxide with added potassium borohydride (*133*).

This type of structural modification was used in experiments that established the relationship of porphyran (from the red alga *Porphyra umbilicalis*) to agarose (Fig. 34) (*84*). Porphyran (*99*) contains a masked repeating unit in which 3-linked residues of β-D-galactopyranose or its 6-methyl ether alternate with those of 4-linked residues of α-L-galactopyranose 6-sulfate or 3,6-anhydro-α-L-galactopyranose. Conversion of the 6-sulfated units to those of the 3,6-anhydride followed by methylation furnished a permethylated polysaccharide that was indistinguishable from permethylated agarose (**100**).

3,6-Anhydrohexopyranose units can be formed in suitably substituted polysaccharides through the introduction of sulfonate substituents at C-6.

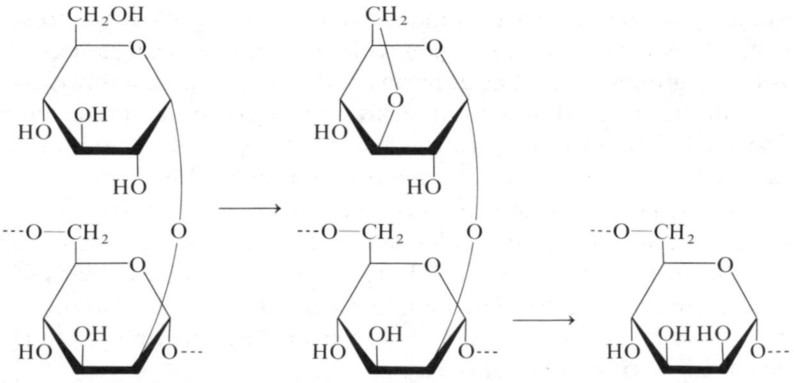

Fig. 35. Structural modification of *Ceratocystis brannea* glucomannan (**101**) involving the formation of 3,6-anhydro-D-glucose residues (in **102**) and their selective acid hydrolysis to give the 6-linked α-D-mannan (**103**).

A notable example of this approach to structural modification (Fig. 35) was provided in studies on the glucomannan (**101**) from *Ceratocistis brannea* in which successive O-tritylation, O-acetylation, detritylation, and *p*-toluene-sulfonylation were followed by treatment with sodium methoxide (*207*). All α-D-glucopyranose residues in side chains were thus modified (**102**), and controlled acid hydrolysis resulted in their selective removal and the isolation of the 6-linked α-D-mannan main chain (**103**).

Rather few examples of polysaccharides carrying secondary sulfate esters with vicinal trans hydroxyl groups have been encountered. Accordingly, epoxide formation as a structural modification leading to new information on the location of the original sulfate esters has been reported only in the case of the complex polysaccharide from the green alga *Ulva lactuca* (*208*). This polysaccharide contains 4-linked D-xylose 2-sulfate residues, which on epoxide formation afford units of 2,3-anhydro-L-lyxose and then on ring opening give units of 2-*O*-methyl-D-xylose (see Chapter 4, Volume II).

D. Location of Removable Substituents

Several polysaccharides with removable substituents occur naturally. The removal of such substituents, notably of *O*-acyl and *O*-sulfate esters and ketals of pyruvic acid, may result in marked alterations of physical properties as well as changes in immunological specificities. The very ease of removal of some of these substituents both provides the basis for their identification and presents difficulties in obtaining unambiguous evidence for their original location. This section deals only with chemical methods for locating these substituents, but reference may be made to the use of ^{13}C-NMR spectroscopy (Chapter 4), especially in conjunction with relevant model substances.

O-Sulfate groups without neighboring hydroxyl groups suitably disposed for epoxide or 3,6-anhydride formation are quite stable to alkaline conditions, so that methylation can be performed without appreciable loss of sulfate. Sulfate esters are at least partially removed during acid hydrolysis, so that methylated sugars are formed in adequate amount, at least for qualitative identification. Since polysaccharides can be desulfated without depolymerization, comparison of the methylated sugars formed from the parent sulfated and the corresponding desulfated polysaccharides gives information on the location of sulfate esters. An example is provided by experiments on chondroitin 4-sulfate (**104** in Fig. 36) (*209*).

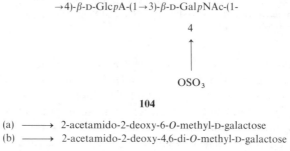

$$\rightarrow 4)\text{-}\beta\text{-}\text{D-Glc}p\text{A-}(1\rightarrow 3)\text{-}\beta\text{-}\text{D-Gal}p\text{NAc-}(1\text{-}$$

4

↑

OSO$_3$

104

(a) ⟶ 2-acetamido-2-deoxy-6-*O*-methyl-D-galactose
(b) ⟶ 2-acetamido-2-deoxy-4,6-di-*O*-methyl-D-galactose

Fig. 36.   Location of sulfate hemiester groups in chondroitin 4-sulfate (**104**) by characterization *inter alia* of methyl ethers of *N*-acetyl-D-glucosamine formed by (a) methylation followed by hydrolysis and (b) desulfation followed by methylation and hydrolysis.

Desulfation of polysaccharides that are relatively insensitive to acid, e.g., chondroitin sulfate, can be performed by treatment with cold methanolic hydrogen chloride (*210*). For polysaccharides containing acid-sensitive groups, e.g., 3,6-anhydrides, solvolytic desulfation without detectable degradation can be achieved by heating as pyridinium salts in DMSO (*211*). In the presence of 5% added water or methanol and under controlled conditions, selective N-desulfation, e.g., of heparin, may be effected (*212*). However, extensive depolymerization occurs when chondroitin sulfate is similarly treated at higher temperature and for a longer period of time (*213*).

A parallel approach to the location of *O*-acetyl (or other *O*-acyl) substituents is impractical since these substituents are either readily removed or undergo migration to other sites during the requisite manipulations. The most satisfactory method for locating *O*-acetyl groups is that of de Belder and Norrman (*214*) and involves the conversion of unsubstituted hydroxyl

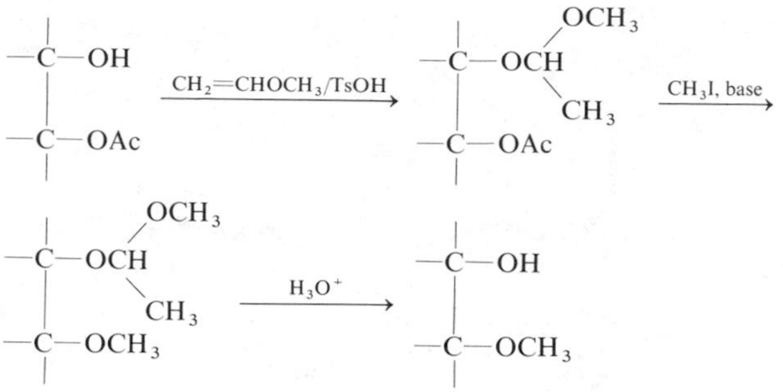

$$
\begin{array}{ccc}
\begin{matrix} | \\ -\text{C}-\text{OH} \\ | \\ -\text{C}-\text{OAc} \\ | \end{matrix}
& \xrightarrow{\text{CH}_2=\text{CHOCH}_3/\text{TsOH}} &
\begin{matrix} | \quad\quad \text{OCH}_3 \\ -\text{C}-\text{OCH} \\ \quad\quad\ \backslash \\ | \quad\quad\ \text{CH}_3 \\ -\text{C}-\text{OAc} \\ | \end{matrix}
\xrightarrow{\text{CH}_3\text{I, base}}
\end{array}
$$

$$
\begin{array}{ccc}
\begin{matrix} | \quad\ \text{OCH}_3 \\ -\text{C}-\text{OCH} \\ \quad\quad \backslash \\ | \quad\quad \text{CH}_3 \\ -\text{C}-\text{OCH}_3 \\ | \end{matrix}
& \xrightarrow{\text{H}_3\text{O}^+} &
\begin{matrix} | \\ -\text{C}-\text{OH} \\ | \\ -\text{C}-\text{OCH}_3 \\ | \end{matrix}
\end{array}
$$

Fig. 37. Location of O-acetyl (and other O-acyl) substituents in glycans.

groups to methyoxyethylacetals on reaction with methyl vinyl ether, followed by base-catalyzed de-O-acetylation and methylation. Hydrolysis of the modified polysaccharide then gives sugar derivatives labeled with O-methyl groups at the sites originally occupied by O-acetyl substituents (Fig. 37). Notwithstanding the above comment concerning the removal/migration of acyl substituents, a recent publication describes the retention of an O-formyl substituent during the carefully controlled methylation of the capsular polysaccharide from Klebsiella type 63 using the Hakomori procedure (215).

Ketals of pyruvic acid are the only common substituents of this type. These groups can be removed by controlled acid hydrolysis, such as "autohydrolysis" of the polysaccharide, but some depolymerization may take place (216). In principle, however, comparison of the methylation analyses of the parent and the depyruvylated polysaccharides may point to the location of ketal function. In many instances pyruvic acid ketals occur as constituents of oligosaccharide repeating units in microbial polysaccharides. Since the ketals are normally stable to methylation* but are removed on acid hydrolysis, the isolation of partially methylated sugars with hydroxyl groups suitably disposed to have carried such a substituent and in amounts suggesting an excess of branch points over end groups is indicative of their probable location. Pyruvic acid ketals are more easily removed by acid hydrolysis after reduction to hydroxyisopropylidene ketals. The preferred method (Fig. 38) from experiments on the polysaccharide from Rhizobium trifollii (105) involves methylation, carboxyl reduction, and remethylation followed by mild acid hydrolysis and trideuteriomethylation in order to label the sugar residue (106) to which the ketal was linked (218).

* The 3,4-O-(1'-carboxyethylidene)-D-galactopyranose residue in the Klebsiella type 33 capsular polysaccharide undergoes loss of the ketal function during Hakomori methylation, apparently in an E2-type anticoplanar elimination of the axially oriented 4-O substituent (127).

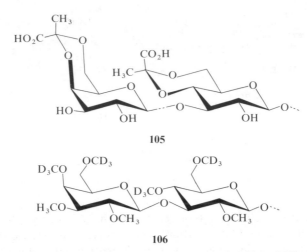

105

106

Fig. 38. Location of pyruvic acid ketals in the terminal residues (**105**) of the capsular polysaccharide from *Rhizobium trifolii* by reduction of the permethylated polysaccharide, followed by selective hydrolysis of the reduced ketal substituents and then trideuteriomethylation to label sites of attachment in the modified polysaccharide (**106**).

VIII. Other Methods for Partial Depolymerization

A. Alkaline Degradations and Other Base-Catalyzed Fragmentations

In general, glycosides are stable under alkaline conditions. However, base-catalyzed β-eliminations can be initiated by strongly electron-withdrawing functional groups with consequent cleavage of glycosidic linkages. Groups that may act in this manner include the masked carbonyl groups of reducing residues, suitably activated hexuronic acid derivatives (especially esters), carbonyl groups introduced by oxidation at selected sites, and sulfone groups formed in structural modifications.

1. Degradations from Reducing Groups

Reducing sugars undergo a number of competing reactions when treated with base, even when precautions are taken to ensure the absence of oxygen and thus to exclude oxidative degradations (*219*). However, extensive studies on partially methylated sugars and appropriately linked oligosaccharides have thrown light on reactions that are of the greatest importance in studies on polysaccharides (*220*). The major degradative pathway for each type of reducing group is determined largely by its substitution pattern, although side reactions nearly always occur. The key reactions, which are summarized in Fig. 39, involve base-catalyzed β-elimination (usually in aqueous solution),

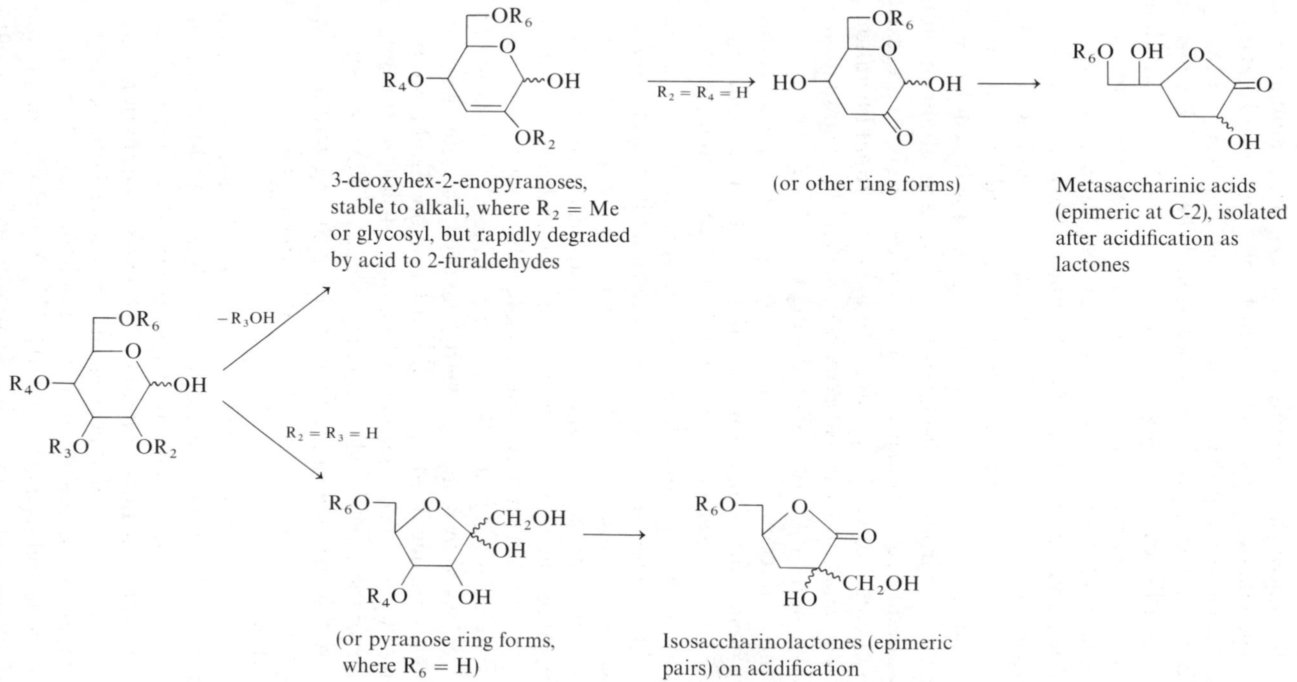

Metasaccharinic acids (epimeric at C-2), isolated after acidification as lactones

(or other ring forms)

$R_2 = R_4 = H$

3-deoxyhex-2-enopyranoses, stable to alkali, where $R_2 = Me$ or glycosyl, but rapidly degraded by acid to 2-furaldehydes

$-R_3OH$

$R_2 = R_3 = H$

Isosaccharinolactones (epimeric pairs) on acidification

(or pyranose ring forms, where $R_6 = H$)

Fig. 39. Some degradation products from variously substituted reducing sugars.

giving initially 3-deoxyhex-2-enopyranoses (*221*). This elimination occurs most easily when there is a 3-O substituent to provide a good leaving group whose departure is not impeded by mass law considerations when degradation is performed typically in aqueous solution. When no 2-O substituent is present, ketonization ensues to give the 3-deoxyglycos-2-ulose, which in turn undergoes a benzilic acid-type rearrangement to give an epimeric pair of 3-deoxyaldonic acids (commonly known as metasaccharinic acids) (*222*). When a reducing sugar carries a 4-O substituent the dominant degradative pathway involves aldose to ketose rearrangement, followed in turn by base-catalyzed β-elimination to give an epimeric pair of 2-*C*-hydroxymethyl-3-deoxyaldonic acids (isosaccharinic acids). 2-O-Substituted 3-deoxyhex-2-enopyranoses, e.g., from 2,3-di-O-substituted sugars, although stable to further base-catalyzed degradations at room temperature, are extremely labile (as enol ethers or enol glycosides) to acid with rapid liberation of the 2-linked substituent and extensive rearrangement or degradation of the residual fragment from the original reducing unit (*221,223*).

Saccharinic acid-forming reactions are of special importance for polysaccharides containing 3-linked main chains, with or without side chains attached, and to a lesser extent for 4-linked glycans. Since β-elimination exposes the next lower polymer homologue as a reducing unit, degradation may proceed in a stepwise manner along the chain in what is known as the "peeling" reaction. However, this reaction rarely if ever proceeds to completion. Like all chain processes, the reaction sequence is particularly susceptible to competing reactions. A notable example of such a competing reaction terminating the erosion process is the so-called stopping reaction, which was first demonstrated during the alkaline degradation of cellulose (*224*) and amylose (*225*). As shown in Fig. 39 the favored reaction when 4-linked reducing sugars are degraded by alkali involves aldose to ketose rearrangement followed by isosaccharinic acid formation. However, the unfavorable β-elimination of the 3-hydroxyl group may occur with lower frequency and will then result in net rearrangement of the reducing terminal unit to a base-stable metasaccharinic acid residue from which no further degradation takes place (Fig. 40).

The alkaline degradation of branched polysaccharides has been less extensively studied. For polysaccharides with 3-linked or 4-linked chains and side-chain units remotely attached at C-6, degradation proceeds as for the linear glycans but with the formation of *O*-glycosylsaccharinic acids from branched residues. Such alkaline degradations have been performed with the galactomannan guaran (4-linked β-D-mannan chain) (*226*) and scleroglucan (3-linked β-D-glucan main chain) (*227*). On the other hand, studies on model compounds containing 3,4-di-O-substituted reducing groups have

$$\rightarrow 4)\text{-}\beta\text{-D-Glc}p\text{-}[(1\rightarrow 4)\text{-}\beta\text{-D-Glc}p\text{-}]_n(1\rightarrow 4)\text{-}D\text{-Glc}p$$

$$\downarrow \text{(a)}$$

$$\rightarrow 4)\text{-}\beta\text{-D-Glc}p\text{-}[(1\rightarrow 4)\text{-}\beta\text{-D-Glc}p\text{-}]_{n\rightarrow 1}(1\rightarrow 4)\text{-}D\text{-Glc}p + \text{iS}$$

$$\downarrow \text{(a)}$$

$$\rightarrow 4)\text{-}\beta\text{-D-Glc}p\text{-}(1\rightarrow 4)\text{-}D\text{-Glc}p + n \text{ iS}$$

$$\downarrow \text{(b)}$$

$$\rightarrow 4)\text{-}\beta\text{-D-Glc}p\text{-}(1\rightarrow 4)\text{-mS}$$

Fig. 40. Alkaline degradation of cellulose with (a) stepwise erosion from the reducing terminus by the "peeling" reaction and (b) formation of an alkali-stable product by the "stopping" reaction (iS, isosaccharinic acid; mS, metasaccharinic acid).

shown that both substituents are eliminated, probably with formation of a different type of degradation product from the reducing residue (228).

In general, however, alkaline degradations from the reducing groups of intact polysaccharides are of limited value in structural studies. Nevertheless, these degradations are potentially of much greater value for studies on otherwise generated oligosaccharides. For such compounds of relatively low molecular weight, degradation reactions diagnostic for linkage type are likely to occur with few detectable side reactions. Specific mention may be made of two situations in which reducing oligosaccharide units are generated under base-catalyzed conditions. Base-catalyzed degradations initiated at hexuronate residues (see the following section) frequently result in the exposure of reducing groups from which further degradation may take place. The extent of such further degradation is dependent on the linkage types of the exposed groups. In one important group of glycoproteins, oligosaccharide chains glycosidically linked to serine or threonine residues are liberated on treatment with base (229). Most commonly, intact chains are trapped as base-stable oligosaccharide alditols when the base-catalyzed β-elimination is performed with sodium hydroxide solutions containing sodium borohydride (Fig. 41) (230). Polysaccharide chains that are O-glycosidically linked to protein in proteoglycans are similarly liberated without appreciable degradation by the "peeling" mechanism (231).

2. Base-Catalyzed β-Elimination from Hexuronic Acid Residues

Base-catalyzed degradations from hexuronic acid residues occur when the uronic acid is (a) esterified (or otherwise activated) and (b) 4-O-substituted

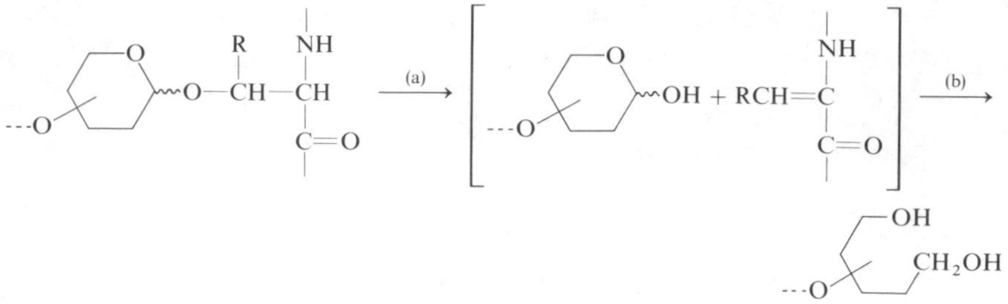

Fig. 41. Alkaline elimination (a) of oligosaccharide chains from base-sensitive O-glycosidic linkages, e.g., to serine or threonine residues in glycoproteins. In the presence of added sodium borohydride (b) "peeling" is avoided by reduction to alkali-stable alditols.

carrying ether or glycosyl groups and result in the β-elimination of the 4-O substituent with the formation of hex-4-enopyranosiduronate residues. The reaction was first postulated as a competing reaction in the saponification of highly esterified pectins in aqueous solution (235). Pectins even undergo extensive depolymerization when heated in phosphate buffer at pH 6.8 (233). The degradation was first used to advantage in structural studies on the bacterial polysaccharide colanic acid (234). Treatment of a suspension of the polysaccharide as its hydroxypropyl ester with methanolic sodium methoxide containing 2,2-dimethoxypropane as a water scavenger resulted in the liberation of 4,6-O-(1-carboxyethylidene)-D-galactose. However, this reaction with an insoluble polysaccharide gave only 50% degradation. Complete degradation is best achieved when the reaction is performed in non-hydroxylic solvents with polysaccharide derivatives carrying base-stable substituents, which may be permanent (e.g., methyl ethers) or temporary (e.g., methoxyethylacetals), and when bases of low nucleophilicity toward ester functions are used, so that the driving force of the carboalkoxy group is maintained.

The most widely used reaction sequence is that developed by Lindberg *et al.* (62,235,236). In this procedure permethylated (fully etherified and esterified) polysaccharide is prepared without appreciable degradation by the Hakomori method in a single operation. However, attempts to repeat the operation involving the addition of fresh base lead to extensive degradation. Treatment of the methylated polysaccharide with sodium methylsulfinylmethanide in DMSO leads to liberation of the 4-linked substituent, with formation of hex-4-enopyranosiduronate residues, which may undergo further degradation. The method is particularly suitable for the study of methylated glycuronans, which are available in only limited quantities, since the course of the reaction may be monitored by examination of the unmodified neutral sugar residues by conversion to partially methylated alditol

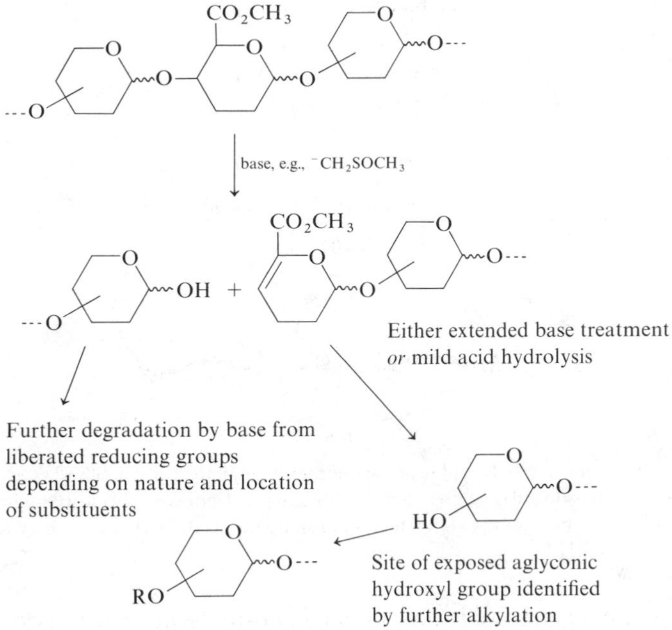

Fig. 42. Base-catalyzed β-eliminations from 4-linked hexuronate residues in permethylated polysaccharides. (For simplicity, substituents not involved in chemical change are omitted.)

acetates and analysis by GLC–mass spectrometry. If necessary, the completeness of degradation of uronic acid residues can be ascertained by methanolysis to give methyl ester methyl glycosides for analysis directly or after acetylation. In contrast to the parent acid-resistant glycosiduronic acid linkages, the hex-4-enopyranosiduronic acid linkages are extremely labile to acid and can be cleaved under very mild conditions without accompanying hydrolysis of normal glycosidic linkages to expose "aglyconic" hydroxyl groups which formerly carried hexuronic acid residues (Fig. 42). These hydroxyl groups are readily recognized after further alkylation with trideuteriomethyl iodide or ethyl iodide by normal compositional analysis. Studies on the methylated capsular polysaccharide from *Klebsiella* type 47 (**107** in Fig. 43) illustrate the application of the method (*235*). Sugar analysis of the methylated polysaccharide **108** after degradation by base showed that terminal rhamnopyranose residues had disappeared, indicating that these residues were present as the 4-O substituents of the glucuronic acid residues and had been lost by further degradation after β-elimination. Further alkylation of the degraded methylated polysaccharide established O-3 of internal rhamnose residues as the site of attachment of uronic acid units.

→3)-β-D-Gal*p*-(1→4)-α-L-Rha*p*-(1-

3

↑

1

α-L-Rha*p*-(1→4)-β-D-Glc*p*A

107

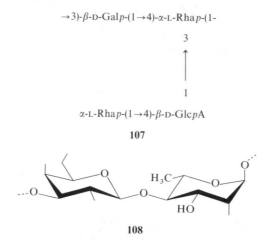

108

Fig. 43. Base-catalyzed degradation of permethylated derivative of *Klebsiella* type 47 capsular polysaccharide (**107**) to give degraded methylated polysaccharide **108** with hydroxyl groups exposed at former sites of attachment of uronic acid residues. (Unless otherwise specified substituents are OCH₃.)

Later experiments showed that, under conditions normally used for base degradations, complete loss of uronic acid residues occurs with direct exposure of aglycone. Consequently, base degradation and further alkylation can be performed successively in a one-pot procedure without intermediate isolation of the product from the first operation (*237*).

In the procedure just described structural information is lost when further degradation of reducing sugars occurs. In an alternative procedure, which has not been widely used because of difficulties in removing excess reagent, exposed reducing groups are simultaneously protected by acetylation if degradation is performed using the organic base 1,5-diazabicyclo[5.4.0]undec-5-ene (DBU) with acetic anhydride. The method is exemplified in studies on methylated degraded leiocarpan A (Fig. 44, **109 → 110**) (*238*).

Two methods have been explored for base-catalyzed degradations of acidic polysaccharides carrying removable protecting groups. In the first method hydroxyl and carboxyl groups are protected by reaction with methyl vinyl ether, and the polysaccharide derivative is then degraded as for methylated polysaccharides with dimsyl base in DMSO. Mild hydrolysis with aqueous acid then removes any remaining unsaturated uronic acid units together with acetal protecting groups. Two treatments were found to be necessary for more or less complete removal of uronic acid units and attendant neutral sugar residues in side chains from the capsular polysaccharide from *Klebsiella* type 47 (K47) (*239*). Heating acetylated polysaccharides at 100°C in acetic anhydride containing triethylamine appears to be a more satisfactory pro-

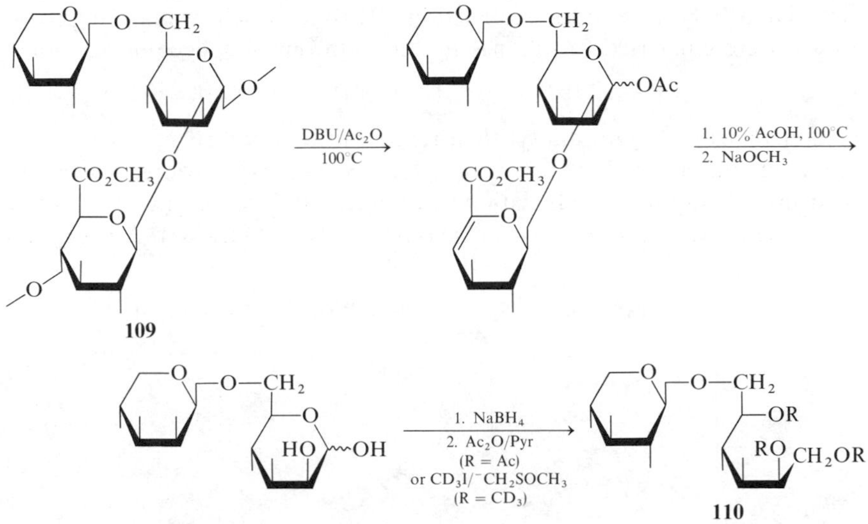

Fig. 44. Base-catalyzed fragmentation of methylated degraded leiocarpan A (**109**) with protection of exposed reducing groups by acetylation followed by processing to furnish oligosaccharide alditol derivatives (**110**).

→2)-α-D-Gal*p*-(1→3)-α-D-Man*p*-(1→2)-α-D-Man*p*-(1→3)-β-D-Glc*p*-(1-

2
↑
1

β-D-Glc*p*-(1→3)-β-D-Glc*p*A

111

cedure, there being virtually complete elimination of terminal or subterminal uronic acid residues in side chains, as shown in the removal of side chains from the capsular polysaccharide (**111**) from *Klebsiella* type 28 (*240*). It is interesting that the K47 polysaccharide was the least degraded. The degradation taking place under these conditions was first observed in studies on saponin glycosides (*241*) and presumably involves a mixed uronic–acetic anhydride as the activating group. The glycosyloxy substituent in the subterminal uronic acid units in K47 may be a relatively poor leaving group.

3. Degradations from Sulfone Derivatives

Sulfone degradation is effected through modification of primary hydroxyl groups. Two examples will illustrate the use of the method in establishing the distribution of such groups. The method was first applied to otherwise methylated polysaccharides (see Section VII,B) in which primary hydroxyl groups were converted to sulfones by the following sequence of reactions:

$$-CH_2OH \rightarrow -CH_2OTs \rightarrow -CH_2I \rightarrow -CH_2-SO_2-C_6H_4-CH_3$$

Treatment with sodium dimsyl then removes the modified residues as unsaturated sulfones with exposure of aglyconic hydroxyl groups, which can be identified after further alkylation (242). Base-catalyzed degradation of the methylated *Streptococcus pneumoniae* type 2 polysaccharide (**112** in Fig. 45)

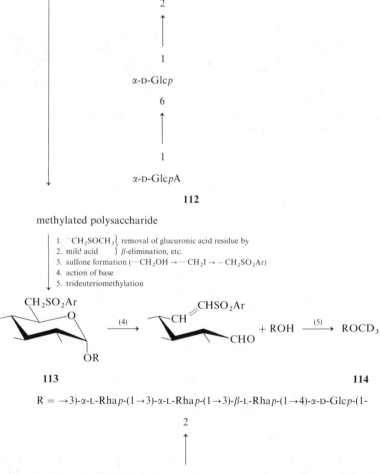

$\rightarrow$3)-α-L-Rha$p$-(1$\rightarrow$3)-α-L-Rha$p$-(1$\rightarrow$3)-β-L-Rha$p$-(1$\rightarrow$4)-α-D-Glc$p$-(1-

2
$\uparrow$
1

α-D-Glc$p$

6
$\uparrow$
1

α-D-Glc$p$A

112

methylated polysaccharide

1. $^-CH_2SOCH_3$ } removal of glucuronic acid residue by
2. mild acid   } β-elimination, etc.
3. sulfone formation ($-CH_2OH \rightarrow -CH_2I \rightarrow -CH_2SO_2Ar$)
4. action of base
5. trideuteriomethylation

CH_2SO_2Ar ⟶(4)⟶ CHSO_2Ar + ROH ⟶(5)⟶ ROCD_3

113   **114**

R = $\rightarrow$3)-α-L-Rha$p$-(1$\rightarrow$3)-α-L-Rha$p$-(1$\rightarrow$3)-β-L-Rha$p$-(1$\rightarrow$4)-α-D-Glc$p$-(1-

2
$\uparrow$

Fig. 45. Stepwise depolymerization of methylated derivative of *S. pneumoniae* type 2 specific polysaccharide (**112**) by removal of uronic acid end groups, introduction and removal of sulfone (**113**), and trideuteriomethylation to locate site of side chains in **114**.

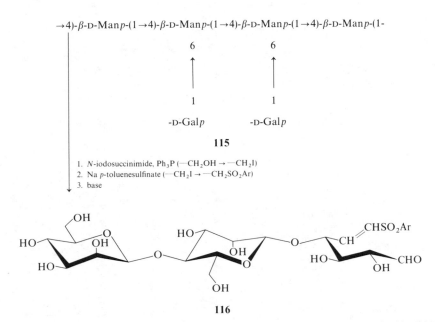

$\rightarrow$4)-β-D-Man$p$-(1$\rightarrow$4)-β-D-Man$p$-(1$\rightarrow$4)-β-D-Man$p$-(1$\rightarrow$4)-β-D-Man$p$-(1-

6   6

↑   ↑

1   1

-D-Gal$p$ -D-Gal$p$

115

1. N-iodosuccinimide, Ph$_3$P ($-$CH$_2$OH $\rightarrow$ $-$CH$_2$I)
2. Na p-toluenesulfinate ($-$CH$_2$I $\rightarrow$ $-$CH$_2$SO$_2$Ar)
3. base

116

Fig. 46. Sulfone degradation of galactomannan (**115**) to furnish oligosaccharide (**116**) containing the previously 6-linked residues.

results in a loss of uronic acid residues and exposure of primary hydroxyl groups, which are converted to sulfones (**113**). Although incomplete removal of terminal glucose residues was achieved, the site of linkage was demonstrated by trideuteriomethylation of exposed hydroxyl groups at O-2 of one of the 3-linked L-rhamnose residues (**114**) (*243*). Subsequently, it was shown that the many laborious series of steps in the previous example can be markedly decreased by direct selective iodination of primary hydroxyl groups in unsubstituted polysaccharides, followed by displacement with sodium *p*-toluenesulfinate (*203*). Degradation of sulfones on modified galactomannans (from **115** in Fig. 46) resulted in the isolation of oligosaccharide fragments (**116**) containing unmodified residues that had carried glycosyl substituents at O-6.

4. Degradations Preceded by Oxidation

Whereas the preceding degradations are dependent on the presence of hexuronic acid residues or primary hydroxyl groups (so far only C-6 of hexoses), the degradative sequence developed by Svensson and collaborators (*244*) can be initiated wherever hydroxyl groups are exposed in methylated polysaccharide derivatives, e.g., by base-catalyzed uronic acid degradations or by selective hydrolysis of acid-labile glycosidic linkages. Specific oxidation is effected with chlorine–DMSO (*245*) and subsequent treatment with base

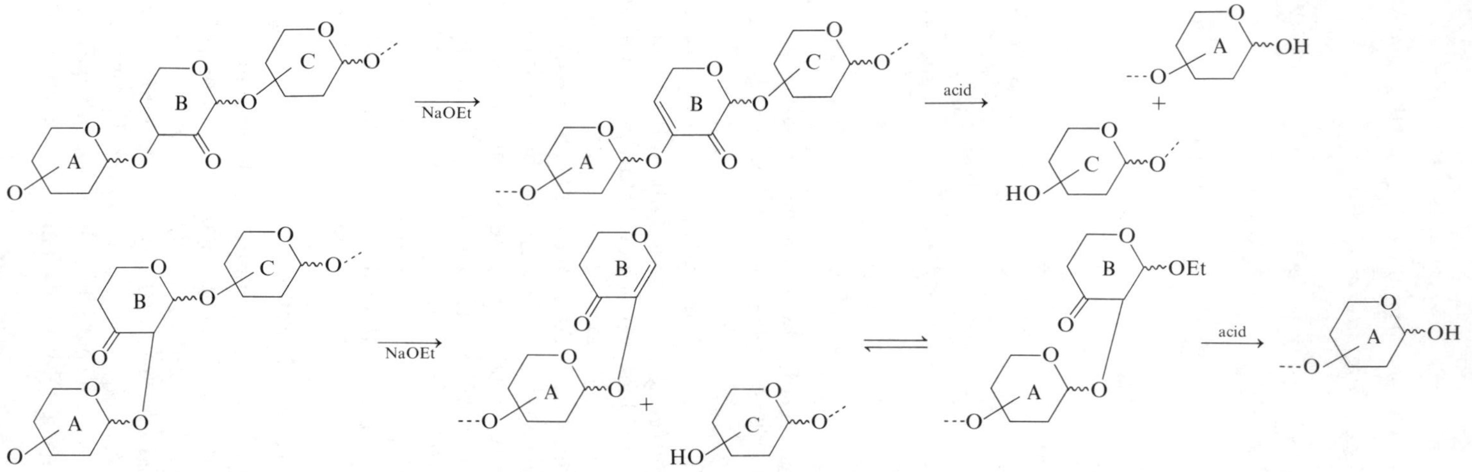

Fig. 47. Base-catalyzed degradation of differently sited glycosidulose residues in polysaccharides.

results directly, or after mild acid hydrolysis of enol glycosidic bonds, in the exposure of new hydroxyl groups from which the whole operation can be repeated (*246–248*). The exact course of this degradation, which can be performed in a stepwise manner, depends on the positions of the hydroxyl groups in the sugar residues and on the overall location of these residues in the polysaccharide structure. Figure 47 shows some general examples, based on studies of model compounds, of degradations initiated from differently sited glycosidulose residues. Figure 48 shows how the degradation has been applied to the degradation of the permethylated derivative of the capsular polysaccharide (**117**) from *Klebsiella* type 59 (*249*). Base-catalyzed degradation leads to the removal of terminal uronic acid residues with exposure of

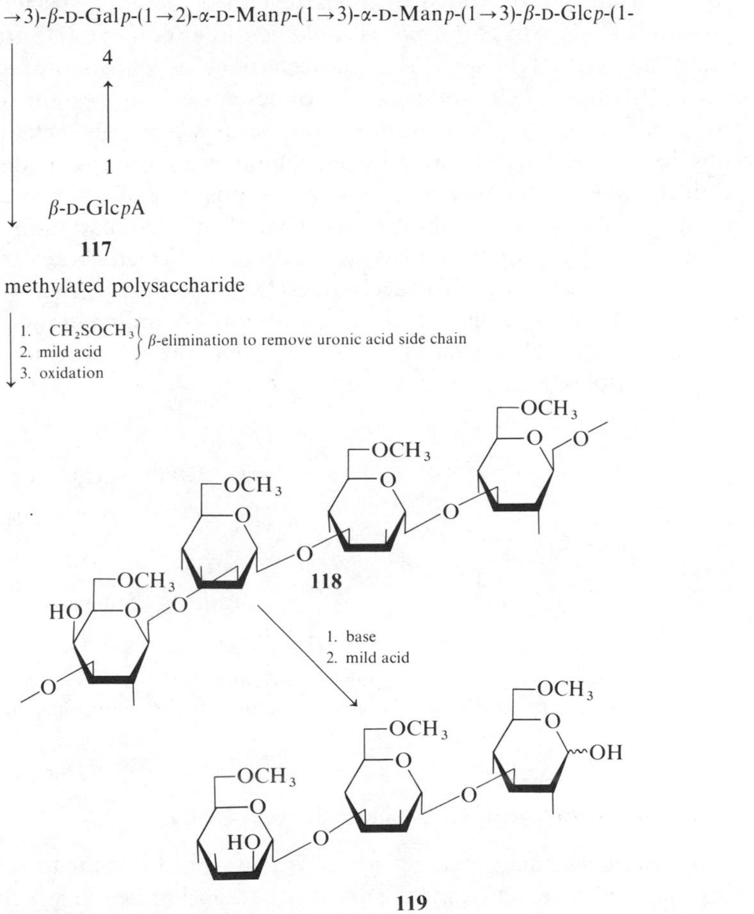

Fig. 48. Stepwise depolymerization of methylated derivative of *Klebsiella* type 59 capsular polysaccharide (**117**) by the Svensson oxidation–alkaline degradation procedure.

hydroxyl groups in the linear portion of the repeating structure of the polysaccharide (118), from which application of the Svensson oxidation–alkaline degradation procedure results in cleavage of interior chains with the formation of oligosaccharide 119.

5. Base-Catalyzed Cleavage of Phosphorodiester Linkages

The susceptibility of ribonucleic acids and other synthetic polyribonu-cleotides to alkali is due to the presence of hydroxyl groups vicinal to phosphorodiester linkages which participate in cyclic phosphate formation and consequent rapid depolymerization. Teichoic acid-like polysaccharides or "capsular teichoic acids" contain oligosaccharide repeating units that are mutually joined through phosphorodiester linkages (250). Polysaccharides of this important group, which includes several of the type-specific capsular substances from *Streptococcus pneumoniae* (see Volume II), are likewise readily degraded by base to oligosaccharides as phosphoromonoesters; enzyme treatment with phosphoromonoesterase removes the phosphate group. Since phosphate migration may occur via the intermediate cyclic phosphates, the location of phosphoromonoester groups in degradation products does not constitute proof of the position of attachment of one terminus of the phosphorodiester linkage in the parent macromolecule. The degradation of the specific substance (120) from *Streptococcus pneumoniae* type 6 to the repeating oligosaccharide (121) is shown in Fig. 49 (251). In several recent investigations carbon–phosphorus coupling in the ^{13}C-NMR spectra of such polysaccharides (e.g., 252) has provided evidence for the positions of phosphorodiester linkages.

$$\text{-2)-}\alpha\text{-D-Gal}p\text{-(1}\rightarrow\text{3)-}\alpha\text{-D-Glc}p\text{-(1}\rightarrow\text{3)-}\alpha\text{-L-Rha}p\text{-(1}\rightarrow\text{3)-D-ribitol-(5-O}\overset{\displaystyle O}{\underset{\displaystyle OH}{\overset{\|}{-}P-}}\text{O-}$$

120

$$\alpha\text{-D-Gal}p\text{-(1}\rightarrow\text{3)-}\alpha\text{-D-Glc}p\text{-(1}\rightarrow\text{3)-}\alpha\text{-L-Rha}p\text{-(1}\rightarrow\text{3)-ribitol}$$

121

Fig. 49. Degradation of the specific substance (120) from *Streptococcus pneumoniae* type 6 with alkali followed by treatment with phosphoromonoesterase affords the otherwise intact oligosaccharide (121).

B. Selective Cleavage of Glycosiduronic Acid Linkages

Glycosiduronic acid linkages are normally very resistant to acid hydro-lysis and related reactions. As a result oligosaccharides in which glycosyl substituents are still linked to nonterminal uronic acid residues have rarely been isolated from polysaccharides as partial hydrolysis products. For-

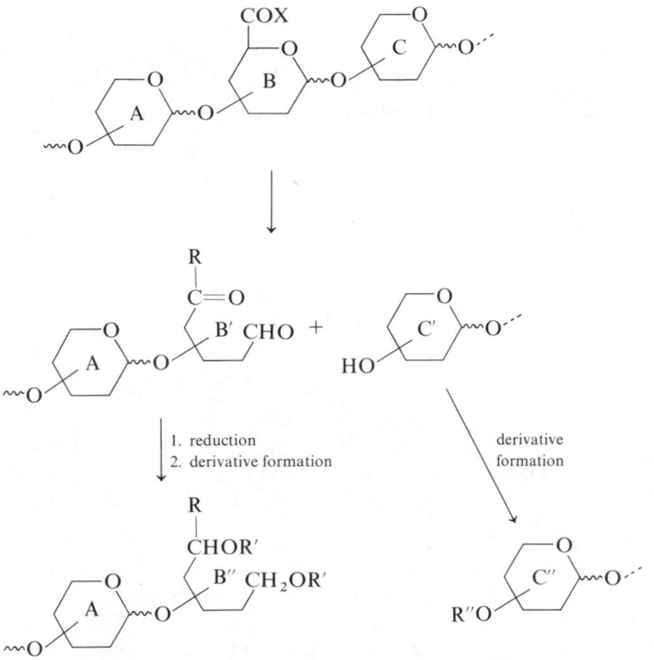

Fig. 50. General scheme for the selective cleavage of glycosidic linkages of hexuronic acid derivatives with formation of dicarbonyl sugars and aglycones followed by derivatization of the exposed residues.

tuitously nonterminal uronic acid residues in polysaccharides are most commonly 4-O-substituted, and these substituents are eliminated on base-catalyzed degradation. Other 2- and 3-linked substituents, if not directly liberated on extended base degradation, would be directly exposed as acid-labile vinyl or allyl glycosidic substituents to the modified hex-4-enopyranosiduronic acid units. Thus, even if such substituents are protected from degradation from their reducing groups during base treatment, they are readily removed during acid work-up, and information is lost on the anomeric configurations of these glycosyl units in the parent polysaccharide. During the past few years three different general approaches have been taken to the problem of selective cleavage of glycosiduronic acid linkages, all leading to the formation of modified units recognizably derived from the uronic acid residues, as shown in the overall scheme in Fig. 50.

1. Degradations by Hofmann, Curtius, and Lossen Rearrangements

Kochetkov *et al.* (*253*) first applied the Hofmann–Weermann degradation to the cleavage of glycosiduronamide linkages. Treatment of glycosiduron-amides with sodium hypochlorite at pH 13 (Fig. 51, pathway 1a) affords a

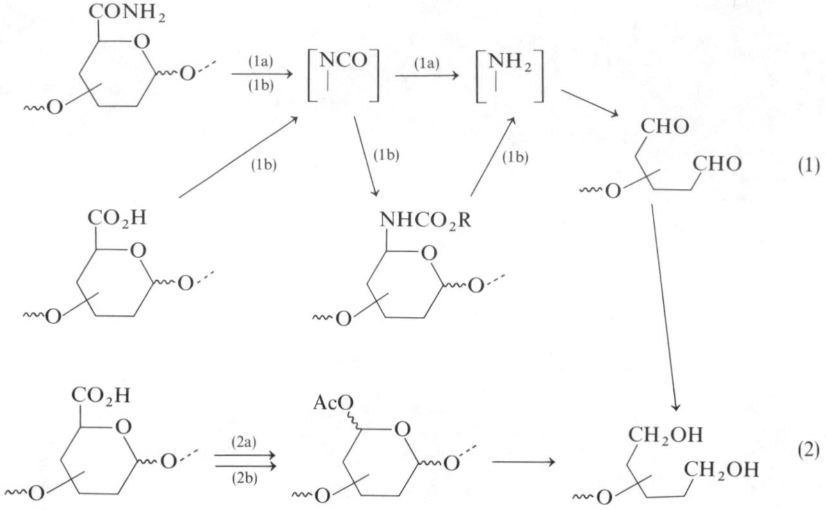

Fig. 51. Selective cleavage of glycosiduronic acid or glycosiduronamide linkages: (1) via intermediate 5-aminopentopyranosides generated (a) directly by the Hofmann amide re-arrangement or (b) from isolated carbamates (R = *tert*-butyl or benzyl) followed by hydrolysis to pentodialdoses and reduction to pentitols; (2) via 5-acetoxypentopyranosides generated by (a) reaction with lead tetraacetate or (b) anodic oxidation in acetic acid followed by direct deacetylation and reduction.

postulated 5-aminopentopyranoside intermediate, which, as a glycosylamine, undergoes rapid hydrolysis at pH 5 to give a pentodialdose with cleavage of the glycosidic linkage. This degradation product is characterized as the corresponding pentitol after reduction with sodium borohydride. The product thus obtained retains structural information on the parent uronic acid unit with respect to attached substituents and stereochemistry other than at C-5. Degradations by this method, therefore, would not differentiate between uronic acid units that are epimeric at C-5, e.g., D-glucuronic and L-iduronic acids. The procedure was shown to result in the removal of 94% of the uronic acid residues in birch 4-*O*-methylglucuronoxylan.

The Hofmann degradation can also be applied to the fragmentation of nonterminal glycosiduronamide linkages in methylated polysaccharides (*254*). Reaction performed on methylated gum arabic amide (**122** in Fig. 52) followed by reduction gave the disaccharide alditol (**123**), from which information was obtained on the anomeric configuration of the attached α-L-rhamnopyranosyl substituent and a degraded methylated polysaccharide in which hydroxyl groups arising from former sites of attachment of glucuronic acid residues were identified after further alkylation.

As a variation of this general approach, and avoiding the strongly alkaline conditions required for the normal Hofmann rearrangement, the same net

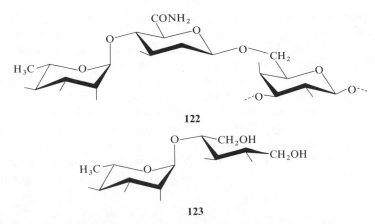

Fig. 52. Hofmann–Weerman degradation of glucopyranosiduronamide linkages in methylated gum arabic amide (**122**) followed by reduction with sodium borohydride affords the disaccharide alditol (**123**).

products can be generated by means of modified Curtius rearrangements. Treatment either of glycopyranosiduronic acids with diphenyl phosphorazidate (*255*) or of glycopyranosiduronamides with lead tetraacetate (*256*) in the presence of a suitable alcohol, e.g., *tert*-butanol or benzyl alcohol, results in the isolation of rearranged 5-aminopentopyranosides as stable carbamates without cleavage of glycosidic linkages (Fig. 51, pathway 1b). Selective cleavage is then carried out, as required, on removal of protecting butyloxycarbonyl (by controlled acid hydrolysis) or benzyloxycarbonyl (by hydrogenolysis) groups and mild acid hydrolysis and reduction to give pentitol residues in place of the original uronic acid units (*257*).

The Lossen hydroxamate rearrangement, which has not yet been applied to glycosiduronic acids, should also give rise to the same net degradation. The reaction has been used to degrade α-glycosyloxycarboxylic acids. Thus, the lipopolysaccharide from *Mycobacterium phlei* contains a glyceric acid terminus from which the glycan chain is liberated on conversion to the hydroxamic acid followed by treatment with a water-soluble carbodiimide (*50*). A similar reaction sequence has been used to remove the acid fragments formed on oxidation of periodate-oxidized residues arising from the core region of a *Salmonella* lipopolysaccharide (*258*).

2. Oxidative Decarboxylation with Lead Tetraacetate or by Anodic Oxidation

Methods that were originally developed by Kitagawa *et al.* (*259*) for the selective removal of uronic acid residues from saponins may be extended to methylated oligo- and polysaccharides, but with modifications in the processing of the initial reaction products. The reaction sequence leads to selective cleavage of glycosiduronic acid linkages with the formation of pentitols, as in the Hofmann and related degradations, and exposure of aglyconic hydroxyl groups. Oxidative decarboxylation of permethylated glycuronans (as carboxylic acids) with lead tetraacetate and accompanying acetylation furnishes epimeric 5-acetoxypentopyranosides (Fig. 51, pathway 2a). Treatment of these derivatives with sodium borohydride results in de-O-acetylation, cleavage of the glycosidic linkage, and reduction of the intermediate pentodialdose derivatives (260). The procedure, which is probably the method of choice for the selective fragmentation of methylated polysaccharides containing β-D-glucosiduronic acid residues, is apparently

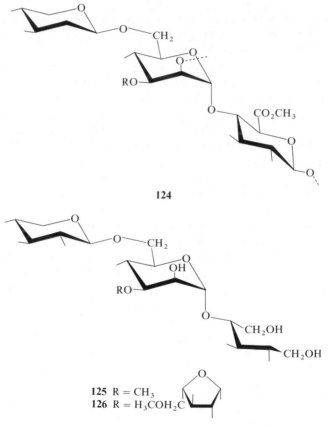

Fig. 53. Selective cleavage of glucopyranosiduronic acid linkages in methylated leiocarpan A (124) followed by reduction furnishes xylitol-terminated oligosaccharides (125 and 126).

sensitive to steric factors since only limited reaction occurs with α-D-galactopyranosiduronic acids. An example of the utility of the method is provided (Fig. 53) by the fragmentation under mild conditions of the glucuronomannan chain of methylated leiocarpan A (124) to yield as major products xylitol-terminated oligosaccharides (125 and 126) in which the more acid-labile side-chain residues are still attached (261).

Anodic oxidation in acetic acid of uronic acid derivatives of low molecular weight results in the formation of the same 5-acetoxypentopyranosides (Fig. 51, pathway 2b) (262), but in experiments carried out so far no reaction has been detected on the electrode surface with polysaccharides (257).

3. Formation and Selective Hydrolysis of Hex-5-Enopyranosides from Uronic Acids

Uronic acid residues in methylated glycuronans are readily reduced with lithium aluminum hydride (71). The resulting primary hydroxyl groups in the otherwise substituted polysaccharide can be converted to 6-deoxy-6-iodo derivatives, and on treatment with DBU these sugar residues give hex-5-enopyranosides (263). Extremely mild conditions are then required for selective hydrolysis to 6-deoxyglycos-5-ulose units. Reduction then affords epimeric pairs of 6-deoxyhexitols (for overall reaction sequence see Fig. 54). In addition to the disadvantage of the formation of two products for each type of glycosiduronic acid residue, the rather lengthy reaction sequence is probably not a preferred procedure. It is satisfactory for glucuronans, but studies on model compounds suggest that the corresponding transformations of galactose (from galacturonic acid) residues are likely to proceed much less

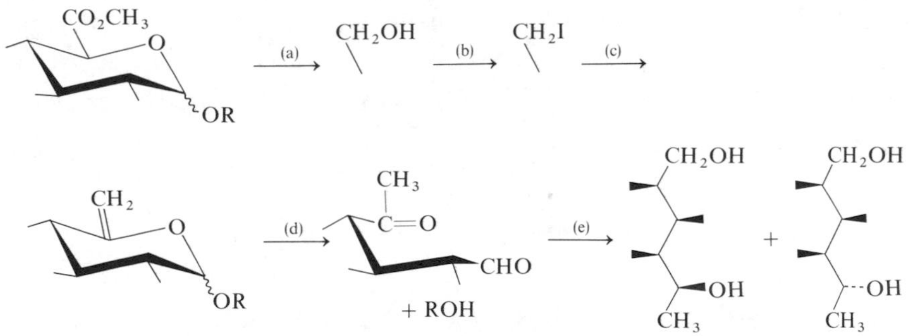

Fig. 54. General procedure for the selective fragmentation of modified methylated hexuronic acid residues in polysaccharides by (a) reduction with LiAlH$_4$, (b) reaction with (PhO)$_3$$\overset{+}{P}$—CH$_3$ I$^-$, (c) dehydroiodination with DBU, (d) hydrolysis of hex-5-enopyranosides, and (e) reduction to give an epimeric pair of 6-deoxyalditols.

readily (*264*). Although not yet demonstrated in practice, this series of reactions could be used as an alternative to those used in the sulfone degradation, in which primary hydroxyl groups are initially immobilized by tritylation.

C. Selective Cleavage of Aminoglycosidic Linkages by Deamination

The selective cleavage of equatorially oriented 2-amino-2-deoxyglycosidic bonds by nitrous acid deamination with the formation of 2,5-anhydro sugars has been extensively studied with derivatives of 2-amino-2-deoxy-D-glucose (D-glucosamine) and D-galactosamine (*265*), which are the most widely occurring amino sugar constituents of polysaccharides and other glycoconjugates. Most frequently, these amino sugars occur as their *N*-acetyl derivatives, and N-deacetylation must precede deamination. This reaction is best accomplished either by hydrazinolysis with anhydrous reagent and a catalytic quantity of hydrazine sulfate (*114*) or by treatment with sodium hydroxide in DMSO (*115*).

Nitrous acid deamination has been used extensively in studies on the glycosaminoglycan heparin. Although this polysaccharide is composed largely of alternating uronic acid and amino sugar residues, the structure is complicated by the fact that the majority of amino sugar residues are present as *N*-sulfate, i.e., as residues of 2-deoxy-2-sulfamino-D-glucose, but in some regions of the structure units of 2-acetamido-2-deoxy-D-glucose predominate, whereas elsewhere the D-glucosamine units are unsubstituted on amino groups. However, modifications of the deamination reaction are available to cleave N-sulfated 2-amino-2-deoxy-D-glucosidic linkages only, N-unsubstituted linkages only, or both. The major repeating segment of heparin

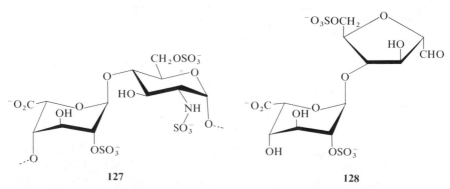

Fig. 55. Selective cleavage by nitrous acid deamination of major segments in heparin (**127**) to give the disulfated disaccharide **128**.

contains alternating units of 4-linked α-L-idopyranosyluronic acid and ‘4-linked 2-deoxy-2-sulfamino-α-D-glucopyranose carrying additional O-sulfate substituents (127 in Fig. 55), and nitrous acid deamination affords as the major product the disulfated 2,5-anhydro-(4-O-α-L-idopyranosyluronic acid)-D-mannose (128) (266, 267).

Deamination of 2-amino-2-deoxy-D-glucopyranosides proceeds mainly by pathway (a) (Fig. 56) with glycoside cleavage and formation of 2,5-anhydro-D-mannose derivatives. Insofar as the stereoelectronic requirements for the loss of nitrogen from the intermediate diazonium ion with concomitant ring contraction involve participation by electrons of the O—C-1 bond, the same geometric requirement is met by participation of the C-3—C-4 bond. Such an alternative ring contraction [pathway (b)] was first demonstrated by Erbing et al. (268), who showed that the reaction, accounting for some 20–25% of the total product, led to the formation of 2-C-formyl pentofuranosides, which readily underwent epimerization at C-2. Important consequences follow from the intermediate formation of oxocarbonium ions. Whereas with the normal ring contraction hydrolysis of this species results in the formation of the 2,5-anhydro-D-mannose and liberation of the aglycone, in the case of the alternative ring contraction hydrolysis of the corresponding intermediate affords the 2-C-formyl pentofuranoside with liberation of the 3-O substituent, if such is present. Thus, in an oligo- or polysaccharide chain a 3-O-glycosyl substituent is liberated as a reducing group. Although this competing reaction that occurs during deamination is a complication, additional structural information may be obtained, provided that the extra products can be separated and identified. Examples have been reported of these reactions and of the competing reactions that occur in the deamination of 2-amino-2-deoxyhexitol residues formed from the reducing units of aminooligosaccharides of natural occurrence or liberated from glycoproteins (67,269).

An example of the products arising from the two degradative pathways is provided by the application of the N-deacetylation–nitrous acid deamination sequence to the O-specific polysaccharide chain of Shigella dysenteriae type 1 lipopolysaccharide (129 in Fig. 57), which affords two oligosaccharides, 130 from the normal ring contraction and 131 from the alternative ring contraction as the 3-O substituent of the N-acetyl-D-glucosamine residue (270). It is probable that the 2-C-formyl pentofuranoside residue, which would be expected to occupy the nonreducing terminus of oligosaccharide 131, is sufficiently acid labile to be liberated even under the extremely mild conditions used for the deamination. Although the liberation of the 3-O-glycosyl substituent is a necessary consequence of the alternative ring contraction, it should be noted that 3-linked 2,5-anhydrohexose residues are quite susceptible to base-catalyzed β-elimination and that this degradation

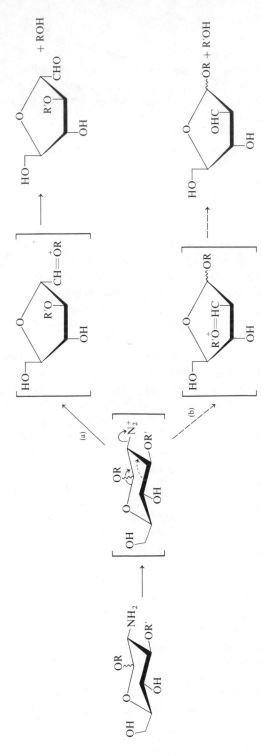

Fig. 56. Alternative ring contractions during the nitrous acid deamination of 3-*O*-substituted 2-amino-2-deoxy-D-glucopyranosides resulting in (a) selective glycoside cleavage with formation of 2,5-anhydro-D-mannose derivatives with liberation of aglycone and (b) formation of 2-*C*-formyl pentofuranosides (this reaction occurs with epimerization at C-2) with liberation of 3-*O*-glycosyl substituents.

→3)-α-L-Rhap-(1→3)-α-L-Rhap-(1→2)-α-D-Galp-(1→3)-α-D-GlcpNAc-(1-

129

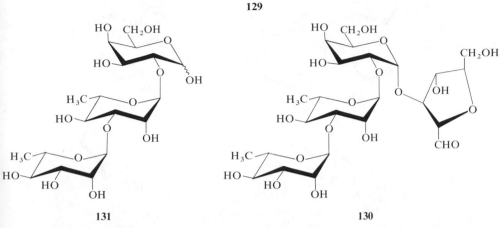

Fig. 57. N-Deacetylation–nitrous acid deamination of the O-specific polysaccharide chain (**129**) from the *Shigella dysenteriae* type 1 lipopolysaccharide leads to the isolation of two oligosaccharides (**130** and **131**).

is often used in the characterization of oligosaccharide fragments formed on deamination (e.g., **132** → **133** in Fig. 58) (*271*).

Deaminations of aminodeoxyglycosides give a variety of products depending on the location and stereochemical orientation of the amino groups in relation to neighboring groups; rarely is a single type of product formed (*265*). Such is the case with axially oriented 2-amino-2-deoxyglycosides, e.g., 2-amino-2-deoxy-D-mannopyranosides. However, Hase and Matsushima (*272*) converted segments of a 2-acetamido-2-dexoy-D-mannurono-D-glucan (**134** in Fig. 59) from the cell walls of *Micrococcus lysodeikticus* to a D-glucan (**135**) by nitrous acid deamination of the N-deacetylated carboxyl-reduced polysaccharide.

2-Acetamido-4-amino-2,4,6-trideoxy-D-galactose is a constituent of the capsular polysaccharide from *Streptococcus pneumoniae* type 1 (*273*), the species-specific cell wall teichoic acid (or C-substance) from *S. pneumoniae* (*274*), and the O-specific side chains of the *Shigella sonnei* phase 1 lipopolysaccharide (*275*). Reaction of the 4-amino group in these residues with nitrous acid proceeds by three competing pathways (Fig. 60): (a) solvolytic displacement at C-4, resulting in the formation of residues of 2-acetamido-2,6-dideoxy-D-glucose without chain scission; (b) a 5 → 4 hydride shift, leading to the formation of a 2-acetamido-2,4,6-trideoxyglycos-5-ulose residue, which is readily hydrolyzed with chain cleavage; and (c) a 3→4 hydride shift, giving an enol glycoside from which the 3-linked substituent is readily

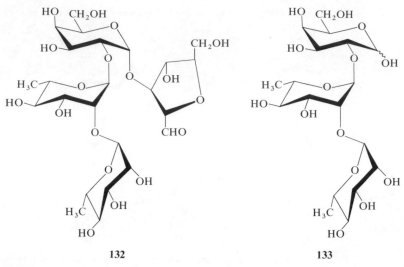

Fig. 58. Tetrasaccharide **132**, formed on N-deacetylation–nitrous acid deamination of the O-specific side chains of the cell wall lipopolysaccharide from *E. coli* O 69 and containing a 3-linked 2,5-anhydro-D-mannose terminus, undergoes ready base-catalyzed β-elimination to give the trisaccharide **133**.

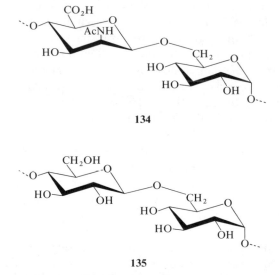

Fig. 59. Segments of 2-acetamido-2-deoxy-D-mannurono-D-glucan (**134**) from *Micrococcus lysodeikticus* are converted to a D-glucan (**135**) on nitrous acid deamination of the N-deacetylated carboxyl-reduced polysaccharide.

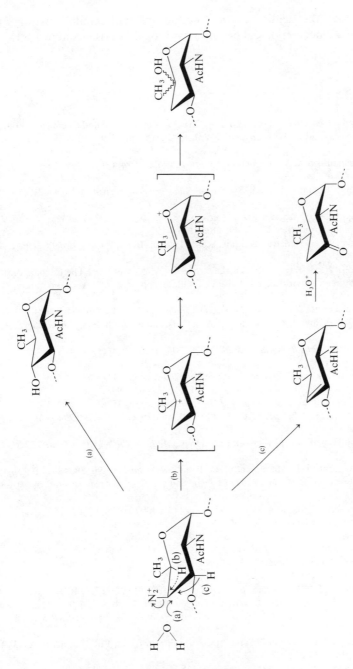

Fig. 60. Competing degradative pathways (see text) on nitrous acid deamination of residues of 2-acetamido-4-amino-2,4,6-trideoxy-D-galactopyranose in polysaccharides.

liberated under the acidic reaction conditions with chain scission and formation of a 2-acetamido-2,4,6-trideoxyglycosid-3-ulose residue.

References

1. G. O. Aspinall, *in* "Elucidation of Organic Structures by Physical and Chemical Methods" (K. W. Bentley and G. W. Kirby, eds.), Vol. IV, Part II, pp. 379–450. Wiley, New York, 1973.
2. G. O. Aspinall and A. M. Stephen, *MTP Int. Rev. Sci.: Org. Chem., Ser. One* 7, 285–313 (1973).
3. B. Lindberg, J. Lönngren, and S. Svensson, *Adv. Carbohydr. Chem. Biochem.* 31, 185–240 (1975).
4. G. O. Aspinall, *Int. Rev. Sci.: Org. Chem., Ser. Two* 7, 201–222 (1976).
5. G. O. Aspinall, *Pure Appl. Chem.* 49, 1105–1134 (1977).
6. For a survey of colorimetric methods, see Z. Dische, *Methods Carbohydr. Chem.* 1, 477 *et seq.* (1962).
7. D. Aminoff, W. W. Binkley, R. Schaeffer, and R. W. Mowery, *in* "The Carbohydrates" (W. Pigman and D. Horton, eds.), 3rd ed., Vol. 2B, pp. 739–807. Academic Press, New York, 1970.
8. J. E. Hodge and B. T. Hofreiter, *Methods Carbohydr. Chem.* 1, 380–394 (1962).
9. M. Dubois, K. Gilles, J. K. Hamilton, P. A. Rebers, and F. Smith, *Anal. Chem.* 28, 350–356 (1956).
10. R. J. Winzler, *Methods Biochem. Anal.* 2, 279–311 (1955).
11. T. Bitter and H. M. Muir, *Anal. Biochem.* 4, 330–334 (1962).
12. N. Blumenkrantz and G. Asboe-Hansen, *Anal. Biochem.* 54, 484–489 (1973).
13. Z. Dische and L. B. Shettles, *J. Biol. Chem.* 175, 595–603 (1948).
14. I. Werner and L. Odin, *Acta Soc. Med. Ups.* 57, 230–235 (1952).
15. G. O. Aspinall, E. L. Hirst, E. G. V. Percival, and R. G. J. Telfer, *J. Chem. Soc.* pp. 337–342 (1953).
16. J. F. Codington, K. B. Linsley, and C. Silber, *Methods Carbohydr. Chem.* 7, 226–232 (1976).
17. R. R. Selvendran, J. F. March, and S. G. Ring, *Anal. Biochem.* 96, 282–292 (1979).
18. P. Albersheim, D. J. Nevins, P. D. English, and A. Karr, *Carbohydr. Res.* 5, 340–345 (1967).
19. R. E. Chambers and J. R. Clamp, *Biochem. J.* 125, 1009–1018 (1971).
20. S. J. Angyal and K. Dawes, *Aust. J. Chem.* 21, 2747–2760 (1968).
21. P. J. Stoffyn and R. W. Jeanloz, *J. Biol. Chem.* 235, 2507–2510 (1960).
22. E. L. Hirst and D. A. Rees, *J. Chem. Soc.* pp. 1182–1187 (1965).
23. D. M. W. Anderson, *Talanta* 2, 73–78 (1959).
24. J. N. C. Whyte and J. R. Englar, *Anal. Biochem.* 59, 426–435 (1974).
25. R. L. Taylor and H. E. Conrad, *Biochemistry* 11, 1383–1388 (1972); R. L. Taylor, J. E. Shively, and H. E. Conrad, *Methods Carbohydr. Chem.* 7, 149–151 (1976).
26. B. Lindberg, J. Lönngren, J. L. Thompson, and W. Nimmich, *Carbohydr. Res.* 25, 49–57 (1972).
27. R. G. Spiro, *in* "Methods in Enzymology" (V. Ginsburg, ed.), Vol. 28, Part B, pp. 3–43. Academic Press, New York, 1973.
28. R. Varma and R. S. Varma, *J. Chromatogr.* 128, 45–52 (1976); 139, 303–310 (1977).

29. Y. C. Lee, *in* "Methods in Enzymology" (V. Ginsburg, ed.), Vol. 28, Part B, pp. 63–73. Academic Press, New York, 1973.
30. J. F. Kennedy and J. E. Fox, *Methods Carbohydr. Chem.* **8**, 3–12 (1980).
31. G. D. McGinnis and P. Fang, *Methods Carbohydr. Chem.* **8**, 33–43 (1980).
32. O. Samuelsson, *Methods Carbohydr. Chem.* **6**, 65–76 (1972).
33. R. A. Laine, W. J. Esselman, and C. C. Sweeley, *in* "Methods in Enzymology" (V. Ginsburg, ed.), Vol. 28, Part B, pp. 159–178. Academic Press, New York, 1973.
34. J. S. Sawardeker, J. H. Sloneker, and A. Jeanes, *Anal. Chem.* **37**, 1602–1604 (1965).
35. B. A. Dmitriev, L. V. Backinowsky, O. S. Chishov, B. M. Zolotarev, and N. K. Kochetkov, *Carbohydr. Res.* **19**, 432–435 (1971).
36. G. M. Bebault, J. M. Berry, Y. M. Choy, G. G. S. Dutton, N. Funnell, L. D. Hayward, and A. M. Stephen, *Can. J. Chem.* **51**, 324–326 (1973).
37. K. Leontein, B. Lindberg, and J. Lönngren, *Carbohydr. Res.* **62**, 359–362 (1978).
38. G. J. Gerwig, J. P. Kamerling, and J. F. G. Vliegenthart, *Carbohydr. Res.* **77**, 1–7 (1979).
39. J. H. Pazur, *in* "Methods in Enzymology" (W. A. Wood, ed.), Vol. 9, pp. 82–87. Academic Press, New York, 1966.
40. P. W. Mobley, R. P. Metzger, and A. N. Wick, *in* "Methods in Enzymology" (W. A. Wood, ed.), Vol. 41, pp. 173–177. Academic Press, New York, 1975.
41. M. Duckworth and W. Yaphe, *Chem. Ind.* (*London*) pp. 747–748 (1970).
42. For an evaluation of various methods, see J. F. Kennedy, "Proteoglycans—Biological and Chemical Aspects in Human Life," pp. 112–113. Elsevier, Amsterdam, 1979.
43. J. F. Kennedy and D. A. Westman, *Anal. Chim. Acta* **55**, 448–449 (1971).
44. T. H. Schultz, *Methods Carbohydr. Chem.* **5**, 189–194 (1965).
45. P. J. Wood and I. R. Siddiqui, *Anal. Biochem.* **39**, 418–428 (1971).
46. G. R. Gray and C. E. Ballou, *J. Biol. Chem.* **246**, 6835–6842 (1971).
47. H. S. Isbell, *Methods Carbohydr. Chem.* **5**, 249–250 (1965).
48. G. W. Hay, B. A. Lewis, F. Smith, and A. M. Unrau, *Methods Carbohydr. Chem.* **5**, 251–253 (1965).
49. S. Peat, W. J. Whelan, and H. G. Lawley, *J. Chem. Soc.* pp. 729–737 (1958).
50. M. H. Saier, Jr. and C. E. Ballou, *J. Biol. Chem.* **243**, 992–1005 (1968).
51. C. T. Greenwood, *Adv. Carbohydr. Chem.* **7**, 289–332 (1952); **11**, 385–393 (1956).
52. R. A. Gibbons, *in* "Glycoproteins" (A. Gottschalk, ed.), 2nd ed., Part A, pp. 31–140. Elsevier, Amsterdam, 1972.
53. J. van Dam and W. Prins, *Methods Carbohydr. Chem.* **5**, 253–261 (1965).
54. C. T. Greenwood, *Methods Carbohydr. Chem.* **5**, 261–265 (1965).
55. G. A. Towle, *Methods Carbohydr. Chem.* **6**, 510–512 (1972).
56. R. St. J. Manley, *Methods Carbohydr. Chem.* **3**, 289–302 (1963).
57. S. C. Churms, *Adv. Carbohydr. Chem. Biochem.* **25**, 13–51 (1970).
58. S. C. Churms, *in* "Chromatography" (E. Heftmann, ed.), 3rd ed., p. 637. Van Nostrand-Reinhold, Princeton, New Jersey, 1975.
59. P. Andrews, *Methods Biochem. Anal.* **18**, 1–53 (1970); see also R. L. Whistler and A. K. M. Anisuzzaman, *Methods Carbohydr. Chem.* **8**, 45–53 (1980).
60. B. Lindberg, *in* "Methods in Enzymology" (V. Ginsburg, ed.), Vol. 28, Part B, pp. 178–195. Academic Press, New York, 1973.
61. P.-E. Jansson, L. Kenne, H. Liedgren, B. Lindberg, and J. Lönngren, *Chem. Commun. Univ. Stockholm* p. 1 (1976); *Chem. Abstr.* **87**, 136–153 (1977).
62. B. Lindberg and J. Lönngren, *in* "Methods in Enzymology" (V. Ginsburg, ed.), Vol. 50, Part C, pp. 3–33. Academic Press, New York, 1978.
63. A. G. Darvill, M. McNeil, and P. Albersheim, *Plant Physiol.* **62**, 418–422 (1978).
64. L. R. Phillips and B. A. Fraser, *Carbohydr. Res.* **90**, 149–151 (1981).

65. E. L. Hirst and E. Percival, *Methods Carbohydr. Chem.* **5**, 287–296 (1965).
66. K. Wallenfels, G. Bechtler, R. Kuhn, H. Trischmann, and H. Egge, *Angew. Chem., Int. Ed. Engl.* **2**, 515–523 (1963); R. Kuhn and H. Egge, *Chem. Ber.* **96**, 3338–3348 (1963).
67. G. O. Aspinall, E. Przybylski, R. G. S. Ritchie, and C. O. Wong, *Carbohydr. Res.* **66**, 225–243 (1978).
68. J. F. Carson and W. D. Maclay, *J. Am. Chem. Soc.* **68**, 1015–1017 (1946).
69. J.-P. Joseleau, G. Chambat, and B. Champitazi-Hermoza, *Carbohydr. Res.* **90**, 339–344 (1981).
70. P. Prehm, *Carbohydr. Res.* **78**, 372–374 (1980).
71. M. Abdel-Akher and F. Smith, *Nature (London)* **166**, 1037–1038 (1950).
72. K. Stellner, H. Saito, and S.-I Hakomori, *Arch. Biochem. Biophys.* **155**, 464–472 (1973).
73. R. L. Whistler and J. N. BeMiller, *Methods Carbohydr. Chem.* **1**, 47–50 (1962).
74. R. L. Whistler and J. N. BeMiller, *Methods Carbohydr. Chem.* **1**, 42–46 (1962).
75. For summary tables of data, see *Methods Carbohydr. Chem.* **5**, 298–357 (1965).
76. T. G. Bonner, E. J. Bourne, and S. McNally, *J. Chem. Soc.* pp. 2929–2934 (1960); T. G. Bonner and E. J. Bourne, *Methods Carbohydr. Chem.* **2**, 206–207 (1963).
77. R. U. Lemieux and H. F. Bauer, *Can. J. Chem.* **31**, 814–820 (1953).
78. G. G. S. Dutton, *Adv. Carbohydr. Chem. Biochem.* **28**, 11–160 (1973); **30**, 9–110 (1974).
79. N. Handa and R. Montgomery, *Carbohydr. Res.* **11**, 467–484 (1969).
80. J. F. Batey and J. R. Turvey, *Carbohydr. Res.* **43**, 133–143 (1975).
81. F. May and H. Weinland, *Hoppe-Seyler's Z. Physiol. Chem.* **305**, 87–96 (1956).
82. B. M. Dmitriev, Y. A. Knirel, N. A. Kocharova, N. K. Kochetkov, E. S. Stranislavsky, and G. M. Mashilova, *Eur. J. Biochem.* **106**, 643–651 (1980).
83. A. K. Bhattacharjee and H. J. Jennings, *Carbohydr. Res.* **51**, 253–261 (1976).
84. N. S. Anderson and D. A. Rees, *J. Chem. Soc.* pp. 5880–5881 (1965).
85. J. Lönngren and S. Svensson, *Adv. Carbohydr. Chem. Biochem.* **29**, 41–106 (1974).
86. M. McNeil and P. Albersheim, *Carbohydr. Res.* **56**, 239–248 (1977).
87. C. G. Wong, S-S. J. Sung, and C. C. Sweeley, *Methods Carbohydr. Chem.* **8**, 55–65 (1980).
88. F. R. Seymour, R. D. Plattner, and M. E. Slodki, *Carbohydr. Res.* **44**, 181–198 (1975).
89. D. P. Sweet, P. Albersheim, and R. H. Shapiro, *Carbohydr. Res.* **40**, 199–216 (1975).
90. D. P. Sweet, R. H. Shapiro, and P. Albersheim, *Biomed. Mass Spectrom.* **1**, 263–268 (1974).
91. T. J. Painter, *Can. J. Chem.* **37**, 497–502 (1959).
92. T. J. Painter, *Methods Carbohydr. Res.* **5**, 280–285 (1965).
93. C. Galanos, O. Lüderitz, and K. Himmelspach, *Eur. J. Biochem.* **8**, 332–336 (1969).
94. G. O. Aspinall, *Adv. Carbohydr. Chem. Biochem.* **24**, 333–379 (1969).
95. C. T. Bishop, M. B. Perry, F. Blank, and F. P. Cooper, *Can. J. Chem.* **43**, 30–39 (1965).
96. C. G. Hellerqvist, B. Lindberg, S. Svensson, T. Holme, and A. A. Lindberg, *Carbohydr. Res.* **8**, 43–55 (1968).
97. G. O. Aspinall and C. C. Whitehead, *Can. J. Chem.* **48**, 3840–3849, 3850–3855 (1970).
98. R. M. Saunders, *Carbohydr. Res.* **7**, 76–79 (1968).
99. L.-Å. Fransson, L. Rodén, and M. L. Spach, *Anal. Biochem.* **23**, 317–330 (1968).
100. G. O. Aspinall, I. W. Cottrell, S. V. Egan, I. M. Morrison, and J. N. C. Whyte, *J. Chem. Soc. C* pp. 1071–1080 (1967).
101. O. Samuelson and L. Wicterin, *Carbohydr. Res.* **4**, 139–144 (1967).
102. R. W. Bailey, "Oligosaccharides." Pergamon, Oxford, 1965.
103. J. Staněk, M. Černy, and J. Pacák, "The Oligosaccharides." Academic Press, New York, 1965.
104. G. O. Aspinall, E. Percival, D. A. Rees, and M. Rennie, *in* "Rodd's Chemistry of Carbon Compounds" (S. Coffey, ed.), 2nd ed., Vol. IF, pp. 596–714. Elsevier, Amsterdam, 1967.
105. R. W. Bailey and J. B. Pridham, *Adv. Carbohydr. Chem.* **17**, 121–167 (1962).

106. R. J. Ferrier, *in* "The Carbohydrates" (W. Pigman and D. Horton, eds.), 3rd ed., Vol. IB, p. 1354, Academic Press, New York, 1980.

107. Y.-T. Li and S.-C. Li, *in* "The Glycoconjugates" (M. I. Horowitz, ed.), Vol. I, pp. 52–68. Academic Press, New York, 1977.

108. A. S. Perlin, *Adv. Carbohydr. Chem.* **14**, 1–61 (1959).

109. A. S. Perlin, *in* "The Carbohydrates" (W. Pigman and D. Horton, eds.), 3rd ed., Vol. IB, pp. 1167–1215. Academic Press, New York, 1980.

110. J. N. BeMiller, *Adv. Carbohydr. Chem.* **22**, 25–108 (1967).

111. U. Lindahl, *Ark. Kemi* **26**, 101–110 (1966).

112. T. Popoff and O. Theander, *Carbohydr. Res.* **22**, 135–149 (1972).

113. R. C. G. Noggridge and A. Neuberger, *J. Chem. Soc.* pp. 745–750 (1938); E. R. B. Graham and A. Neuberger, *J. Chem. Soc. C* pp. 1638–1641 (1968).

114. Z. Yosizawa, T. Sato, and K. Schmid, *Biochim. Biophys. Acta* **121**, 417–420 (1966).

115. L. Kenne and B. Lindberg, *Methods Carbohydr. Chem.* **8**, 295–296 (1980).

116. M. L. Wolfrom and B. O. Juliano, *J. Am. Chem. Soc.* **82**, 1673–1677 (1960).

117. K. Onodera, T. Komano, and S. Hirano, *Biochim. Biophys. Acta* **83**, 20–26 (1964).

118. Y. C. Lee and C. E. Ballou, *Biochemistry* **4**, 287–298 (1965).

119. C. J. Lawson and D. A. Rees, *J. Chem. Soc. C* pp. 1301–1304 (1968).

120. G. O. Aspinall and J. Baillie, *J. Chem. Soc.* pp. 1702–1714 (1963).

121. G. O. Aspinall, A. J. Charlson, E. L. Hirst, and R. Young, *J. Chem. Soc.* pp. 1696–1702 (1963).

122. B. Bayard and J. Montreuil, *Carbohydr. Res.* **24**, 427–443 (1972).

123. R. Kuhn and H. Wiegandt, *Chem. Ber.* **96**, 866–880 (1963).

124. G. Spik, B. Bayard, B. Fournet, G. Strecker, S. Bouquelet, and J. Montreuil, *FEBS Lett.* **50**, 296–299 (1975).

125. R. D. Guthrie and J. F. McCarthy, *Adv. Carbohydr. Chem. Biochem.* **22**, 11–23 (1967).

126. G. O. Aspinall and J. P. McKenna, *Carbohydr. Res.* **7**, 244–254 (1968).

127. B. Nilsson and S. Svensson, *Carbohydr. Res.* **72**, 183–190 (1979).

128. A. Lundblad, S. Svensson, B. Löw, L. Messeter, and B. Cedergren, *Eur. J. Biochem.* **104**, 323–330 (1980).

129. B. Nilsson and S. Svensson, *Carbohydr. Res.* **65**, 169–171 (1978).

130. B. Nilsson and S. Svensson, *Carbohydr. Res.* **69**, 292–296 (1979).

131. B. Nilsson and S. Svensson, *Carbohydr. Res.* **62**, 377–380 (1978).

132. S. Hirase and C. Araki, *Bull. Chem. Soc. Jpn.* **27**, 105–109 (1954).

133. D. A. Rees, *J. Chem. Soc.* pp. 1821–1832 (1963).

134. S. Hirase, *Bull. Chem. Soc. Jpn.* **30**, 75–79 (1957).

135. N. S. Anderson, T. C. S. Dolan, and D. A. Rees, *J. Chem. Soc. C* pp. 596–601 (1968).

136. J. Hoffman, B. Lindberg, and S. Svensson, *Acta Chem. Scand.* **26**, 661–666 (1972); J. Hoffman and B. Lindberg, *Methods Carbohydr. Chem.* **8**, 117–122 (1980).

137. F. W. Parrish, A. S. Perlin, and E. T. Reese, *Can. J. Chem.* **38**, 2094–2104 (1960).

138. H. R. Goldschmidt and A. S. Perlin, *Can. J. Chem.* **41**, 2272–2277 (1963).

139. C. Araki and K. Arai, *Bull. Chem. Soc. Jpn.* **30**, 287–293 (1957).

140. H. Niemann, H. Beilharz, and S. Stirm, *Carbohydr. Res.* **60**, 353–366 (1978).

141. G. G. S. Dutton, K. L. Mackie, A. V. Savage, D. Riegler-Hug, and S. Stirm, *Carbohydr. Res.* **84**, 161–170 (1980).

142. G. G. S. Dutton, A. V. Savage, and M. Vignon, *Can. J. Chem.* **58**, 2588–2591 (1980).

143. B. Lindberg, K. Samuelsson, and W. Nimmich, *Carbohydr. Res.* **30**, 63–70 (1973).

144. B. Valent, A. G. Darvill, M. McNeil, B. K. Robertson, and P. Albersheim, *Carbohydr. Res.* **79**, 165–192 (1980).

145. A. G. Darvill, M. McNeil, and P. Albersheim, *Carbohydr. Res.* **86**, 309–315 (1980).

145a. D. Rolf and G. R. Gray, *J. Am. Chem. Soc.* **104**, 3539–3541 (1982).

146. G. R. Waller, ed., "Biochemical Applications of Mass Spectrometry." Wiley, New York, 1972.
147. G. R. Waller and O. C. Dermer, eds., "Biochemical Applications of Mass Spectrometry," Suppl. Vol. Wiley, New York, 1980.
148. O. S. Chizhov and N. K. Kochetkov, *Adv. Carbohydr. Chem. Biochem.* **21**, 29–93 (1966).
149. N. K. Kochetkov, O. S. Chizhov, and A. F. Bochkov, *MTP Int. Rev. Sci.: Org. Chem., Ser. One* 7, p. 147–190 (1973).
150. D. C. DeJongh, *in* "The Carbohydrates" (W. Pigman and D. Horton, eds.), 3rd ed., Vol. IB, pp. 1327–1353. Academic Press, New York, 1980.
151. O. S. Chizhov, V. I. Kadentsev, A. A. Solov'yov, P. F. Levonowich, and R. C. Dougherty, *J. Org. Chem.* **41**, 3425–3428 (1976).
152. S. Ando, K. Kon, Y. Nagai, and T. Murata, *J. Biochem. (Tokyo)* **82**, 1623–1631 (1977).
153. J. Moor and E. S. Waight, *Org. Mass Spectrom.* **9**, 903–912 (1974).
154. K.-A. Karlsson, *Prog. Chem. Fats Other Lipids* **16**, 207–230 (1977).
155. K.-A. Karlsson, I. Pascher, B. E. Samuelsson, J. Finne, T. Krusius, and H. Rauvala, *FEBS Lett.* **94**, 413–417 (1978).
156. H. Yamada, R. E. Cohen, and C. E. Ballou, *J. Biol. Chem.* **254**, 1972–1979 (1979).
157. M. Linscheid, J. D'Angona, A. L. Burlingame, A. Dell, and C. E. Ballou, *Proc. Natl. Acad. Sci. U.S.A.* **78**, 1471–1475 (1981).
158. V. Kováčik, Š. Bauer, J. Rosík, and P. Kováč, *Carbohydr. Res.* **8**, 282–290 (1968).
159. V. Kováčik, Š. Bauer, and J. Rosík, *Carbohydr. Res.* **8**, 291–294 (1968).
160. P. Hallgren and A. Lundblad, *J. Biol. Chem.* **252**, 1014–1022, 1023–1033 (1977).
161. B. Fournet, J.-M. Dhalluin, G. Strecker, J. Montreuil, C. Bosso, and J. Defaye, *Anal. Biochem.* **108**, 35–56 (1980).
162. G. W. Hay, B. A. Lewis, and F. Smith, *Methods Carbohydr. Chem.* **5**, 357–361 (1965).
163. G. W. Hay, B. A. Lewis, and F. Smith, *Methods Carbohydr. Chem.* **5**, 377–380 (1965).
164. J. R. Dixon, J. G. Buchanan, and J. Baddiley, *Biochem. J.* **100**, 507–511 (1966).
165. W. Dröge, O. Lüderitz, and O. Westphal, *Eur. J. Biochem.* **4**, 126–133 (1968).
166. A. Jeanes and F. R. Seymour, *Carbohydr. Res.* **74**, 31–40 (1979).
167. L. Hough, *Methods Carbohydr. Chem.* **5**, 370–377 (1965).
168. T. J. Painter and B. Larsen, *Acta Chem. Scand.* **24**, 813–833 (1970).
169. T. J. Painter and B. Larsen, *Acta. Chem. Scand.* **24**, 2366–2378 (1970).
170. T. J. Painter and B. Larsen, *Acta Chem. Scand.* **24**, 2724–2736 (1970).
171. M. Fahmy Ishak and T. J. Painter, *Acta Chem. Scand.* **25**, 3875–3877 (1971).
172. J. E. Scott and M. J. Tigwell, *Biochem. J.* **173**, 103–114 (1978).
173. L.-Å. Fransson, *Carbohydr. Res.* **36**, 339–348 (1974).
174. L.-Å. Fransson, *Carbohydr. Res.* **62**, 235–244 (1978).
175. L.-Å. Fransson, A. Malmström, I. Sjöberg, and T. N. Huckerby, *Carbohydr. Res.* **80**, 131–145 (1980); I. Sjöberg and L.-Å. Fransson, *Biochem. J.* **191**, 103–110 (1980).
176. L.-Å. Fransson and I. Carlstedt, *Carbohydr. Res.* **36**, 349–358 (1974).
177. V. Zitko and C. T. Bishop, *Can. J. Chem.* **44**, 1749–1756 (1966).
178. I. J. Goldstein, G. W. Hay, B. A. Lewis, and F. Smith, *Methods Carbohydr. Chem.* **5**, 361–370 (1965).
179. J. Johnson, S. Kirkwood, A. Misaki, T. E. Nelson, J. V. Scaletti, and F. Smith, *Chem. Ind. (London)* pp. 820–822 (1963).
180. G. G. S. Dutton and K. B. Gibney, *Carbohydr. Res.* **25**, 99–105 (1972).
181. S. C. Churms and A. M. Stephen, *Carbohydr. Res.* **19**, 211–221 (1971).
182. P. A. J. Gorin and J. F. T. Spencer, *Can. J. Chem.* **43**, 2978–2984 (1965).
183. B. Lindberg, J. Lönngren, W. Nimmich, and U. Rudén, *Acta Chem. Scand.* **27**, 3787–3790 (1973).

184. P. S. O'Colla, *Methods Carbohydr. Chem.* **5**, 382–392 (1965).
185. G. O. Aspinall, *Methods Carbohydr. Chem.* **5**, 397–400 (1965).
186. D. A. Rees and J. W. B. Samuel, *Chem. Ind. (London)* pp. 2008–2009 (1965).
187. G. O. Aspinall and J. M. McNab, *J. Chem. Soc. C* pp. 845–851 (1969).
188. K. Heyns and H. Paulsen, *Adv. Carbohydr. Chem.* **17**, 169–221 (1962).
189. G. O. Aspinall and I. M. Cairncross, *J. Chem. Soc.* pp. 3998–4000 (1960).
190. D. Abbott, E. J. Bourne, and H. Weigel, *J. Chem. Soc. C* pp. 827–831 (1966).
191. B. Lindberg and S. Svensson, *Acta Chem. Scand.* **22**, 1907–1912 (1968).
192. H. Miyaji, A. Misaki, and M. Torii, *Carbohydr. Res.* **31**, 277–287 (1973).
193. J. K. Rogers and N. S. Thompson, *Carbohydr. Res.* **7**, 66–75 (1968).
194. B. Lindberg, J. Lönngren, and D. A. Powell, *Carbohydr. Res.* **58**, 177–186 (1977).
195. G. O. Aspinall, R. M. Fairweather, and T. M. Wood, *J. Chem. Soc. C* pp. 2174–2179 (1968).
196. C. A. Dekker and L. Goodman, *in* "The Carbohydrates", (W. Pigman and D. Horton, eds.), 3rd ed., Vol. IA, pp. 1–68. Academic Press, New York, 1970.
197. S. Umezawa, *Pure Appl. Chem.* **50**, 1453–1476 (1978).
198. S. Hanessian, *Acc. Chem. Res.* **12**, 159–165 (1979); S. Hanessian, D. M. Dixit, and T. J. Liam, *Pure Appl. Chem.* **53**, 129–148 (1981).
199. J. P. K. Verheyden, A. C. Richardson, R. S. Bhatt, B. D. Grant, W. L. Fitch, and J. G. Moffatt, *Pure Appl. Chem.* **50**, 1363–1383 (1978).
200. B. Fraser-Reid and R. C. Anderson, *Fortschr. Chem. Org. Naturst.* **39**, 1–61 (1980).
201. J. Yoshimura, *Pure Appl. Chem.* **53**, 113–128 (1981).
202. R. L. Whistler and S. Hirase, *J. Org. Chem.* **26**, 4600–4605 (1961).
203. C. W. Baker and R. L. Whistler, *Carbohydr. Res.* **45**, 237–243 (1975); *Methods Carbohydr. Chem.* **7**, 152–156 (1976).
204. K. Takeo, T. Sumimoto, and T. Kuge, *Staerke* **26**, 111–118 (1974).
205. C. J. Lawson and D. A. Rees, *Nature (London)* **227**, 392–393 (1970).
206. D. A. Rees, *Biochem. J.* **81**, 347–352 (1961).
207. P. A. J. Gorin and J. F. T. Spencer, *Carbohydr. Res.* **13**, 339–349 (1970).
208. E. E. Percival and J. K. Wold, *J. Chem. Soc.* pp. 5459–5468 (1963).
209. R. W. Jeanloz and P. J. Stoffyn, unpublished results; quoted by R. W. Jeanloz, *in* "The Carbohydrates", (W. Pigman and D. Horton, eds.), 3rd ed., Vol. IIB, pp. 589–625. Academic Press, New York, 1970.
210. E. Percival, *Methods Carbohydr. Chem.* **8**, 281–285 (1980).
211. A. I. Usov, K. S. Adamyants, L. I. Miroshnikova, A. A. Shaposhnikova, and N. K. Kochetkov, *Carbohydr. Res.* **18**, 336–338 (1971).
212. Y. Inoue and K. Nagasawa, *Carbohydr. Res.* **46**, 87–95 (1976).
213. Y. Inoue and K. Nagasawa, *Carbohydr. Res.* **85**, 107–119 (1980); see also K. Nagasawa, Y. Inoue, and T. Tokuyasu, *J. Biochem. (Tokyo)* **86**, 1323–1329 (1979).
214. A. N. de Belder and B. Norrman, *Carbohydr. Res.* **8**, 1–6 (1968).
215. G. G. S. Dutton and E. H. Merrifield, *Carbohydr. Res.* **85**, C13–C14 (1980).
216. G. G. S. Dutton and K. L. Mackie, *Can. J. Chem.* **55**, 49–63 (1977).
217. B. Lindberg, F. Lindh, J. Lönngren, and W. Nimmich, *Carbohydr. Res.* **70**, 135–144 (1979).
218. P.-E. Jansson, B. Lindberg, and H. Ljunggren, *Carbohydr. Res.* **75**, 207–220 (1979).
219. J. W. Green, *in* "The Carbohydrates", (W. Pigman and D. Horton, eds.), 3rd ed., Vol. IB, pp. 1101–1166. Academic Press, New York, 1980.
220. R. L. Whistler and J. N. BeMiller, *Adv. Carbohydr. Chem.* **13**, 289–329 (1958).
221. E. F. L. J. Anet, *Adv. Carbohydr. Chem.* **19**, 181–218 (1964).
222. J. C. Sowden, *Adv. Carbohydr. Chem.* **12**, 35–79 (1957).
223. G. O. Aspinall, R. Khan, R. R. King, and Z. Pawlak, *Can. J. Chem.* **51**, 1359–1362 (1973).

224. G. Machell and G. N. Richards, *J. Chem. Soc.* pp. 4500–4506 (1957).
225. G. Machell and G. N. Richards, *J. Chem. Soc.* pp. 1199–1204 (1958).
226. R. L. Whistler and J. N. BeMiller, *J. Org. Chem.* **26**, 2886–2892 (1961).
227. G. O. Aspinall, T. N. Krishnamurthy, I. Furda, and R. Khan, *Can. J. Chem.* **53**, 2171–2177 (1975).
228. G. O. Aspinall and K. M. Ross, *J. Chem. Soc.* pp. 3674–3677 (1961).
229. K. O. Lloyd, *Int. Rev. Sci.: Org. Chem., Ser. Two,* **1**, 251–281 (1976).
230. R. N. Iyer and D. M. Carlson, *Arch. Biochem. Biophys.* **142**, 101–105 (1971).
231. U. Lindahl and L. Rodén, in "Glycoproteins" (A. Gottschalk, ed.), 2nd ed., Part A, pp. 491–517 Elsevier, Amsterdam, 1972.
232. B. Vollmert, *Makromol. Chem.* **5**, 110–127 (1950).
233. P. Albersheim, H. Neukom, and H. Deuel, *Arch. Biochem. Biophys.* **90**, 46–51 (1960); A. J. Barrett and D. H. Northcote, *Biochem. J.* **94**, 617–627 (1965).
234. C. J. Lawson, C. W. McCleary, H. I. Nakada, D. A. Rees, I. W. Sutherland, and J. F. Wilkinson, *Biochem. J.* **115**, 947–958 (1969).
235. B. Lindberg, J. Lönngren, and J. L. Thompson, *Carbohydr. Res.* **28**, 351–357 (1973).
236. B. Lindberg and J. Lönngren, *Methods Carbohydr. Chem.* **7**, 142–148 (1976).
237. G. O. Aspinall and K.-G. Rosell, *Carbohydr. Res.* **57**, C23–C27 (1977).
238. G. O. Aspinall and A. S. Chaudhari, *Can. J. Chem.* **53**, 2189–2193 (1975).
239. M. Curvall, B. Lindberg, and J. Lönngren, *Carbohydr. Res.* **41**, 235–239 (1975).
240. B. Lindberg, F. Lindh, and J. Lönngren, *Carbohydr. Res.* **61**, 81–87 (1978).
241. I. Kitagawa, Y. Ikenishi, M. Yoshikawa, and K. S. Im, *Chem. Pharm. Bull.* **25**, 1408–1416 (1977).
242. H. Björndal and B. Wagström, *Acta Chem. Scand.* **23**, 3313–3320 (1969).
243. L. Kenne, B. Lindberg, and S. Svensson, *Carbohydr. Res.* **40**, 69–75 (1975).
244. L. Kenne, J. Lönngren, and S. Svensson, *Acta Chem. Scand.* **27**, 3692–3698 (1973).
245. E. J. Corey and C. U. Kim, *Tetrahedron Lett.* pp. 919–922 (1973).
246. For a stepwise degradation, see P.-E. Jansson, L. Kenne, B. Lindberg, H. Ljunggren, J. Lönngren, U. Rudén, and S. Svensson, *J. Am. Chem. Soc.* **99**, 3812–3815 (1977).
247. S. Svensson, in "Methods in Enzymology" (V. Ginsburg, ed.), Vol. 50, Part C, pp. 33–38. Academic Press, New York, 1978.
248. S. Svensson and L. Kenne, *Methods Carbohydr. Chem.* **8**, 67–71 (1980).
249. B. Lindberg, J. Lönngren, U. Rudén, and W. Nimmich, *Carbohydr. Res.* **42**, 83–94 (1975).
250. J. Baddiley, *Acc. Chem. Res.* **3**, 98–105 (1970).
251. P. A. Rebers and M. Heidelberger, *J. Am. Chem. Soc.* **81**, 2415–2419 (1959); **83**, 3056–3059 (1961).
252. D. R. Bundle, I. C. P. Smith, and H. J. Jennings, *J. Biol. Chem.* **249**, 2275–2281 (1974); P. Branefors-Helander, B. Classon, L. Kenne, and B. Lindberg, *Carbohydr. Res.* **76**, 197–202 (1979).
253. N. K. Kochetkov, O. S. Chizhov, and A. F. Sviridov, *Carbohydr. Res.* **14**, 277–285 (1970).
254. G. O. Aspinall and K.-G. Rosell, *Can. J. Chem.* **56**, 685–690 (1978).
255. K. Ninomiya, T. Shioiri, and S. Yamada, *Tetrahedron* **30**, 2151–2157 (1974); *Chem. Pharm. Bull.* **22**, 1398–1404 (1974).
256. H. E. Baumgarten, H. L. Smith, and A. Staklis, *J. Org. Chem.* **40**, 3554–3561 (1975).
257. G. O. Aspinall, H. K. Fanous, N. S. Kumar, and V. Puvanesarajah, unpublished results.
258. V. Lehmann, O. Lüderitz, and O. Westphal, *Eur. J. Biochem.* **21**, 339–347 (1971).
259. I. Kitagawa, M. Yoshikawa, and A. Kadota, *Chem. Pharm. Bull.* **26**, 484–496 (1978).
260. G. O. Aspinall, H. K. Fanous, N. S. Kumar, and V. Puvanesarajah, *Can. J. Chem.* **59**, 935–940 (1981).
261. G. O. Aspinall and V. Puvanesarajah, unpublished results.

262. I. Kitagawa, T. Kamigauchi, H. Ohmori, and M. Yoshikawa, *Chem. Pharm. Bull.* **28**, 3078–3086 (1980).
263. G. O. Aspinall and K.-G. Rosell, *Can. J. Chem.* **56**, 680–684 (1978).
264. G. O. Aspinall, O. Igarashi, T. N. Krishnamurthy, W. Mitura, and M. Funabashi, *Can. J. Chem.* **54**, 1708–1713 (1976).
265. J. M. Williams, *Adv. Carbohydr. Chem. Biochem.* **31**, 9–79 (1975).
266. U. Lindahl and O. Axelsson, *J. Biol. Chem.* **246**, 74–82 (1971).
267. A. S. Perlin, N. M. Ng Ying Kin, S. S. Bhattacharjee, and L. F. Johnson, *Can. J. Chem.* **50**, 2437–2441 (1972).
268. C. Erbing, B. Lindberg, and S. Svensson, *Acta Chem. Scand.* **27**, 3699–3704 (1973).
269. G. O. Aspinall, M. M. Gharia, R. G. S. Ritchie, and C. O. Wong, *Carbohydr. Res.* **85**, 73–92 (1980).
270. B. A. Dmitriev, Y. A. Knirel, N. K. Kochetkov, and I. L. Hofman, *Eur. J. Biochem.* **66**, 559–566 (1976).
271. C. Erbing, L. Kenne, B. Lindberg, G. Naumann, and W. Nimmich, *Carbohydr. Res.* **56**, 371–376 (1977).
272. S. Hase and J. Matsushima, *J. Biochem.* (*Tokyo*) **72**, 1117–1128 (1972).
273. B. Lindberg, B. Lindqvist, J. Lönngren, and D. A. Powell, *Carbohydr. Res.* **78**, 111–117 (1980).
274. H. J. Jennings, C. Lugowski, and N. M. Young, *Biochemistry* **19**, 4712–4719 (1980).
275. L. Kenne, B. Lindberg, K. Petersson, E. Katzenellenbogen, and E. Romanowska, *Carbohydr. Res.* **78**, 119–126 (1980).

Spectroscopic Methods

ARTHUR S. PERLIN AND BENITO CASU

I. Introduction	133
II. Nuclear Magnetic Resonance Spectroscopy	135
A. Methods and Instrumentation	135
B. Characteristics of NMR Spectra of Polysaccharides	145
C. Studies on Chemical Structure	151
D. Other Applications	169
III. Infrared–Raman Spectroscopy	172
A. Methodology: Instrumentation and Techniques	172
B. Functional Group Detection and Configuration	176
C. Analytical Aspects	180
D. Hydrogen Bonding and Hydration	182
IV. Comparative Evaluation of Spectroscopic Methods	184
References	186

I. Introduction

The earliest spectroscopic observations on polysaccharides were concerned with classic color reactions (*1*)—the formation of a blue starch–iodine aggregate or the change in the wavelength of absorption of a dye (metachromasia) induced by the staining of polysaccharides in tissues. Observations such as these were subsequently translated into important analytical methods, for example, the spectrophotometric determination of amylose as

THE POLYSACCHARIDES, VOL. 1

its iodine complex (2) or of heparin in the presence of azure A (3). An example of much more recent origin (4) is energy-dispersive X-ray microanalysis of proteoglycan in cartilage stained with brominated toluene blue, although here X rays given off by the bromine atom are assayed rather than the color transmitted. With developments in spectropolarimetry, the visible and ultra-violet regions have come to be used extensively for investigations on higher molecular order in polysaccharides by optical rotatory dispersion and cir-cular dichroism techniques (5,6). Applications in this specific area are de-scribed in Chapter 5.

Mass spectrometry is normally included in any discussion of spectro-scopic methods. However, although advances in this field are rapidly ex-tending the upper limits of molecular size that can be accommodated, applications to polysaccharides are confined mainly to the characterization of fragmentation products, e.g., by combined gas chromatography and mass spectrometry of the hydrolysate of methylated polysaccharides. Accordingly, that topic is reserved for Chapter 3 in conjunction with chemical methods of structural analysis.

This chapter is concerned with applications of infrared (IR)–Raman and nuclear magnetic resonance (NMR) spectroscopy. Initially, in the early 1940's, IR studies dealt with hydroxyl absorption bands of cellulosic ma-terials and evidence for hydrogen bonding (7). The technique was then em-ployed (7, 7a) to obtain such structural information about complex polysaccharides as ratios of uronic acid and aminodeoxy sugar in mucopoly-saccharides or the position and configuration of glycosidic linkages in dextran. As discussed later, applications of this kind are now supported and enhanced by laser Raman spectroscopy. In recent years, however, NMR spectroscopy has become far more important as a source of structural information. Carbon-13 NMR, in particular, is now used so extensively in polysaccharide studies that a large percentage of current publications dealing with structure make use of ^{13}C-NMR data.

It is evident that with large, complex molecules no single spectroscopic technique is sufficiently sensitive for all parameters of interest. An NMR spectrum, for example, may easily divulge a great deal about the number and nature of constituent residues present in a polysaccharide, but relatively little may be apparent as to its macromolecular conformation. Information about the latter is more likely to be gained from chiroptical and fiber diffrac-tion methods. Hence, it is important to recognize the specific advantages of each spectroscopic technique and of equal importance to appreciate that they are complementary. Studies on polysaccharides are greatly facilitated, therefore, by combining data from a range of spectroscopic sources, as well as by utilizing information furnished by chemical and enzymatic methods.

II. Nuclear Magnetic Resonance Spectroscopy

A. Methods and Instrumentation

Nuclear magnetic resonance spectroscopy has a number of characteristics that make this technique especially advantageous for studies on polysaccharides. Several texts (*8–11*) and review articles (*12–23*) deal specifically with these features as they apply to polymers in general or/and to carbohydrates.

In common with other physical methods, NMR spectroscopy is nondestructive; that is, it is possible to examine a polymer without modifying or degrading it and then recover the material intact. Until very recently, high-resolution NMR studies were restricted to solutions, so that it was impractical to deal directly with polysaccharides of low solubility other than by wide-line NMR techniques. These limitations have been substantially overcome (*24*) by new methods and instrumentation that give high-resolution ^{13}C spectra with solid samples. Consequently, the nondestructive qualification attributed above now includes the preservation of the physical nature of a solid polymer and, indeed, permits NMR spectroscopy to probe intact tissue. Although few solid-state NMR spectra of polysaccharides have yet been reported, they will likely be relatively commonplace within a few years.

This development in the spectroscopy of solid materials is one example of how advances in NMR techniques facilitate studies on macromolecules. A dramatic improvement in resolution was realized (*8,9*) with the introduction of superconducting magnets, when it became feasible to increase operating frequencies from the 40–100 MHz range of early instruments using electromagnets to over 200 MHz. Similarly, the pulsed Fourier transform (FT) technique (*25*) afforded a large enhancement in sensitivity and also made possible the development of instruments for routine ^{13}C-NMR spectroscopy. Of necessity, a dedicated computer is an integral part of these instruments, and its presence has stimulated many additional advances in technique (*9–11*), including improved methods for the measurement of spin–lattice and spin–spin relaxation times (T_1 and T_2, respectively) and numerous possibilities for the enhancement of spectral quality by data manipulation.

1. ^1H-NMR Spectra

Two factors dominate the acquisition of high-resolution ^1H-NMR spectra of polysaccharides, i.e., interference by exchangeable protons (O—H, N—H) and line broadening of signals. The preparation of aqueous solutions of polysaccharides involves a prior exchange treatment with deuterium oxide and the use of good-quality deuterium oxide (preferably 99.95 atom %) as solvent. Nevertheless, a strong peak due to residual water (HOD signal, Fig. 1), as well as substantial side bands of the peak, are often obtained, especially if the concentration of polysaccharide is less than 1 mM. The chemical shift of the HOD signal at room temperature ($\delta \sim 4.8$) is such as to interfere with the vitally important anomeric region; e.g., it is close to the H-1 signal of β-glucopyranosyl residues. By raising the temperature, one can displace the HOD signal upfield [$\delta \sim 4.5$ at 70°C (Fig. 1b)], thereby exposing more of the spectrum; spinning side bands are shifted as well. When spectra are recorded in the FT mode, a strong solvent resonance may exceed the dynamic range of the system, creating intense spectral artifacts and preventing the detection of weak resonances. Fortunately, there are a num-

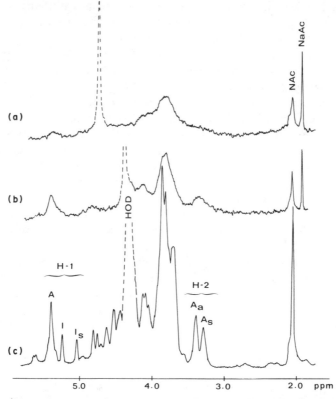

Fig. 1. ^{1}H-NMR spectra of heparan sulfate in D_2O, illustrating the effects of changes in the probe temperature and operating frequency (a) at a temperature of 30°C and frequency of 100 MHz; (b) at 70°C and 100 MHz; (c) at 70°C and 400 MHz (HOD, residual water; NAc, acetamido methyl; NaAc, sodium acetate; A, aminodeoxyglucose; I, iduronic acid; a, acetylated; s, sulfated; signal H-2 of β-glucuronic acid residue is expected to appear at $\sim\delta$ 3.4).

ber of FT techniques (26,27) for minimizing interference by the HOD signal. "Saturation decoupling," for example, takes advantage of the fact that the solvent protons relax much more slowly (T_1 of H_2O is >2 s) than most of those of a polymer ($T_1 < 0.5$ s). Hence, the solvent signal can be largely eliminated by the choice of a suitable pulse sequence, although care is necessary to avoid the introduction of spurious lines. This operation is illustrated in Fig. 2 by the effect of "saturation decoupling" on a strong HOD signal that overlaps several other signals. Correlation spectroscopy (28) may offer advantages for experiments in which there are strong resonances from the solvent, because the frequency range can be adjusted to exclude such resonances. Also, the use of H_2O as solvent rather than D_2O seems to be more feasible with this technique than with pulsed FT. However, "two-dimensional" FT ^1H-NMR spectroscopy offers (29,30) a particularly effective approach to the examination of polymers in H_2O while minimizing interference from the DOH peak in D_2O solution.

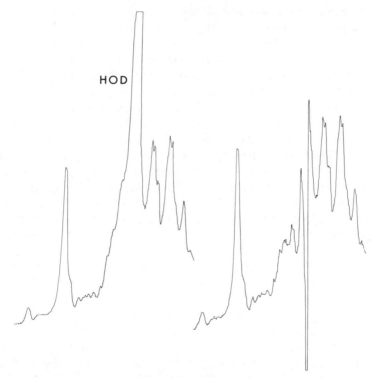

Fig. 2. Portion of a 200 MHz ^1H-NMR spectrum (4.3–4.9 ppm region) of heparin in D_2O at 60°C, showing (left) the presence of a strong peak (HOD) due to residual water and (right) after suppression of the HOD resonance by "saturation decoupling."

The problem of signal broadening ($8-11$) is attributable largely to the fact that polymer protons have short relaxation times; i.e., the linewidth ($\Delta v_{1/2}$) is proportional to the rate of spin–spin lattice relaxation (T_2). Not atypical are the extremely wide signals that comprise the 100 MHz spectrum of heparan sulfate at 25°C (Fig. 1a); the degree of broadening becomes more apparent from a comparison of the two relatively narrow signals upfield, which represent the acetamido methyl substituent of the polymer (left, $\Delta v_{1/2}$ 5.5 Hz) and the methyl group of added acetate anion (right, $\Delta v_{1/2}$ 1.5 Hz). A substantial enhancement in the quality of most polysaccharide ^{1}H spectra can be realized by using elevated temperatures, as shown by the appreciable increase in signal intensities in the spectrum at 70°C (Fig. 1b). More importantly, when the spectrum is recorded at high field (e.g., at 220 MHz) (31), there is a corresponding enhancement in signal separation, which helps to counteract much of the line broadening effect and to expose details in fine structure as are apparent in the ^{1}H spectrum of heparan sulfate at 400 MHz (Fig. 1c).

Comparable advances in the quality of ^{1}H spectra of heparins, a related group of glycosaminoglycan, were observed over a period of many years. The spectra were recorded sequentially at 60 ($32-34$), 100 ($35,36$), 220 ($31,37-39$), 270 ($40,41$), and 400 (42) MHz. As the progressive improvements in resolution lead to increasingly more complete analyses of the spectra, these heparin studies serve to emphasize the advantages of high-field NMR spectroscopy for polysaccharides. They also furnish an illustration of the great improvement in resolution that can be obtained ($9,10$) by computer manipulation of the lineshapes of resonance signals. Thus, a convolution difference FT ^{1}H-NMR spectrum (43) at 270 MHz (Fig. 3b), obtained (40) by appropriate operations on the free-induction decay (FID) signal, subtracted away much of the envelope that partially masked resonance signals in the unmodified spectrum (Fig. 3a). Even relatively small spin–spin splittings were then observed (40). The same coupling information was obtained ($43a$) more recently by two-dimensional FT ^{1}H-NMR spectroscopy ($29,30$), which effectively affords a marked enhancement in resolution. This is illustrated in Fig. 4a, which depicts the low-field portion of the spectrum shown in Fig. 3a, and in Fig. 4b by the display of the corresponding cross section of the two-dimensional spectrum.

The advantages of high-field FT spectrometers for ^{1}H spectra of polymers are such as to have generated continuous pressure over the years for the development of more and more powerful instruments. At present, spectrometers operating at 360–400 MHz (83–92 kG) have become standard research instruments, and it appears that 500–600 MHz spectrometers will soon be far from rare. Because of the enormous expenditure entailed in their purchase and maintenance and the high level of expertise they require, many of these

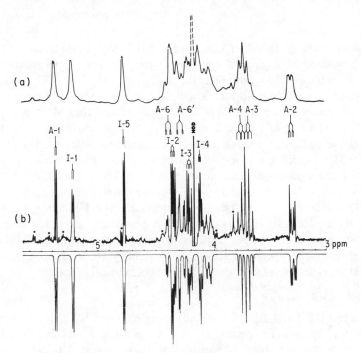

Fig. 3. (a) ^{1}H-NMR spectrum (270 MHz) of beef lung heparin in D_2O at 90°C; (b) spectrum after convolution difference processing (A, aminodeoxyglucose; I, iduronic acid). Reproduced with permission from (*40*), *Macromolecules* **12**, 1001–1007. Copyright (1979) The American Chemical Society.

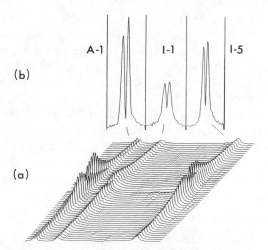

Fig. 4. (a) Partial 270 MHz ^{1}H-NMR "δ-J-resolved" two-dimensional spectrum of beef lung heparin in D_2O at 90°C, equivalent to the 4.7–5.4 ppm region in Fig. 3; (b) the corresponding cross section of the stacked plots in spectrum (a) (A, aminodeoxyglucose; I, iduronic acid) (*43a*).

instruments are installed in major regional NMR laboratories. Through these laboratories, high-field spectra (as well as their interpretation) are often available at relatively moderate cost.

It should be emphasized that low-field spectrometers are satisfactory for acquiring much of the ^{1}H-NMR data required and that the use of a high-field instrument should be reserved mainly for key experiments to supplement those data. Information useful for many purposes is available from 60–100 MHz spectra, e.g., for rapid monitoring of the isolation of a polysaccharide, for distinguishing among samples that differ substantially in composition, or for the detection of such distinctive substituent groups as O- or N-acetyl. In these types of applications, the FT mode, by offering markedly higher sensitivity, is compatible with such isolation and purification procedures as gel permeation chromatography or electrophoresis. Fourier transform spectrometers in this category are likely to be accessible to most workers engaged in chemical or biochemical research.

2. ^{13}C-NMR Spectra

Whereas the FT mode may be advantageous for ^{1}H spectra, it is mandatory for ^{13}C spectra owing to the low natural abundance (1.1%) and intrinsically low sensitivity of the ^{13}C nucleus. Indeed, ^{13}C-NMR spectroscopy provided the major impetus for the development of FT spectrometers (44). At 15–25 MHz, operating frequencies corresponding to 60–100 MHz for protons, a useful separation of ^{13}C resonance signals is commonly attainable because of the wide range of chemical shifts exhibited (16,17) by ^{13}C nuclei (> 200 ppm). Mainly for this reason, there is far less need for, or interest in, high-field instruments for carbon than for protons, although advantages can easily be demonstrated in some applications. For example, a comparison of ^{13}C-NMR spectra of the β-D-glucan of barley at 22.6 and 100 MHz (Fig. 5) shows (44a) that there is a marked improvement in overall quality at the higher frequency.

For convenience, several other aspects of ^{13}C-NMR methodology and instrumentation are discussed in succeeding sections. As yet, there are few spectrometers in operation that are suitable for the acquisition of ^{13}C-NMR spectra of solids,* and spectral service in this area will likely be available only from a limited number of specialized laboratories for some time to come because of the cost and technical complexity involved.

3. Other Nuclei: ESR Spectroscopy

Aside from ^{1}H and ^{13}C, few other elements have elicited interest as a part of studies on polysaccharides. Phosphate groups, present in some classes of

---

* This technique makes use of ^{1}H–^{13}C cross-polarization and magic-angle spinning and hence is referred to as CP/MAS (24).

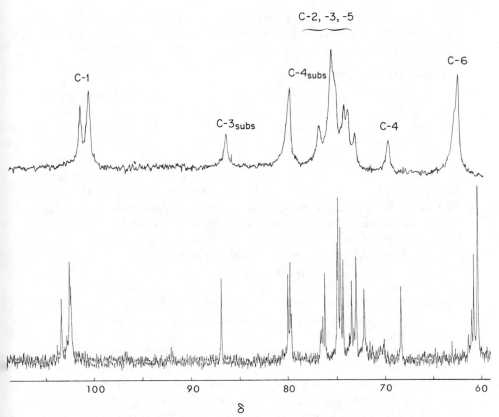

Fig. 5 Proton-decoupled ^{13}C-NMR spectra of β-D-glucan from barley in DMSO-d_6 at 90°C: upper, at 22.6 MHz, and lower, at 100 MHz. Advantages of high-field operation for promoting improved resolution and increased sensitivity are shown by such details as the presence at 100 MHz of three distinct signals involving C-4 of the glycosidic linkage (C-4$_{subs}$), as well as many other major and minor signals not clearly evident at 22.6 MHz (*44a*).

microbial origin, can be examined (*45*) by ^{31}P-NMR spectroscopy, and the solution characteristics of sodium salts of acidic polysaccharides can be investigated (*46*) by ^{23}Na-NMR. In addition, one can consider materials isotopically enriched, either biosynthetically or chemically, with deuterium (*47,48*), tritium, ^{15}N, ^{17}O, as well, of course, as ^{13}C (*49,50*). Such NMR-active nuclei can then be utilized as probes for specific purposes. Hence, access to multinuclear spectrometers, which now are relatively widespread, could be a substantial asset.

A related area involves (*51*) the labeling of polysaccharides with radical species amenable to ready detection by electron spin resonance (ESR) spectroscopy, an approach widely employed in protein studies. By no means are ESR spectrometers rare in chemistry departments, and the extremely high sensitivity of this technique could make demands on instrument time relatively modest.

4. Chemical Shift Standards: Solvents

An important question associated with NMR spectra of aqueous solutions concerns the choice of a suitable reference for chemical shifts. External tetramethylsilane (TMS) in a coaxial capillary is frequently employed for ^{1}H spectra, although its use entails a large bulk susceptibility correction (-0.4 to -0.5 ppm). Tetramethyltin is a more satisfactory capillary fluid (18), especially above 60°C because of its lower volatility, and its chemical shift with a D_2O solution is only 2–3 Hz downfield of that of TMS in chloroform. Water-soluble derivatives—sodium 2,2-dimethyl-2-silapentane 5-sulfonate or 3-(trimethylsilyl) propionate-2,2,3,3-d_4 (TSP)—are suitable as internal standards for both ^{1}H and ^{13}C. However, their presence complicates recovery of the sample. The use of D_2O as a solvent for ^{13}C-NMR spectra is convenient because FT instruments usually employ a deuterium "lock," although mixtures of H_2O and D_2O or H_2O with an external lock (e.g., C_6D_6) may also be used. It is common practice to introduce a drop or two of an easily removable internal reference, such as methanol or acetone. However, since the chemical shift of these compounds with respect to TMS or TSP is influenced by temperature, concentration, and possibly the nature of the polymer, many ^{13}C data on polysaccharides are not directly comparable because of variations in experimental conditions. There remains a serious need for a consistent, universally accepted basis for the referencing of ^{1}H and ^{13}C chemical shifts of polysaccharides in aqueous solution. Indeed, this need also applies to the monosaccharides and oligosaccharides that furnish reference data.

Since dimethyl sulfoxide (DMSO) is a good solvent for many polysaccharides, DMSO-d_6 is often employed in NMR studies. Because of slow exchange rates in this medium, hydroxyl and amino proton resonances may be clearly observable (52) in the spectra and thus provide valuable structural information that is lost with aqueous solutions.

Derivatized polysaccharides, e.g., peracetates or fully methylated material, are readily handled as solutions in such common deuteriated organic solvents as chloroform-d_6 and acetone-d_6. Chemical shifts are then referenced with respect to internal TMS.

5. Sensitivity Enhancement: Quantitative Measurements

The ability to detect a spectral line is determined partly by the shape of the signal. Since a broad signal is less easily observed than a narrow one against a background of random noise, it can be a great deal more difficult to acquire a satisfactory spectrum of a polysaccharide than of low molecular weight compounds. Also, high viscosity, chain stiffness, and poor solubility may make it impractical to employ a concentrated solution of a polysaccharide in order to increase the strength of its resonance signals. Pulsed FT NMR spectroscopy, referred to earlier, is the principle technique available for improving the ratio of signal to noise (S/N) (the net gain is equal to the square root of the number of consecutive pulses applied), a factor that is particularly important (44) in ^{13}C-NMR spectroscopy.

Quantitative measurements of signal intensity in the FT mode require special attention, especially for ^{13}C (10,16,21,44), to ensure that all nuclei produce a uniform signal response despite variations in their relaxation rates and nuclear Overhauser enhancement. To compensate for differences in these properties, it is advisable that a delay of at least five times that of the longest T_1 be used during gated decoupling and that relative intensities of signals be measured for analogous types of nuclei to ensure similar nuclear Overhauser enhancement effects, e.g., that comparisons be made between anomeric carbons.

A significant enhancement in sensitivity can be obtained simply by raising the operating temperature. At 60°–90°C, the ^{1}H signals of many polysaccharides become appreciably sharper (Fig. 1), and dramatic improvement is possible with very viscous solutions and gels. When the quantity of available material is an additional limitation, the use of a specially designed microprobe can reduce the sample size needed by a factor of about 10. With FT NMR spectroscopy under optimum conditions, it is feasible to obtain ^{1}H spectra with microgram quantities of polysaccharides and ^{13}C spectra with a few milligrams in a reasonable time period (< 20 h).

6. Proton-Coupled and Proton-Decoupled ^{13}C-NMR Spectra

Most ^{13}C-NMR spectra are acquired in the ^{1}H-decoupled mode, which involves noise-modulated irradiation of all of the protons simultaneously. This reduces the spectrum to a group of singlets and results in a marked improvement in S/N because the intensity of each signal is increased in proportion to the ^{1}H splitting eliminated, as well as by nuclear Overhauser enhancement.

The visual simplicity of this type of spectrum is represented by the ^{1}H-decoupled ^{13}C spectrum of a preparation of chondroitin 4-sulfate (Fig. 6a) (53), which exhibits 13 individual lines (one, of relatively greater intensity, accounting for two nuclei) in accord with the fact that there are 14 nonequivalent carbons in the disaccharide repeating sequence (1) of this polymer. Consequently, this spectrum is easily distinguished from that of a structurally less homogeneous preparation of chondroitin 4-sulfate (Fig. 6b) (53), owing to the additional signals produced by the latter and the associated line

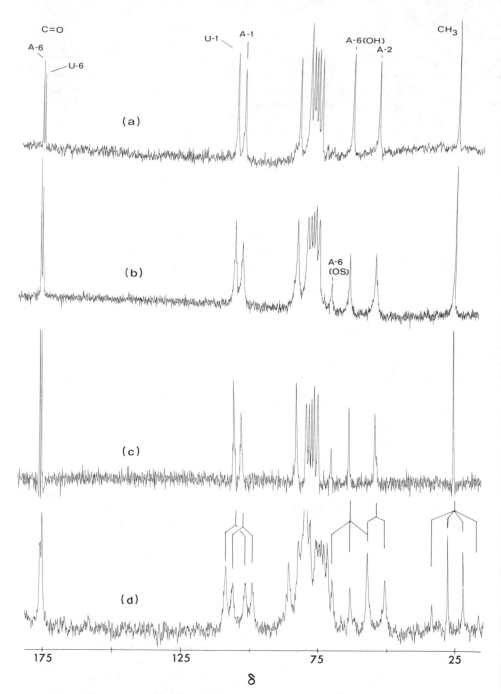

Fig. 6. Proton-decoupled ^{13}C-NMR spectra (22.6 MHz) of chondroitin 4-sulfate in D_2O at 70°C. (a) Sample having a high degree of structural homogeneity; (b) sample containing minor component structurally akin to chondroitin 6-sulfate; (c) resolution-enhanced form of spectrum (b) by convolution difference. Tracing (d) is the fully proton-coupled spectrum corresponding to proton-decoupled spectrum (a), showing splitting patterns (singlet, doublet, triplet, or quartet) associated with the number of protons (0, 1, 2, or 3) appended to various carbons (*53*). U, uronic acid; A, aminodeoxyhexose; S, sulfate.

broadening caused by the contribution of minor sequences other than **1**. Resolution of spectra of this complexity can be enhanced by such techniques as convolution difference [equivalent to that applied to protons (Fig. 3)], as shown by the narrower signals and more clearly defined shoulders in Fig. 6c.

$$\beta\text{-D-Glc}p\text{A-}(1\rightarrow 3)\text{-}\beta\text{-D-Gal}p\text{NAc-4-SO}_3^-\text{-}(1\rightarrow 4)$$

1

To obtain ^{13}C–^{1}H coupling information without too large a sacrifice in S/N, irradiation of the protons is gated (9,10), so that most of the nuclear Overhauser enhancement contribution is preserved. ^{1}H-Coupled spectra are usually complex (17,19), however, because the magnitude of one-bond coupling is large enough ($^{1}J_{CH} \sim 150$–200 Hz) to cause extensive overlap among many of the signals in typical spectra of carbohydrates. An indication of the degree of complexity introduced can be gained by comparing the ^{1}H-decoupled and ^{1}H-coupled spectra in Figs. 6a and 6d, respectively, especially in the δ 60–80 region. One effective approach to the solution of this problem is two-dimensional FT ^{13}C-NMR spectroscopy (54). A spectrum is obtained in two frequency dimensions, one of which constitutes a conventional ^{13}C–^{1}H-decoupled subspectrum and gives chemical shift data. The other subspectrum, which is displayed in the second dimension, depicts the ^{13}C–^{1}H coupling information for each individual signal, within the limits of resolution. On the basis of recent experiments with oligosaccharides (55), this technique promises to be an excellent procedure for obtaining chemical shift and ^{13}C–^{1}H coupling data simultaneously for polysaccharides.

Partial ^{1}H decoupling of a ^{13}C spectrum—"off-resonance coherent" decoupling—can afford a less cluttered group of signals that clearly exhibit reduced one-bond splitting ($^{1}J_{CH}$) as compared with those found in Fig. 6d. In this manner, a methyl signal is displayed as a narrow quartet, a methylene as a triplet, etc., providing a useful method for verifying the number of protons appended to a carbon atom.

B. Characteristics of NMR Spectra of Polysaccharides

One of the most simple and generally useful applications of NMR spectroscopy to polysaccharides is in assessing the degree of molecular complexity. Thus, a measure of the different kinds of constituent residues of a polysaccharide and their ratios is obtained from the number and relative intensities of ^{1}H and ^{13}C signals in the spectra. Carbon-13 NMR spectroscopy is particularly valuable for this purpose, because the singlets that constitute the ^{1}H-decoupled spectrum (e.g., Fig. 6) are often far better dispersed than the signals of the corresponding ^{1}H spectrum.

Another type of information that often is directly available concerns the purity of the polysaccharide. For example, the detection of upfield ^1H or ^{13}C signals attributable to alkyl groups or of downfield signals attributable to aromatic moieties may be indicative (9,10) of residual protein, or it may show (56) the presence of a contaminant introduced during a chromatographic step in the isolation. If uronic acid-containing polysaccharides sequester paramagnetic impurities during processing, severe line broadening of their ^1H and ^{13}C spectra is observed (56).

In general, the magnetic resonance characteristics of polysaccharides are closely analogous to those of the monosaccharides from which they are constituted. Ranges of characteristic ^1H and ^{13}C chemical shifts for sugar

TABLE I

**Representative ^1H and ^{13}C Chemical Shifts
for Nuclei of Polysaccharides[a]**

^1H	δ (ppm)	^{13}C	δ (ppm)
$\underline{C}H_3C$	~1.5	$\underline{C}H_3C$	~15
$\underline{C}H_3CON$	1.8–2.1	$\underline{C}H_3COH$	
$\underline{C}H_3CO_2$	2.0–2.2	$\underline{C}H_3CO_2$	20–23
$C\underline{H}(NH)$	3.0–3.2	$\underline{C}H_2C$	38
$\underline{C}H_3O$	3.3–3.5	$\underline{C}H_3O$	55–61
H-2 to H-6′	3.5–4.5	$\underline{C}H(NH)$	58–61
H-5	4.5–4.6	$\underline{C}H_2OH$	60–65
H-1 (ax)	4.5–4.8	C-2 to C-5	65–75
$\underline{H}$—$C(OH)_2$	5.2	C—X[b]	80–87
H$\underline{O}$	5.0–5.4	C-1 (ax-O, red)	90–95
H-1 (eq)	5.3–5.8	C-1 (eq-O, red)	95–98
HCO_2	5.9	C-1 (ax-O, glyc)	98–103
		C-1 (eq-O, glyc)	103–106
		C-1 (fur)	106–109
		COOH	174–175
		C=O	175–180

Substituent effects on α-^1H and α-^{31}C (ppm)[c]

	O-Alkyl	O-Acyl	O-Sulfate	O-Phosphate
^1H	− 0.2–0.3	+ 0.3–0.5	+ 0.3–0.6	+ 0.3–0.5
^{13}C	+ 7–10	+ <3	+ 6–10	+ 2–3

[a] Abbreviations: ax, axial; eq, equatorial; red, reducing; glyc, glycosidic; fur, furanosyl.

[b] Nonanomeric ^{13}C involved in glycosidic linkage.

[c] Downfield, +; upfield, −.

residues are given in Table I. Also included are data for various substituents, e.g., methyl resonances due to acetamido groups of many polymers containing amino sugars, to pyruvic acid acetals found in a number of microbial polysaccharides, or to O-methyl groups that may occur naturally or through alkylation.

Anomeric resonances in both ^{1}H and ^{13}C spectra are well separated from signals produced by most of the other nuclei (Table I). This fact helps greatly in determining the number of different kinds of residues in a polysaccharide and in estimating their relative proportions. It also makes it easier to obtain from these signals the chemical shift and coupling parameters necessary for assigning anomeric configuration by NMR spectroscopy. An unsubstituted primary alcohol group produces (57–60) a ^{13}C resonance signal distinctively upfield (at δ 62.5 in Fig. 6) and hence is easily recognized, although its ^{1}H signal is less distinctive since it may overlap signals of secondary protons. Secondary positions are characterized by a relatively narrow range of chemical shifts (Fig. 6), which often makes a group of resonance signals in this category difficult to correlate with specific ^{1}H or ^{13}C nuclei in the molecule or in a particular residue.

A further distinction among anomeric carbons is the chemical shift difference between C-1 of internal residues in a polymer (61) and that of the reducing glycose residue. Carbon 1 of the latter resonates some 5–10 ppm upfield of the glycosidic carbons, in the region of δ 90–100 characteristic of C-1 of monosaccharides (57–60). Hence, for molecules of degree of polymerization (DP) of less than 10–15, C-1 signals due to the reducing-end moiety should be readily detected. Examples of this kind are commonly encountered during enzymatic degradation studies on polysaccharides.

Although there is a generally close correspondence between chemical shifts of polysaccharides and those of their monomeric or oligomeric analogues, it is hazardous to attempt to identify monosaccharide constituents of polysaccharides from spectra of the latter without supporting chemical evidence. Rather, it is far more appropriate to employ NMR spectroscopy for determining the *proportions* of the constituent residues. However, a related application for which NMR spectroscopy appears to be well suited is the identification and determination of the component sugars in the chemical or enzymatic hydrolysate of a polysaccharide. More specifically, ^{13}C chemical shift patterns (57–64) for each of the various sugars known to be constituents of polysaccharides are probably sufficiently distinctive for this purpose, as well as to allow for the detection of a new constituent residue. Nevertheless, as with chromatography, it is not normally possible to designate enantiomeric configurations of sugars on the basis of NMR evidence.

1. Chemical Shifts of Carbons of the Glycosidic Linkage

In ^{13}C spectra, the signals of both carbons engaged in the glycosidic linkage (C-1—O—C-X) are strongly displaced downfield (61), commonly by 6–9 ppm, of the signals of the corresponding carbons in related monosaccharides. These displacements are attributed (61) to inductively deshielding β-substituent effects. Hence, C-1 signals of polysaccharides are located (16,17,61,62) in the region of δ 100–110, well separated from signals due to the other carbons (Table I; see Figs. 5 and 6).

If C-X is a primary carbon, its resonance signal will no longer be located upfield of those of secondary carbons (Table I) because of the deshielding β effect of the glycosidic bond. Then, the absence of a signal around δ 60–65 and the presence of an *extra* signal in the intermediate region associated with secondary carbons serve to identify C-X as primary. These characteristics are seen in the stick diagram (Fig. 7) relating α-glucose to dextran.

When secondary, C-X often gives (16,17,61,62) the most strongly downfield of all signals of secondary carbons, and, in principle, it may be identified by reference to the chemical shifts of appropriate low molecular weight compounds. As shown, for example, in relating α-glucose to amylose (Fig. 7) the major difference among carbons 2 to 5 is the downfield location of the C-4 signal of the polymer. Lesser effects are observed on adjacent positions, frequently amounting to a 2–3 ppm shift upfield; this may be attributed (67) to a γ steric effect. However, exact values for these β and γ effects can vary

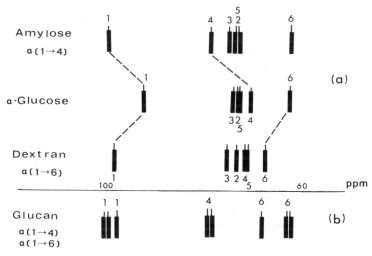

Fig. 7. Stick diagrams (a) correlating the ^{13}C chemical shifts for α-glucopyranose (57,63) with the chemical shifts of (1→4)- and (1→6)-linked α-glucans (65,66), (b) representing the ^{13}C-NMR spectrum of a (1→4)- and (1→6)-linked α-glucan from *T. mesenterica* (66).

by several parts per million, depending on the position of the glycosidic linkage, its configuration, and possibly the individual sugars involved. This makes it difficult to identify a C-X signal unequivocally by reference to model compounds, particularly when a polysaccharide contains several different constituent sugars and/or kinds of linkage. Selective ^{1}H decoupling (e.g., 40) is one practical technique for confirming the origin of the C-X signal, provided, however, that the ^{1}H spectrum itself has been adequately resolved.

Ongoing developments in methodology, nevertheless, markedly enhance prospects in this direction. Of great promise is a technique [double quantum coherence (68)] whereby the sequence of C—C linkages in a molecule can be determined unambiguously, at natural ^{13}C abundance, from the pattern of spin–spin couplings between adjacent carbons [$^1J \simeq 40$–45 Hz (57)]. This should give the *position* of the glycosidic bond, because it furnishes an unequivocal assignment for the *only* secondary carbon (C-X) that is strongly deshielded relative to the corresponding carbons in its monosaccharide analogue, as illustrated in Fig. 7. This information should be available also from the two-bond coupling (68) between C-1 and C-X across the glycosidic bond. Although at present limited in application by low sensitivity, this technique will likely become important for establishing positions of linkage, as well as sequences of residues, in higher saccharides.

2. Chemical Shifts of Protons of the Glycosidic Bond

The influence of a glycosidic bond on the chemical shift of the anomeric proton (H-1) and the proton (H-X) at the other point of attachment of the linkage is relatively much smaller than observed with carbon, since chemical shifts of H-1 of glycosyl residues are in the same range (69,70) as those of reducing sugars. There can be no deshielding β effect analogous to that with carbon, because H-1 is γ with respect to the point at which substitution (glycosidation) occurs. As shown (71) with D-glucopyranose disaccharides, the main impact is on H-X, for which downfield shifts average 0.2 ppm, and protons adjacent to it are deshielded by $\sim$0.15 ppm, depending on the position of the linkage. Because linewidths in the ^{1}H spectra of many polysaccharides are at least 10 Hz, shielding effects of that magnitude are unlikely to be of help in the assignment of H-X and thereby the position of the glycosidic bond.

3. Configuration of Glycosidic Linkages: Conformation

As in monosaccharides, the chemical shift of H-1 or C-1 of a glycosyl residue in a polysaccharide is intimately related to configuration and conformation. An equatorial C-1—H-1 bond of an α-gluco or α-galacto residue, for example, is normally represented by an H-1 signal in the region of δ 5–5.5, whereas axial H-1 of the β-anomer should resonate upfield, closer to δ 4.5 (Table I). Characteristically, the chemical shift of C-1 of an α-anomer in these series is expected to be $\delta \sim 100$ and 3–4 ppm upfield of that of the β-anomer (Table I). It is anticipated (72,73) as well that similar differences in ^{13}C chemical shifts will be found with anomeric furanosyl residues of polysaccharides and, furthermore, that the C-1 (and other) signals will be located several parts per million downfield of those of the configurationally related pyranoses. However, variations in chemical shift may occasionally be large enough to position the signal of one anomer within the chemical shift range of the other, or, as in the manno- and idopyranose series, H-1 and C-1 chemical shifts of the two anomers may differ by relatively small values. Consequently, in determining anomeric configuration it is best to employ chemical shift data in conjunction with spin–spin coupling information.

Coupling between C-1 and H-1 ($^{1}J_{CH}$) appears to be a generally reliable criterion of configuration. That is, the difference of about 10 Hz in ^{1}J between the equatorial and axial C-1—H-1 bonds of α- and β-glucose (57,74), as well as of isomeric sugars (74a,b) appears to be maintained in the polysaccharide series (53,84). Some representative examples are given in Table II (also see Fig. 6d).

Spin–spin splitting is often difficult to observe in ^{1}H spectra because of line broadening. Even the large, characteristic (85) difference in coupling between gauche and anti vicinal protons (3–4 Hz versus 7–8 Hz) (Table II) may be completely obscured. However, the enhancement in lineshape attainable with such new techniques as convolution difference and two-dimensional spectroscopy (Figs. 3 and 4, respectively) makes measurements of coupling much more feasible. Since this applies not only to anomeric protons, ^{1}H–^{1}H coupling data for other positions become accessible as well, permitting a more complete analysis of conformations of glycosyl residues.

Nuclear magnetic resonance spectroscopy holds promise (19) as a source of information about angles ϕ and ψ describing the conformation of a glycosidic linkage. For example (86,87), the magnitude of coupling between ^{13}C-1 and H-X and between ^{13}C-X and H-1 ($^{3}J_{CH}$) is a reflection of these angles given by the Karplus curve relationship for ^{13}C–O–C–^{1}H arrays of nuclei. As yet, measurements of this type have been carried out (86,87) only with disaccharides and cyclohexaamylose. With ^{13}C-enriched material, observations on coupling between carbons ($^{3}J_{CC}$) across the glycosidic linkage are also feasible (49). Another approach (88) consists of evaluating the influence of H-X on T_1 of H-1, or vice versa, as a measure of interatomic distances and hence of the relative orientations of adjacent glycosyl residues. Chemical shifts are very sensitive to changes in polysaccharide conformation,

TABLE II

**Spin–Spin Coupling Data for ^{1}H-1 and ^{13}C-1 Nuclei of
Polysaccharides and Anomeric Configuration**

Residue/polysaccharide	$^3J_{H\text{-}1,H\text{-}2}$ (Hz)		$^1J_{C\text{-}1,H\text{-}1}$ (Hz)		Ref.
	α	β	α	β	
Glucopyranosyl					
Amylose	~3				75
Dextran	~3		171		76,77
Lichenan		7.5		160	78,79
Pustulan				160	77
2-Deoxy-2-sulfaminoglucopyranosyl					
Heparin	3.6		170		31,40
Galactopyranosyl					
Klebsiella	~3		172		55,80
2-Acetamido-2-deoxygalactopyranosyl					
Chondroitins		~7		161–162	31,53
Rhamnopyranosyl					
Klebsiella		<1		162	55,80,81
Glucopyranosyluronic acid					
Chondroitins		8.0		162	82
Mannopyranosyluronic acid					
Alginate				162	83
Idopyranosyluronic acid					
Heparin	2.6		172		40,53
Dermatan	3.0		168		31,82

as when the pH or temperature is altered, although the patterns of displacements in chemical shifts such as these have yet to be correlated with the direction and magnitude of changes in angles ϕ and ψ.

C. Studies on Chemical Structure

1. Early Investigations

Many well-known polysaccharides were examined (*52,69,76,89,90*) initially to determine the degree to which the spectral characteristics of polysaccharides may be correlated with those of mono- and oligosaccharides. Although linkage positions and configurations in most of these polymers were already known, an important consideration was the possibility of using ^{1}H-NMR spectroscopy for determining the solution conformations of individual types of residues present. Good correlation was found (*69*) between the chemical shifts and coupling constants for H-1 of α- and β-D-glucose in D$_2$O and anomeric protons of a number of D-glucose oligosaccharides, which was consistent with a $^4C_1(D)$ conformation for residues in the larger molecules. Although resolution was much poorer with D_2O solutions of dextrans, partially hydrolyzed starch, and other glucans, the observed chemical shifts and (to a lesser extent) spacings indicated (89,90) that glucopyranose residues bonded by α-(1→4), α-(1→6), β-(1→2), or β-(1→3) linkages possess $^4C_1(D)$ conformations in water. A comparison of H-1 chemical shifts and spacings in the spectra of chitin and partially hydrolyzed chitin with those of 2-acetamido- and 2-amino-2-deoxy-α,β-D-glucose provided evidence (91) for the β-D configuration and $^4C_1(D)$ conformation of residues in this polymer.

An ^1H-NMR examination of heparin demonstrated (31,36) the usefulness of this technique for detecting structural features that were sometimes overlooked by the chemical methods then available. That is, the ^1H spectrum was incompatible with the generally accepted view that D-glucuronic acid is a major constituent of the polymer and, with support from chemical data, showed instead that the principal acid component is L-iduronic acid.

With the introduction of ^{13}C-NMR spectroscopy to polysaccharide studies came the prospect of far better signal dispersion than usually available in ^1H spectra. Detailed studies (61,62,65,66,92,93) soon followed on characteristics of the ^{13}C chemical shifts of many known polysaccharides, and much attention was devoted to correlations between ^{13}C data for the polymers and those of low molecular weight model compounds. Consequently, extensive data are available (16,17,22,62,66,93) for virtually all possible homopolymers of D-glucose: β-(1→4), β-(1→6), β-(1→2), β-(1→3), α-(1→4), α-(1→6), α-(1→2), and α-(1→3). These data have provided an invaluable source of reference material for studies on less well characterized polysaccharides.

2. Details of Chemical Structure: Sequences and Chemical Heterogeneity

One very effective way in which NMR spectroscopy can be applied is in examining the degree of complexity of a polysaccharide. Information about the number and relative proportions of constituent residues, i.e., different linkage configurations and positions in homopolymers, as well as different kinds of sugars in heteropolymers, is readily apparent from a well-resolved spectrum. Proton and ^{13}C spectra reinforce each other in this regard, although the latter often are more informative because of superior signal separation. In favorable circumstances, it is possible to account for all nonequivalent carbons (e.g., Fig. 6) and hence for the number of distinct types of residues present. Corresponding information in ^1H spectra comes mainly from the number of anomeric protons detected, because so many of the other signals may overlap heavily, although two-dimensional spectra (Fig. 4) can greatly ease this problem.

Extensive studies on dextrans (*66,76,77,94–104*) illustrate these points. In one of the earliest NMR studies on polysaccharides, it was shown (*76*) that there are two H-1 signals in the spectrum of dextran B-742 (at δ 4.9 and 5.2) representing, respectively, the major 6-linked α-D-glucopyranosyl residues and minor residues bearing a $(1\rightarrow3)$ or $(1\rightarrow4)$ branch (**2**). Supporting observations included the rationale (*90*) that the difference in shift between these

$$\alpha\text{-}D\text{-}Glc}p\text{-}(1\rightarrow6)\text{-}\alpha\text{-}D\text{-}Glc}p\text{-}(1\rightarrow6)\text{-}\alpha\text{-}D\text{-}Glc}p\text{-}(1\rightarrow6)$$

$$m$$

$$\uparrow$$

$$1$$

$$\alpha\text{-}D\text{-}Glc}p$$

2 (*m* = 3 or 4)

two signals had its origin in electronegativity differences. The ratio of the H-1 integrals afforded (*76*) a measure of the degree of branching of the dextran (in good agreement with chemical data). In broad surveys of a large number of dextrans, in which H-1 signals were analyzed to determine the nature and degree of branching, it was found (*94–97*) that the detection of minor linkage types was not always successful, based on comparisons with chemical analysis. Also, because of overlapping signals, quantitative measurements usually involved an appreciable level of uncertainty. An evaluation of integrals of FT ^{1}H-NMR spectra at 90°C suggested (*97*) that some anomeric protons do not contribute proportionally and that errors of perhaps 15% entered into the linkage analysis. Some of these measurements were aided (*97*) by suppression of residual water peaks (as in Fig. 2) by the "WEFT" technique.

Carbon-13 spectra of dextrans have been subjected to more thorough analysis (*66,77,98–104*) because many signals other than those of the anomeric region could be examined. In the spectrum of dextran B-742, for example, the signals of C-1 and several other carbons concurred in demonstrating the presence of a minor proportion ($\sim9\%$) of α-$(1\rightarrow4)$ linkages, in addition to the α-$(1\rightarrow6)$ linkages (57%) and α-$(1\rightarrow3)$ linkages (34%). [By contrast, chemical analysis did not clearly differentiate the $(1\rightarrow4)$ assignment from the possibility of $(1\rightarrow2)$.] Similarly, the chemical shifts of secondary carbons in the region of δ 70–85 were shown (*100*) to be diagnostic of the presence of α-$(1\rightarrow2)$, α-$(1\rightarrow3)$, and α-$(1\rightarrow4)$ linkages, provided that the dextran spectra were taken at constant temperature (*99*), preferably near 90°C. Although there was concern as to the accuracy of ^{13}C integral values, comparisons with methylation structural analysis indicated (*104*) that the two methods

TABLE III

Comparative Analyses of Dextrans by
^{13}C-NMR Spectroscopy and Methylation-GLC

Relationship between $(1\rightarrow6)$ and $(1\rightarrow3)$ linkages[a]		
NRRL dextran	^{13}C	Methylation
B-1355	0.7	0.7
B-1498	1.4	1.5
B-1501	3.9	2.7

Degree of linearity[b]		
NRRL dextran	^{13}C	Methylation
B-1299	0.87	0.67
B-1396	10.7	7.9
B-1402	2.0	1.98
B-1422	4.6	3.0

[a] Seymour *et al.* (*102*)
[b] Seymour *et al.* (*101*).

agree within 10% and hence that ^{13}C-NMR spectroscopy provides a reliable description of branching in many dextrans (Table III).

Mannans of yeast constitute another highly diverse series of branched homopolymers that have been very profitably investigated (*105–113*) by ^{1}H- and ^{13}C-NMR spectroscopy. In these polymers (**3**), the main chain of $(1\rightarrow6)$-linked residues of α-D-mannopyranose bears side chains of differing lengths containing mainly α-$(1\rightarrow2)$ linkages and occasionally α-$(1\rightarrow3)$ linkages. Residues of β-D-mannopyranose are found in some instances. Most

α-D-Man*p*-$(1\rightarrow6)$-α-D-Man*p*-$(1\rightarrow6)$-α-D-Man*p*-$(1\rightarrow6)$

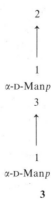

2

1
α-D-Man*p*
3

1
α-D-Man*p*

3

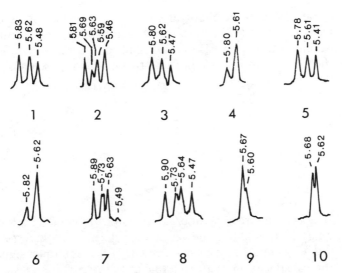

Fig. 8. Signals due to anomeric protons of mannans in D_2O at 100 MHz from various yeasts, showing characteristic differences in patterns of chemical shift and relative intensities. From (*108*), with permission.

members of this class of polysaccharide gave ^1H spectra (at 100 MHz and $\sim 70°C$) that were well resolved in the anomeric region (*105–108*). Appreciable chemical shift differences were observed between the H-1 signals of unsubstituted and 2-O-substituted residues of the main chain and between the H-1 signals of residues of the side chains. Among the latter, there were differences depending on whether the residue was appended to the main chain, was internal, or was terminal. Since there were wide variations as well in the ratios of residue types, the yeast mannans afford an extensive series of highly distinctive spectra (Fig. 8).*

The anomeric regions of ^{13}C spectra of yeast mannans are equally distinctive (*109*). As in the dextran series, chemical shifts of many of the other nuclei have been correlated with the structures proposed. Throughout, these investigations made elaborate use of oligosaccharides prepared from the mannans and provide an excellent illustration of the value of partial fragmentation as a complement to NMR studies on the polymer itself. It is especially noteworthy that the patterns of H-1 or C-1 resonance signals for many of these yeast mannans are sufficiently distinctive to be of use for taxanomic purposes (*108,110*).

---

* There are a number of galactomannans of related structure that exhibit (*108*) analogous patterns of H-1 resonances, supplemented by the contribution of α-galactopyranosyl groups that terminate the side chains.

An example of nearest-neighbor analysis in a branched polysaccharide is found (*114*) in ^{13}C-NMR studies on galactomannans of legume seeds. The main chain of these consists of (1→4)-linked β-D-mannopyranose residues and the side chains of (1→6)-linked α-D-galactopyranose units (**4**). Both

$$\beta\text{-D-Man}p\text{-}(1\rightarrow4)\text{-}\beta\text{-D-Man}p\text{-}(1\rightarrow4)\text{-}\beta\text{-D-Man}p\text{-}(1\rightarrow4)$$

6

↑

1

α-D-Gal*p*

4

^{1}H and ^{13}C spectra afforded manno/galacto ratios in good agreement with chemical analysis. In addition, however, the influence of nearest-neighbor residues on ^{13}C chemical shifts was reflected (*114*) in multiplet signals. For example, three distinct signals attributable to C-4 of mannose residues represented diads in which both residues were unbranched, one was branched, or both were branched. Relative intensities showed how these diad sequences may vary according to the plant source.

Carbon-13 NMR spectroscopy is an excellent method for examining the degree of regularity of sequences in linear polysaccharides containing different types of linkages or/and sugar residues. A consistent alternating sequence of (1→3) and (1→4) linkages in the linear β-D-mannan of *Rhodotorula glutinis* was clearly evident (*48*) from the 12 resonances that comprise its ^{13}C spectrum. Similarly, the absence of multiplicity in the C-1 signals of de-O-acetylated meningococcal polysaccharides as compared with the native polymers was indicative (*20,21*) of uniformly linked sialic acid residues in these materials.

The high sensitivity of the ^{13}C-NMR method is especially well illustrated (*66*) in the example of a glucan from *Tremella mesenterica*. The glucan was shown to consist of a uniform, linear, trisaccharide sequence of α-glucopyranose residues joined by two consecutive (1→4) linkages and one (1→6) linkage (**5**) from the fact that the ^{13}C spectrum (Fig. 7) contained two signals

$$\alpha\text{-D-Glc}p\text{-}(1\rightarrow4)\text{-}\alpha\text{-D-Glc}p\text{-}(1\rightarrow6)\text{-}\alpha\text{-D-Glc}p\text{-}(1\rightarrow4)$$

5

for each of the C-1, C-4, and C-6 resonances due to the 4-substituted residues. Hence, the latter were in either of two nonequivalent environments, whereas the 6-substituted residues exhibited only one type of relationship with respect to neighboring residues. Similar characteristics were evident (*44a,115,115a*) in the ^{13}C spectrum of lichenan from Icelandic moss, which contains approximately two $(1\rightarrow4)$ linkages per $(1\rightarrow3)$ linkage. Since all of the 3-substituted residues produced only single resonances for C-1 and C-3, whereas there were two pairs of C-1 and C-4 signals for the other residues, the polymer must consist mainly of a repeat trisaccharide sequence (6) (which is in agreement with

$$\beta\text{-D-Glc}p\text{-}(1\rightarrow3)\text{-}\beta\text{-D-Glc}p\text{-}(1\rightarrow4)\text{-}\beta\text{-D-Glc}p\text{-}(1\rightarrow4)$$

6

other analyses). Two related glucans, one from oats and the other from barley, gave ^{13}C spectra that were virtually indistinguishable from that of lichenan (44a). Although other studies had suggested that these cereal glucans are slightly richer in $(1\rightarrow4)$ linkages, the ^{13}C data are probably more reliable, having been recorded (44a) under optimum conditions for quantitative measurements (see Fig. 5 for the spectrum of the barley glucan).

Both ^{1}H and ^{13}C data (18,19,31,39,116) have contributed to the overall representation of glycosaminoglycans (mucopolysaccharides) as linear heteroglycans consisting of alternate residues of an aldopyranuronic acid and a 2-amino-2-deoxyaldopyranose (7). Carbon-13 spectra of heparin, hyaluronic acid, chondroitin 4- and 6-sulfate, and dermatan have all exhibited

$$\text{Gly}p\text{A-}(1\rightarrow x)\text{-Gly}p\text{N-}(1\rightarrow4)$$

7

fundamental patterns based on the 12 sugar carbons required for sequence **7**. However, lesser groups of signals showed that there are minor irregularities in these patterns, or chemical heterogeneity. Thus, the ^{13}C spectra of chondroitin 4- and 6-sulfate showed (53) that each polymer contains structural elements of the other, to the extent of 20–30% (see Fig. 6b). From ^{1}H and ^{13}C spectra it was evident (31,36,39,40) that hog mucosal heparin, although consisting mainly of **8**, contains a higher proportion of β-D-glucuronic acid

$$\alpha\text{-L-Ido}p\text{A2SO}_3^-\text{-}(1\rightarrow4)\text{-}\alpha\text{-D-Glc}p\text{NSO}_3^-\,6\text{SO}_3^-\text{-}(1\rightarrow4)$$

8

and 2-acetamido-2-deoxy-α-D-glucose than does heparin from beef lung.* Nevertheless, well-resolved ^{13}C signals of the mucosal heparin were not multiplets, suggesting either that the effect of nearest neighbors was atypically slight or that the minor types of sequences were relatively highly concentrated in certain molecules. The latter possibility was more consistent with evidence available (42) from the NMR spectra of fractions of mucosal heparin.

---

* This distinction was expressed graphically (19) in the convenient form of a "difference" ^{13}C spectrum obtained by subtraction with a computer of the spectrum of beef lung heparin from that of mucosal heparin.

For molecules of alginic acid, the influence of neighboring residues on ^{13}C chemical shifts is pronounced (83). These heteropolymers of marine origin have a nonregular primary structure in which residues of β-D-manno-pyranosyluronic acid (M) and α-L-gulopyranosyluronic acid (G) are linked through (1→4) bonds in blockwise fashion. This was clearly reflected in the form of multiplet signals in their ^{13}C spectra: four C-1 signals corresponded to diad sequences MM, MG, GM, and GG, whereas C-5 of mannosyl residues was apparently sensitive to both neighbors and provided a means of identifying triplet sequences (MMM, MMG, etc.). In this series also, ^{13}C spectra recorded under appropriate experimental conditions gave (83) a reliable measure of M/G ratios.

Agars and carrageenans, closely related families of linear polysaccharides obtained from red seaweeds, provide a unique example of structural differences that are distinguishable by ^{13}C-NMR spectroscopy (117,118). Although both polymers consist of D-galactose and 3,6-anhydrogalactose linked (1→3) and (1→4), respectively, the anhydrogalactose in carrageenan has the D configuration, whereas that in agarose is L. For each polysaccharide the C-1 signal of galactose was found to resonate at δ 103, whereas C-1 signals of the anhydro sugars occurred at δ 99.2 (agarose) and δ 96.2 (κ-carrageenan), a difference of 3 ppm. This difference in chemical shift was a clear demonstration (118) that, although the anhydro sugars are enantiomeric, their comparable environments in the polymers are diastereomeric and hence nonequivalent.

3. Anomeric Configuration in Multiresidue Types of Polysaccharides

With polysaccharides made up of several kinds of sugar residues, NMR spectroscopy can be of great advantage despite the high degree of complexity inherent in their spectra. The use of a high-field spectrometer is especially beneficial for ^1H applications in this area. By examining both ^1H and ^{13}C spectra, it should be possible to detect signals characteristic of each species represented and to estimate relative proportions. With supporting data from chemical analysis, signals of anomeric protons and carbons are relatively easily identified and analyzed to determine the configurations of the glycosidic linkages. Indeed, since information about anomeric configuration in complex materials is extremely difficult to obtain in other ways, this is perhaps the most important role for NMR spectroscopy.

Numerous examples of this type of application are found (80,81,119–124) in studies on capsular polysaccharides of *Klebsiella*. These polysaccharides and a variety of fragments prepared from them (by partial acid or enzymatic hydrolysis, Smith degradation) were examined extensively by ^1H- and ^{13}C-NMR spectroscopy in combination with chemical methods. The oligosaccharides, of course, were more amenable to spectroscopic analysis. Primarily, the spectra were used to determine anomeric configurations, although they also furnished information about such matters as the presence and location of O-carboxyethylidene substituents (*119,121,123*), as well as broad support for chemical analysis. With individual oligosaccharides fully characterized, their sequences were then translated to the polymeric level, and unique structures were determined thereby. The level of complexity can be appreciated, given that these polysaccharides (e.g., **9** is the sequence elaborated for *Klebsiella* K56) may contain five different sugars linked in various combinations of anomeric configuration and position and sometimes bearing a

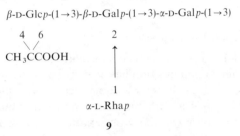

β-D-Glc*p*-(1→3)-β-D-Gal*p*-(1→3)-α-D-Gal*p*-(1→3)

```
    4   6           2
     \ /            ↑
    CH₃CCOOH        |
                    |
                    1
```

α-L-Rha*p*

9

4,6- or 2,3-O-(1-carboxyethylidene) acetal function. To this it may be added that there are some 80 serologically distinct strains of *Klebsiella*. In structural studies of this scope, as mentioned earlier in conjunction with fungal mannans, NMR spectroscopy is obviously of invaluable assistance both for determining structural details and for distinguishing among polysaccharides from varied sources.

Studies on other polysaccharides of great complexity provide additional examples of the utility of NMR spectroscopy for determining the configuration at anomeric centers, as well as other structural features. Among these may be cited investigations (*125*) on the C-teichoic acid of *Streptococcus pneumoniae* and an extensive series of ^{13}C measurements (*126*) on capsular polysaccharide antigens from *Hemophilus influenzae* and *Neisseria meningitidis*.

High-field ^{1}H-NMR spectroscopy has been utilized extensively (*127–129a*) in determining the structures of glycoproteins. Large oligosaccharide segments (e.g. **10**), isolated from proteolytic digests and terminated at the

α-D-Man*p*-(1→6)-β-D-Man*p*-(1→4)-β-D-Glc*p*NAc-(1→4)-β-GlcpNAc

```
            3
            ↑
            |
            |
            1
```

α-D-Man*p*

10

reducing end by an amino acid residue, were found to give very well resolved spectra at 360 MHz. Analyses of coupling constants of the H-1 signals in the spectra, aided by complementary data on smaller model compounds, provided detailed information about anomeric configurations in many such oligosaccharides. Also, H-1 chemical shifts often were found to be indicative of positions of attachment of glycosidic bonds, which afforded support for other structural data obtained by chemical and enzymatic means. Because the quantities of oligosaccharide available in glycoprotein studies are frequently very small, ^{13}C-NMR spectroscopy has not been widely used, owing to its inherently lower sensitivity.

4. Conformation of Individual Residues

The value of H-1 and C-1 chemical shift and coupling data for the determination of anomeric configuration has already been stressed, and a number of examples have been cited. For pyranosyl residues, configuration is often assigned on the basis of an assumed conformation. Thus, a D-galactopyranosyl residue may be given the α designation because its NMR characteristics are consistent with those of an equatorial H-1 and an axial ^{13}C-1—O-1 bond in the $^4C_1$(D) conformation; frequently, the remainder of the spectrum is not amenable to a straightforward analysis. With most of the constituent sugars of polysaccharides, it is undoubtedly satisfactory to rely on signals due to the anomeric nuclei because, as is well known, one particular conformation of each of these sugars is highly favored.

Less certainty, however, is associated with sugars of the ido or gulo configuration, which has given rise to conformational studies on L-iduronic acid residues in heparin, dermatan (chondroitin B), and heparan sulfate and on L-guluronic acid residues of alginic acid. The convolution difference spectrum of heparin, referred to earlier (Fig. 3), afforded (40) a complete set of $^3J_{HH}$ parameters for the L-iduronosyl 2-O-sulfate moiety, consistent with a $^1C_4$(L) conformation and an α-L configuration (8). Other NMR sources of support for these designations were provided (39,40) by ^{13}C chemical shift and ^{13}C–^{1}H coupling data, by ^{1}H coupling data (130) on an unsaturated disaccharide prepared enzymatically from heparin, as well as by selective line broadening effects (131) on the ^{13}C–^{1}H spectra promoted by Gd^{3+} ion. Proton–proton coupling data, afforded by a full analysis of the convolution difference spectrum of dermatan sulfate and supplemented by other NMR data as above, showed (82) that in this polymer also the L-iduronic acid residues have the α configuration and favor the $^1C_4$(L) conformation.

Studies on homopolymeric regions of the alginate molecule and on model glycosides furnished ^{1}H–^{1}H coupling data indicating (132) that L-guluronic acid residues are α-linked and in the $^1C_4$(L) conformation.

5. Polysaccharide Conformations: Relaxation Characteristics

Nuclear magnetic resonance spectroscopy provides several kinds of information related to the shapes of polysaccharides in solution. In an early study (75) on the ^{1}H spectrum of amylose in DMSO it was found that signals attributable to OH-2 and OH-3 resonate at substantially lower field ($\delta > 5$) than would normally be expected.* This observation, supported by analogous data for model compounds (maltose, cyclohexaamylose) and related IR measurements, was advanced (75) as evidence that OH-2 and OH-3 are engaged in intramolecular hydrogen (bonding between contiguous residues

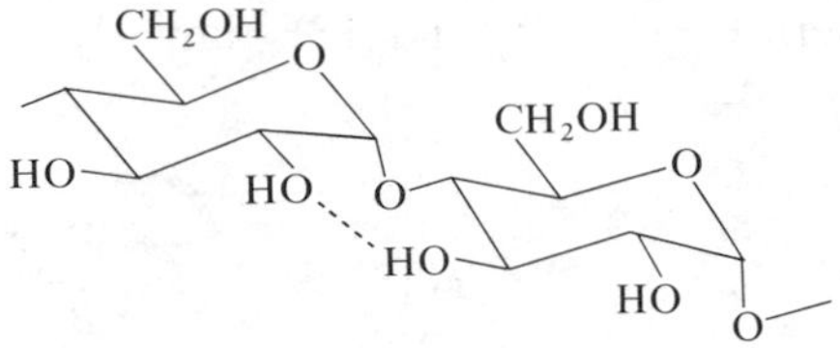

11

(as in **11**). The patterns of upfield shifts exhibited by these two signals as the temperature of the solution was increased indicated (134) that OH-3 donates its proton in the interaction with OH-2; the same conclusion was reached with cyclohexaamylose. These findings, together with coupling constant data ($^3J_{HOH}$), are consistent (134) with right-handed helical character for the amylose chain in DMSO solution.

A similar illustration of intraresidue hydrogen bonding was described (79) for solutions of lichenan in DMSO. In the spectrum of this polymer, signals attributable to OH-3 and OH-4 were found at δ 5.36, whereas those due to OH-2 and OH-6 were upfield ($\delta \sim 4.7$). Furthermore, the former two signals migrated upfield with an increase in temperature more rapidly than did the latter, which is characteristic of intramolecular hydrogen bonding. For this polymer, it is likely (79) that OH-3 and OH-4 engage in hydrogen bonding to O-5 of adjacent residues (as in **12**), by analogy with cellulose in

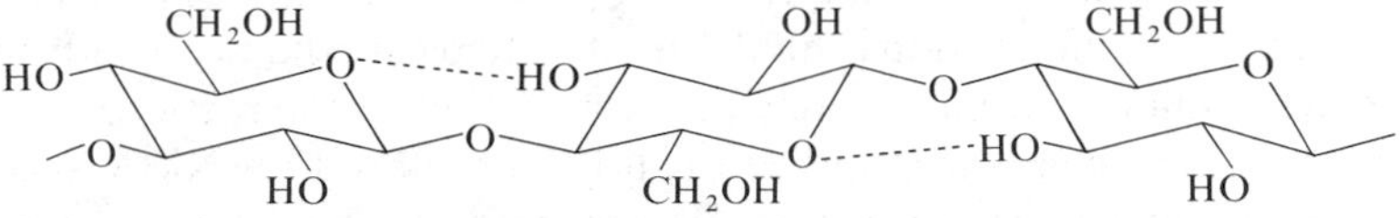

12

---

* Much larger downfield shifts (δ 13–14) are observed (133) for H-bonded systems of the type C=N $\cdots$ H—N, which occur in base-paired segments of RNA.

the solid state. Comparable data were obtained (*135*) for cellotetraose and cellobiose in DMSO solution.

Coupling between ^{13}C and ^{31}P was utilized (*45*) to examine the conformations of meningococcal polysaccharides in which 2-acetamido-2-deoxy-α-D-glucopyranose or 2-acetamido-2-deoxy-α-mannopyranose units are linked together by phosphate bridges. Couplings between the phosphorus atom and C-2 were found to be generally large, i.e., $^{3}J_{CP}$ was 8.0–8.6 Hz. Based on the known dihedral angle dependence of three-bond ^{13}C–^{31}P coupling, this showed that the phosphorus atom is preferentially trans with respect to C-2 and gauche with respect to the ring oxygen atom in rotamers involving the anomeric C—O—P bond (**13**). Similarly, values of 6.0–7.6 Hz

13

for coupling between phosphorus and C-5 corresponded to a P—O bond trans to the C-5—C-6 bond. Hence, these data were consistent (*45*) with highly extended backbone conformations for the polysaccharides. The ^{13}C–^{31}P coupling observed (*136*) in the ^{13}C spectrum of the capsular antigen from *Pneumococcus* type 26 has been utilized in a similar fashion to describe the conformation of the polysaccharide.

Changes in chemical shift induced by pH should provide a measure of how near the nuclei affected are to sites of dissociation. The ^{1}H spectra of heparin recorded over a range of pH showed (*40*) that the H-5 signal of the aminodeoxyglucose residue (in **8**) shifted markedly upfield and became broader when the pH was lowered. Since H-5 in related mono- and oligosaccharides shows no analogous pH effect, this macromolecular property was interpreted to mean that H-5 is so positioned in the polymer as to sense changes in the ionization of the carboxyl group of the iduronosyl moiety. Accordingly, a conformational model for the heparin chain was advanced (*40*) to accommodate appropriately positioned sites of interaction.

Helix–random coil transitions are readily monitored by NMR spectroscopy. Early observations (*137*) on ^{1}H spectra of polyuridylic acid, polyadenylic acid, and transfer RNA of *Escherichia coli* illustrate this. The tRNA molecule exhibited extreme line broadening at 31°C and only slightly less at 72°C, consistent with very high degrees of order. By contrast, the structure of polyuridylic acid was random at both temperatures, since its spectra were of high-resolution quality, whereas the characteristics of polyadenylic acid were intermediate. Carbon-13 NMR spectroscopy may be somewhat more sensitive. An examination of polyuridylic acid revealed (*138*) a helix–coil transformation over the range 12°–18°C, as evidenced by a marked sharpening of all ^{13}C resonance lines as the temperature was increased. This change from an ordered to a random conformation over so narrow a temperature span suggested that the process was a cooperative one.

Proton NMR spectroscopic studies on solutions of sodium hyaluronate in D_2O clearly showed (*139*) that only a portion of the polymer was mobile enough to produce a high-resolution spectrum, whereas the remainder gave extremely broad signals. Even at 104°C, the linewidths of the acetamido methyl signals for the two portions were 3 and ~ 40 Hz, respectively.* The stiff chain segments were estimated to account for 55–70% of the polymer under different conditions of ionic strength, temperature, and moderate changes in pH. Spin–spin relaxation times, measured by pulsed NMR, also were consistent (*139*) with the mobility characteristics of the polymer, and qualitative support was furnished by ^{13}C-NMR spectroscopy.

Extensive line broadening of ^{13}C resonances, attributable to gel formation, was observed (*140–143*) with $(1\rightarrow 3)$-β-D-glucans of the curdlan type. Integrated peak intensities (with nuclear Overhauser enhancement suppression) of the C-1 and C-5 signals of the gel in D_2O were 20–30% of those for the random coil (in NaOD). Greater internal rotation was exhibited by the hydroxymethyl group, however, since the C-6 signal of the gel accounted for 60% of the theoretical intensity.† The fact that the backbone carbons of the gelled polymer gave clearly visible signals at all, suggested that there is a considerable proportion of single helical regions in curdlan. A double helical conformation, by contrast, would have been expected (*140,141*) to exhibit the reduced mobility of double-stranded carrageenan, the ^{13}C spectrum of which contains virtually no detectable resonances. From an examination of the dependence of linewidth on molecular weight, it was concluded (*143*) that the relatively flexible regions of curdlan correspond to chain segments of $\overline{DP} \geq 49$.

Curdlan and other β-D-glucans also have been examined (*144,145*) in the solid state by CP/MAS ^{13}C-NMR spectroscopy, representative spectra being shown in Fig. 9. The observed dispersion of chemical shifts indicated the presence of both helical and random coil forms in these solid materials. Thus, splitting of the C-3 peak of curdlan or laminaran was ascribed to contributions by both forms, whereas the shape of the C-3 signal of lentinan

---

* For comparison, linewidths of the acetamido methyl signals in Figs. 1a and 1b are 5.5 and 3.0 Hz, respectively.
† By contrast, and as yet unaccounted for, the C-6 signal of the aminodeoxyglucose residue in heparin (**8**) is substantially broader than the signals of other carbons (*116*).

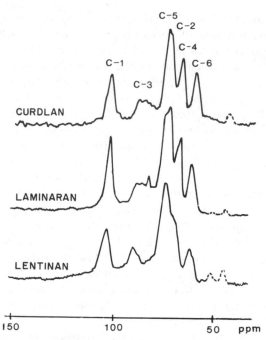

Fig. 9. CP/MAS ^{13}C-NMR spectra (75.46 MHz) of powder specimens of (1→3)-linked β-D-glucans: curdlan (DP 20), laminaran, and lentinan. Samples were contained in a rotor machined from perdeuteriated poly(methyl methacrylate). Peaks at 44.4 and 51.0 ppm are from the rotor. From (*144*), with permission.

suggested that most molecules in this specimen adopt a single helix conformation.

Another notable feature (*142*) of curdlan solution spectra is a downfield displacement by 2–3 ppm of the C-1 and C-3 signals relative to the corresponding signals from partially degraded curdlan or short-chain laminaran. These displacements were attributed to the effect of restricted rotation about the glycosidic bonds in the single helical conformation, by analogy with earlier observations. That is, deshielding influences amounting to 2–5 ppm had been found (*21,92*) in comparing the chemical shifts of C-1 and C-4 of (1→4)-α-D-glucans with those of structurally related cyclohexaamylose, and the population of dominant rotamers was expected to be more limited in the latter due to geometric restraints. In this context, it is worth noting that there are substantial differences (*86,87*) between cyclohexaamylose and maltose in the magnitude of angles ψ and ϕ of their respective glycosidic linkages, as shown by measurements of coupling between C-1 and H-4 and C-4 and H-1′.

A number of characteristics, some analogous to those described above, were exhibited (146) by ι-carrageenan in lithium iodide solution. Whereas at 90°C a high-resolution spectrum was observed for most of the polymer, broad signals appeared at 25°C, originating from C-1 of both monomeric residues and C-3 of the galactosyl residue. Furthermore, these three broad peaks experienced downfield shifts of 2–4 ppm. Consequently, it appeared (146) that an ordered conformation existed within a random coil at the lower temperature and that the glycosidic linkages were constrained to relatively fixed orientations within this more rigid framework.

In branched polysaccharides, very large differences in mobility can exist between side-chain residues and those of the backbone. Nuclear magnetic resonance spectroscopic measurements with gel-forming fungal glucans consisting of a $(1 \rightarrow 3)$-β-linked main chain and $(1 \rightarrow 4)$-α-linked side chains have shown (142,143) that they may differ by several orders of magnitude. Carbon-13 signals due to the $(1 \rightarrow 3)$-linked units had linewidths (at 25.2 MHz) as large as ~ 1000 Hz, which corresponded to a correlation time of $> 10^{-6}$ s for backbone motion. By contrast, measurements on the narrower signals due to carbons of the side chains, of T_1 and nuclear Overhauser enhancement values, as well as linewidths, gave correlation times in the region of 10^{-8}–10^{-9} s. Consequently, the side chains of these polysaccharide gels reorient at least a hundred times as rapidly as does the backbone. Analogous, although smaller, differences were found (147) in T_1 values between ^{13}C nuclei of side-chain and main-chain residues of several mannans. For example, T_1 values for C-1 of a nonreducing end group of a side chain, of an adjacent side-chain unit, and of a main-chain residue were 0.20, 0.13, and 0.09 s, respectively. Measurements on dextrans gave (103) T_1 values of the same magnitude, e.g., 0.22 s for C-1 of a 6-mono-O-substituted residue located in a side chain as compared with values of 0.15 and 0.11 s, respectively, for C-1 of 3- and 6-O-mono-substituted residues of the backbone chain.

Relaxation data for a completely different type of macromolecular system were furnished (148) by measurements on the glycoprotein, glycoamylase. Its sugar constituents are not part of a polysaccharide structure but are single glycosyl groups or in very short chains appended to the protein. Since many of the ^{13}C linewidths for these carbohydrate residues were found to be less than 40 Hz, it was evident that internal rotation was rapid, corresponding to correlation times of $\sim 10^{-8}$ s. The measured T_1 values (0.13–0.25 s) were comparable to those of side chains of branched polysaccharides, suggesting (148) that the influence imparted by the protein matrix is not unlike that of a polymeric carbohydrate backbone. Also noteworthy was the fact that the relaxation times were not significantly changed upon denaturation.

Substantial variations in segmental motion also have been observed (*149*) for residues of linear oligosaccharides. Thus, ^{13}C nuclei of the terminal D-galactose group in the tetrasaccharide stachyose were readily distinguished from those of the internal D-galactose residue because of larger T_1 values. Analogous results were obtained with other oligosaccharides.

An additional advantage of this capacity to differentiate between resonance signals of side chains and main chains, or between those of inner residues and end groups of oligosaccharides, is that it may be used in appropriate cases for the assignment of ^{13}C resonance signals that are difficult to identify from chemical shifts. Applications to glucans (*140,141*), mannans (*147*), and dextrans (*103*) illustrate the value of T_1 measurements for this purpose.

Differences in the tertiary structure of various preparations of cellulose fibers were detected (*150,151*) by CP/MAS ^{13}C-NMR, which afforded distinctive spectra for cellulose I, cellulose II, and amorphous material. In the spectrum of highly crystalline cellulose I (*150*), the C-1 and C-4 signals were split, indicating that glucose residues are in two magnetically nonequivalent environments. By contrast, the spectrum of largely amorphous material (*151*) was similar to that of the high-resolution ^{13}C spectrum of low molecular weight cellulose in DMSO. Values of T_1 were found (*151*) to be in a range between those observed for glassy synthetic polymers and crystalline polymers.

Broad-line NMR observations (*152,153*) on water molecules associated with the surface of cellulose have shown the presence of two signals: a doublet (maximum separation of 350 mG) and a narrow, central singlet. The doublet was attributed to water species tightly bonded to the macromolecule, rather than to rapid tumbling in an anisotropic environment. However, the singlet appeared to be due not to free water, but to protons of the macromolecule; after exchange with D_2O, the singlet remained, whereas the doublet was no longer evident. A second singlet, detected at about 65°C, was considered to be due to energetic water species tumbling rapidly in the macromolecular environment, but less mobile than bulk water.

Observations (*154*) on surface relationships between solid starches and water illustrate a different type of NMR measurement. They involve deuterium NMR linewidths and intensities for starch samples exchanged with D_2O. There appeared to be three fractions of water represented in the resonance signals examined over a range of temperature: one forming a monolayer, one less firmly bound, and one free, characteristics analogous to those exhibited by other types of natural polymers.

6. Polysaccharide Derivatives

The industrial importance of polysaccharide derivatives and the valuable role they play in the purification and characterization of many polysaccharides have resulted in many NMR spectroscopic studies (14) on materials in this category. Not surprisingly, the most extensive group of investigations deal with industrial derivatives of cellulose. Proton NMR spectra of cellulose triacetate (155–158) as a dilute solution in $CDCl_3$, particularly at 220 or 250 MHz, clearly demonstrated that the glucopyranosyl residues retain the $^4C_1$(D) conformation when esterified. Also, the finding (157) that signals of the ring protons and those of the 2-, 3-, and 6-O-acetyl signals (at δ 2.05, 1.95, and 2.17, respectively) were well resolved raised the possibility of using NMR spectroscopy for the characterization of partially acetylated materials (159). Carbon-13 NMR spectroscopy also offered promise in this direction and, in fact, was employed (160) to show that OH-2 was selectively esterified by acetic anhydride in the presence of pyridine.

O-Methyl derivatives may also be considered in this context in view of the fact (161) that the 6-O-methyl group of tri-O-methylcellulose resonated at δ 3.33 as compared with δ 3.52 and 3.48 for the substituents at the secondary positions. In the ^{13}C spectrum of O-methylcellulose of DS (degree of substitution) $\sim$ 2.8 (162), there were signals at δ 61.4, 61.0, and 60.1 due to the 2, 3, and 6 substituents, respectively. However, a derivative of DS $\sim$ 0.7 gave (162) signals attributable only to 2- and 6-O-methyl groups and exhibited concomitant downfield displacements of (a proportion of) the C-2 and C-6 signals. These displacements were due to the strong deshielding β effect (by $\sim$ 7–9 ppm) introduced (59,60) by the conversion of ^{13}C—O—H to ^{13}C—O—alkyl.

A problem common to these investigations is the high viscosity of solutions of cellulose derivatives. Consequently, limited hydrolytic degradation (with acid or cellulase) has been employed (see, e.g., 162) to furnish sufficiently concentrated solutions for satisfactory ^{13}C spectra of some derivatives. However, when direct NMR analysis of the intact cellulose derivative is not feasible, it may be more satisfactory (162) to use a total acid hydrolysate. This should offer a sufficiently large improvement in resolution and in the measurement of integrals to compensate for errors introduced in the handling of hydrolysates. Analyses of carboxymethylcellulose by ^{13}C NMR (Fig. 10) (162), as well as by ^1H-NMR spectroscopy (163), are examples of the latter approach. In the spectra of the polymer and hydrolysate (Fig. 10) specific signals were easily attributable to the O-$^{13}CH_2$—COOH substituents and to carbons bearing these ether substituents, but structural detail was more evident in the spectrum of the hydrolysate. Furthermore, the latter was found to be readily amenable to quantitative analysis (163) by ^1H-NMR spectroscopy because the methylene protons of the substituents were well resolved in the δ 4.0–4.5 region.

As well as serving as sources of data on the average number and location of hydroxyl groups substituted per glucose residue, both ^1H- and

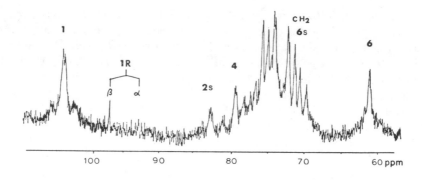

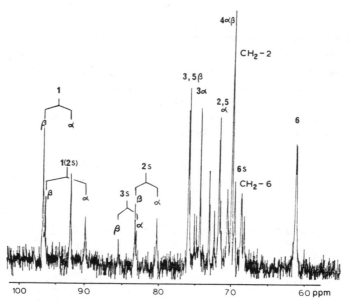

Fig. 10. ¹³C-NMR spectra at 22.6 MHz of (upper) *O*-(carboxymethyl)cellulose (DS 0.7, partially degraded by cellulase) in D_2O at 30°C and (lower) of acid hydrolysate of *O*-(carboxymethyl)cellulose (DS 0.7) in D_2O at 30°C (R, signal of reducing-end residue; s, carbon is bonded to alkoxyl group). From (*162*), with permission.

¹³C-NMR spectroscopy have furnished information about molar substitution per residue (MS), for such derivatives as hydroxyethyl- (*162,164*) and hydroxypropyl cellulose (*165,166*). Substituents of the latter (A and B, respectively) have various degrees of chain length (*n*), depending on the number of moles of ethylene oxide or propylene oxide, respectively, that

become attached to the polymer:

$$R-O-(CH_2-CH_2-O)_nH \qquad R-O-(CH_2-\overset{\overset{\textstyle CH_3}{|}}{CH}-O)_n-CH_2-\overset{\overset{\textstyle CH_3}{|}}{CH}-OH$$

A   B

It was shown (162,164) that the average chain length of hydroxyethyl substituents (A) can be easily estimated by ^{13}C-NMR, since the chemical shifts of CH_2 groups within the chain are typically much larger than that of the terminal CH_2OH groups; for spectral integrations, appropriate precautions were taken (164) to minimize possible differences in relaxation time and nuclear Overhauser effects. For hydroxypropyl substituents (B), the ^{13}C nuclei of methyl groups of inner units resonated at ~ 3 ppm upfield (γ effect) of the CH_3 of the terminal unit, and, as would be expected, the CH_3 signals were much sharper than the signals of the cellulosic backbone (166).

Substituent effects evident in the ^{13}C spectra of nitrocellulose served (167) to indicate the positions and degrees of substitution. For example, chemical shift changes for C-1 and C-6 due to replacement of a hydroxyl proton with a nitro group β, γ, or δ with respect to the carbon of interest amount to 10, -5, and 1.5 ppm, respectively. These values, which are of the same order as observed in alkylation, were utilized as well (167) in studying selective denitration reactions.

Proton NMR spectroscopy was used (168) in an early study to show that, on permethylation, the glucose residues of amylose retain the $^4C_1$(D) conformation and also (52) that OH-3 is relatively slowly alkylated. Among other methyl ether derivatives examined were those of dextrans (169) and pullulan (170). With a $(1 \rightarrow 4)$-linked glucan (171) from red algae and a $(1 \rightarrow 6)$-linked glucan (172) from *Clostridium botulinum*, H-1 signals of the methylated polysaccharides were more easily observed than those of the unmodified polysaccharides in D_2O (or of the peracetates), which was helpful for determining anomeric configurations.

D. Other Applications

Nuclear magnetic resonance spectroscopy has been utilized in a number of investigations on chemical and enzymatic reactions, as well as for a few other purposes. Several examples of applications in these categories may be cited from studies on glycosaminoglycans. Proton NMR spectroscopy proved to be useful for monitoring the N-desulfation of heparin with dilute acid (36) and deaminative degradation of the polymer with nitrous acid (36), whereas its oxidation with periodic acid (173) and hydride reduction of uronide residues (174) were examined conveniently by ^{13}C-NMR spectroscopy. A distinctive indicator (53,130) of the action of eliminase enzymes on heparin and chondroitins was the downfield signal of the olefinic proton generated as fragments terminated by α,β-unsaturated uronic acid residues (e.g., 14) were formed.

14

Information concerning the stereochemistry of the action of several glycosidases on polysaccharide substrates was furnished by ^{1}H-NMR spectroscopy. For example, it was shown (Fig. 11) (175) that the hydrolysis of laminaran by an exo-(1→3)-β-D-glucanase resulted in the release of α-glucose and hence that the reaction involved an inversion of anomeric configuration, in contrast to a retention of configuration in the attack of an endo-β-(1→3)-D-glucanase on the polysaccharide. Similarly, an ^{1}H-NMR kinetic study of hyaluronic acid by testicular hyaluronidase showed (176) that the release of the (1→4)-linked residue of 2-acetamido-2-deoxy-β-D-glucopyranose was accompanied by retention of anomeric configuration. During the hydrolysis of microbial cell wall oligosaccharides by lysozyme, a decrease in the value of $J_{1,2}$ for the liberated α-muramic acid residue was observed (177), which indicated that this residue was substantially distorted toward a half-chair conformation when bound to the enzyme.

Marked changes are induced in ^{1}H and ^{13}C chemical shifts of uronic acid moieties by varying the pH. For instance, downfield shifts of ~ 0.4 ppm by the H-5 signal was observed (40) upon lowering the pH from 7 to 2, whereas the carboxyl carbon became more shielded by ~ 2 ppm. Accordingly, it was feasible (40) to plot NMR titration curves for variations in δ with changes in pH (pD for D_2O solutions), from which pK_a data for the carboxyl group of L-iduronic acid residues in heparin were obtained.

Nuclear magnetic resonance spectroscopy provides a sensitive probe for examining the stoichiometry of complex formation between polysaccharides and various ionic species, as well as for localizing the site of such interactions. The binding of Ca^{2+} by heparin was found (178,179) to be characterized by displacements of several ^{1}H signals. Most notable was strong deshielding of the iduronic H-1 and H-5 signals, implying that the carboxyl groups, rather than sulfate, are the main sites of binding. Evidence that the carboxyl groups are also the sites for attachment of Na^+ was obtained (46) by measurements of ^{23}Na linewidths of sodium heparin solution; i.e., the linewidths increased with an increase in pH while exhibiting an equivalence point at pH 4.5. In

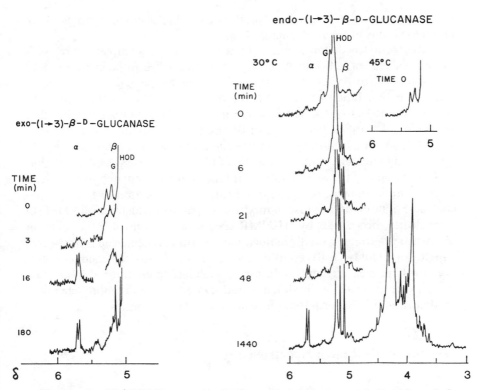

Fig. 11.   Partial ^{1}H-NMR spectra in D_2O recorded during assay of glycosidases. Left: exo-$(1\rightarrow3)$-β-D-glucanase on laminaran; α-glucose is produced, and only later is it seen to mutarotate to the β-anomer. Right: endo-$(1\rightarrow3)$-β-D-glucanase on laminaran; the β-D-anomers are seen to be the initial products of the reaction. From (*175*), with permission.

addition, ^{23}Na linewidth measurements were employed in competition experiments to measure the capacity of a large number of cations to bind to heparin, which showed (*46*) that Ca^{2+} forms the strongest ion complex with this polymer. Carbon-13 NMR spectroscopy is effective (*109,180*) for monitoring complex formation between polysaccharides and borate ion or diphenyl boronate.

In the presence of Gd^{3+}, there was pronounced broadening of the C-1 and C-6 signals of the α-L-iduronic residues in heparin and in dermatan sulfate and also of the H-1 and H-5 signals of these residues (*131*). Since the β-D-glucuronosyl residues of chondroitin 6-sulfate were unaffected, the transition metal ion promoted selective relaxation of ^{1}H and ^{13}C nuclei of anomers in which the anomeric carbon–oxygen bond was axial, in agreement with data for low molecular weight model compounds. According to these observations

Gd^{3+} is a sensitive diagnostic NMR reagent for distinguishing between anomeric forms of hexuronic acids (131).

Copper–ethylenediamine and related metal–amine complexes are well-known solvents of cellulose. Although chelation between the metal ion and the 2- and 3-hydroxyl groups of the cellulose had appeared to be a feasible step in the dissolution process, NMR observations (181) on the analogous solvent properties of Cd–ethylenediamine were not consistent with that possibility. Advantage was taken of the high sensitivity of ^{113}Cd chemical shifts to its ligand environment and of the utility of ^{13}C chemical shifts for determining sites of complexation. Overall, the data suggested (181) that hydrogen bonding between cellulose and the diamine, mediated by the metal through steric and electronic factors, brings about solution of the polysaccharide. Solutions of cellulose in mixtures of paraformaldehyde and DMSO were clearly shown (182) by ^1H-NMR spectroscopy to involve the reaction of hydroxyl groups with formaldehyde to form a hemiacetal derivative ("methylol cellulose"). By contrast (183), according to ^{13}C chemical shift and lineshape measurements, solutions of cellulose in mixtures of N-methyl-morpholine N-oxide or hydrazine and DMSO do not entail chemical modification but the formation of relatively stable complexes.

III. Infrared–Raman Spectroscopy

A. Methodology: Instrumentation and Techniques

Infrared and Raman are two related forms of vibrational spectroscopy. Infrared spectra originate from the absorption of IR frequencies by vibrating chemical bonds. Raman spectra have their origin in the perturbation of these bonds by high-frequency (visible) radiation, leading to—among other scattering effects—the emission of radiation whose frequency differs from that of the incident light by an increment corresponding to vibrational bond frequencies (184,185).

Infrared and Raman spectra are plots of absorption and emission, respectively, as a function of frequency in the $\bar{v}$ range 4000–50 cm^{-1} typical of IR radiation.* Whereas IR radiation interacts only with molecular vibrations

---

* Strictly speaking, $\bar{v}$ values are not *frequencies* but wavenumbers (i.e., the number of waves per centimeter, corresponding to the reciprocal of the wavelength λ in centimeters). Since Raman bands represent frequency differences, they are also called Raman shifts and are designated as $\Delta\bar{v}$ in reciprocal centimeters. Infrared *intensities* are expressed in terms of absorbance ($A = \log(I_0/I)$, where I_0 and I are the intensities of the light before and after passage through the sample) or as percent transmittance ($T\% = I/I_0 \times 100$) with a logarithmic absorbance scale. Ordinates in Raman spectra are usually linear in arbitrary intensity units.

associated with a change in dipole moment, Raman scattering is associated only with vibrations involving a change in polarizability. Consequently, vibrations symmetric with respect to a center of molecular symmetry do not give rise to oscillating dipoles. Although "Raman active," they are "IR inactive"; i.e., they give rise to weak bands, if any, in the IR spectrum. The converse is true for asymmetric vibrations. Therefore, IR and Raman spectra are complementary, and both are required to describe the vibrational systems of molecules (*184,185*).

Infrared spectrometers consist basically of a source of IR radiation, a dispersive system (a prism or a grating) for separating the component frequencies of the radiation, and a detector. A double-beam system permits the absorbance of the sample to be measured with respect to a reference beam (*186*). To maximize energy output, the transmission components are reduced to a minimum number and are made of such material as alkali halides that are relatively transparent to IR radiation. Signal-to-noise ratios can be increased by several orders of magnitude by storing the signals from multiple scans and processing these data with a computer.

A high level of sensitivity can be attained with FT IR spectrometers, a new generation of instruments in which the dispersive system is substituted by an interferometer. The interferogram is Fourier-transformed by a dedicated computer into a conventional IR spectrum. Although the main advantage of an FT IR spectrometer over the usual dispersion instrument is convenient computer manipulation of spectra, the feasibility of obtaining spectra of small samples in real time, or of accumulating weak signals much faster than with conventional instruments, is an additional advantage of FT IR spectrometry (*187*).

Raman spectrometers are usually of the dispersive type. The source of monochromatic radiation is most commonly a laser, and the optics are those typical of ultraviolet–visible spectrophotometers.

1. Sampling Techniques in IR Spectroscopy (*185,186,188*)

Infrared and Raman spectra can be obtained from samples in the solid state as well as in solution. However, the sampling requirements are somewhat different for the two types of spectroscopy.

Infrared spectra of polysaccharides are most frequently obtained from *solid samples*. To minimize scattering of the IR radiation, which decreases the energy available and also distorts absorption bands, the solid specimen should approximate a "pseudomolecular dispersion"; i.e., it should consist of small, evenly distributed particles. (Ideally, the particles should not exceed a size corresponding to the wavelength of the incident light, i.e., 2–3 μms for the high-frequency region of the IR spectrum.) Further reduction of scattering can be achieved by dispersing the solid granules in a medium whose refractive index is intermediate between that of air and the solid. Typical dispersing media are paraffin oil (Nujol) and fluorocarbons, which, however, are not completely transparent over all regions of the IR spectrum. A more popular technique for solid sampling is that of the pressed disk, obtained by grinding the sample with dry KBr (or other halide) and pressing the mixture into a disk. Good samples can generally also be obtained from polysaccharide films, cast from solutions or spread from gels on a disk of AgCl or other suitable water-insoluble material, and dried under an infrared lamp.

Alternatively, IR spectra of solid samples and pastes can be obtained by the use of reflectance methods such as the attenuated total reflection (ATR), a technique especially useful for characterizing surfaces of paper or textiles. This is illustrated (Fig. 12) by the ATR spectra of cellulose, gum arabic, and starch obtained from a base paper and the adhesive sides of a postage stamp and a gummed label (*189*).

Infrared spectra of solid polysaccharides usually require 1–10 mg of sample, although as little as 10 μg will suffice if microtechniques, which

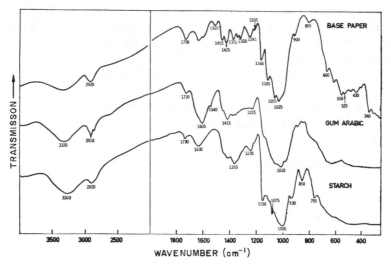

Fig. 12. ATR–IR spectra of a base paper (essentially cellulose), gum arabic, and starch (adhesive sides of a postage stamp and a gummed label). The spectrum of the base paper also shows peaks at 1730 and 1245 cm^{-1}, indicative of residual *O*-acetyl and uronic acid groups of hemicelluloses, bands at 1595 and 1505 cm^{-1} of residual lignin, and a band at 805 cm^{-1} derived from a softwood component of the furnish. The curved background in the high-frequency region is typically due to scattering of low-wavelength radiation by solid samples, whereas the decreasing background from 800 to 200 cm^{-1} is due to absorption by the supporting crystal (KRS-5). From (*189*), with permission.

require special beam condensers, are used. Spectral accumulation by repetitive scanning further increases sensitivity.

A special technique for investigating polysaccharide fibers or oriented films involves the use of IR polarizers. Spectra are obtained with the sample oriented at different angles with respect to the plane of plane-polarized light (generally, parallel and perpendicular). Since absorption occurs only for groups whose oscillating dipoles are oriented parallel with respect to the polarized radiation, the measurement of "dichroic ratios," i.e., the ratio of band intensities for parallel and perpendicular orientations of the sample, affords information about the orientation of bonds with respect to the axis of the polysaccharide chain (7a).

The frequency, intensity, and shape of IR bands (and also the number of observed bands) are also affected by the lateral order of polysaccharide chains. Although this effect permits different polymorphs of a given polysaccharide to be characterized and the degree of crystallinity of each polymorph to be evaluated, it is a complicating factor when spectra are compared for identification purposes, especially in view of the fact that crystallinity can be altered or even destroyed by mechanical grinding.

Although most polysaccharide solvents absorb strongly in the IR region, useful spectra can often be obtained for *solutions*, especially in those regions where the solvent is relatively transparent. Water has transmittance "windows" in the regions 2900–1800 and 1500–950 cm^{-1}, and D_2O at 2050–1300 and 1150–700 cm^{-1} (190).* Dimethyl sulfoxide and DMSO-d_6 also may be used in complementary regions (191), whereas chloroform and other organic solvents are suitable for acetates and other derivatives of polysaccharides.

In order to minimize solvent absorption and, accordingly, to obtain spectra over as wide a spectral region as possible, the use of a thin cell (0.025–0.1 mm) and a relatively high concentration of polysaccharide (1–10%) is usually required. Although the sample requirement with a normal cell is 10–100 mg, this limitation is greatly reduced with computer-equipped spectrometers, because they can accurately subtract strong backgrounds and accumulate data from numerous scans.

2. Sampling Techniques in Raman Spectroscopy (192)

Raman spectra of *solid samples* are usually obtained directly from powders in glass capillaries. Sample requirements seldom exceed a few milligrams. A fluorescence background due to trace impurities is often superimposed on Raman spectra of polysaccharides and should be subtracted (preferentially by a computerized baseline correction procedure). In favorable cases,

* Since solutes in D_2O consist largely of species in which exchangeable hydrogens are substituted with deuterium, vibrational spectra in D_2O are different from those in H_2O, especially for those bands involving vibrations of exchangeable H (or D).

fluorescence can be photochemically quenched by exposing the sample to the laser beam for a period before obtaining the spectrum.

Since H_2O and D_2O have weak Raman bands, they can be used to obtain Raman spectra in *solution* much more easily than IR spectra. Normal cells require a substantial amount of sample (up to 100 mg), whereas only a droplet of the solution is needed with a capillary cell. In addition, because fluorescence is much more efficiently quenched in the capillary (*193*), the microsampling technique often has the advantage of producing spectra of better quality.

B. Functional Group Detection and Configuration

The vibrational frequencies of bonds (and therefore those of the corresponding IR and Raman bands) are a function of k/μ, where k is the force constant of the bond and μ the "reduced mass" of the vibrating atoms. Hence, double bonds absorb (or emit) at frequencies higher than single bonds consisting of the same atoms, and for bonds having similar force constants the vibrational frequency decreases with increasing masses of the atoms (*184,185*). Although these are useful generalizations, it is important to realize that vibrations in a molecular system are not isolated from each other. This is amply evident for such complex molecules as carbohydrates and polysaccharides, which contain many C—C and C—O bonds that have similar force constants and masses. Thus, normal coordinate analysis has shown (*194,195*) that most of the vibrations of glucose and glucans are highly "coupled," i.e., are not localized in specific bonds.

In spite of these limitations, "group frequencies" can readily be associated with the presence of certain functional groups. The most generally useful group frequencies are those arising from stretching vibrations of multiple bonds (typically, C=O) or single bonds formed between atoms of greatly different masses (O—H, N—H, C—H). These vibrations are essentially unaffected by coupling with vibrations of the C—C and C—O bonds of the polysaccharide backbone and give rise to IR bands in characteristic regions. Also, modifications in bond character caused by inductive and resonance effects of substituents may bring about sufficiently large second-order shifts of the IR group frequencies to differentiate between different functional groups containing the same type of bond (e.g., C=O).

Typical frequency ranges for some functional groups of polysaccharides and their most common derivatives are given in Table IV. It is worth emphasizing that, whereas the absence of a characteristic absorption band indicates the absence of the appropriate functional group, for the detection of this group one cannot rely solely on the presence of an absorption band in its typical region. Additional criteria of characteristic band intensities and

TABLE IV

Typical IR Frequencies of Functional Groups Commonly
Encountered in Polysaccharides

Group	Form	v (cm^{-1})	Vibrational mode[a]
O—H	Free (monomeric)	3650–3600	v O—H[b]
	H-Bonded	3600–3200	
		1100–1050	δ O—H-related
C—H		2900–2800	v C—H
		1470–1380	δ C—H
COOH	Monomer	$\sim$1760	v C=O
	Dimer	$\sim$1710	v C=O
COO$^-$		1600–1550	v_{as} C=O
		1450–1400	v_{as} C=O
CONH (Acetamido)		3300–3250	Band "A," v N—H
		3100–3070	Band "B" (12)
		$\sim$1650	Amide I, v C=O-related
		$\sim$1550	Amide II, δ N—H-related
COOR		$\sim$1735	v C=O
		$\sim$1250	v C—O—C
S=O (Sulfates)		$\sim$1240	v_{as} S=O (broad, complex)
(Sulfonates)		1200–1150	v_s S=O (broad, complex)
NO$_2$ (Nitrates)		1640–1620	v_{as} N=O
		1285–1270	v_s N=O

[a] v, stretching; δ, deformation.
[b] Band frequency, intensity, and shape strongly affected by H bonding.

shapes assist assignments by ruling out such alternative possibilities as overtones or combination bands associated with strong "fundamentals" at higher frequency or absorption by impurities or residual solvents. A few functional groups give rise to more than one characteristic band. This is true of the carboxylate ion, which constitutes a "coupled oscillator," its components having significantly different frequencies due to assymetric and symmetric stretching of the C=O bonds. The amide linkage is another coupled system, with two characteristic bands (amide I and amide II) associated with vibrations not strictly localized in individual bonds but nevertheless very useful for diagnostic purposes (195). Absorption patterns composed of weak, but characteristic bands also may be useful in assignments, e.g., the series of O—H bands between 2800 and 2000 cm^{-1}, which, in conjunction with the strong C=O band at $\sim$1710 cm^{-1}, are indicative of the presence of COOH groups.

Although functionalization alters the vibrational pattern of a polysaccharide backbone, bands arising from polar functional groups usually make such a strong contribution to an IR spectrum as to appear essentially as if

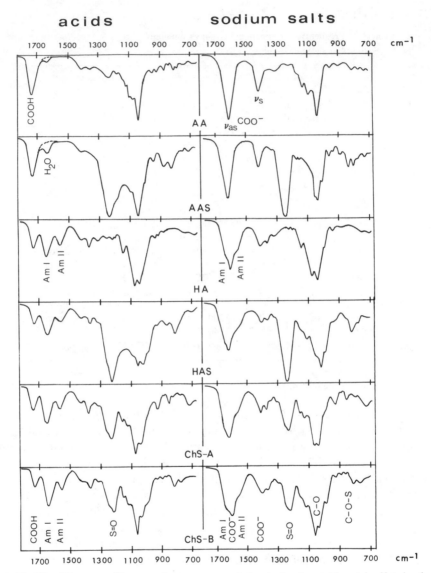

Fig. 13. IR spectra (films cast from aqueous solutions) of polyuronides and sulfated poly-uronides (AA, alginic acid; AAS, sulfated alginic acid; HA, hyaluronic acid; HAS, sulfated hyaluronic acid; ChS-A, chondroitin 4-sulfate; ChS-B, dermatan sulfate; Am. I, amide I; Am. II, amide II). Redrawn from (*196*), with permission.

superimposed on the background absorption of a neutral polysaccharide. Such an effect is illustrated in Fig. 13 by the prominent absorptions due to carboxyl and sulfate groups of polyuronides and sulfated polyuronides (*196*).

Deuterium oxide is a useful solvent for the observation of carboxyl and acetamido bands (*197–199*). As illustrated in Fig. 14a for heparin, in neutral solution the carboxylate ($v_{as}COO^-$) and N-acetyl (amide I or ND—CO) bands absorb at frequencies very close to each other; however, the COOD and ND—CO bands can be resolved in acidic solution (*199*). As expected for vibrations of polar groups, bands due to stretching of COOH (COO$^-$) and amido groups are much weaker (and less diagnostic) in Raman spectra than in IR spectra. Conversely, as already noted, vibrational modes associated with small changes in dipole moment give rise to stronger bands in Raman than in IR spectra. This complementary nature of IR and Raman spectroscopy is illustrated by spectra of heparin in H_2O (Fig. 14b). That is, the strong IR band centered at ~ 1240 cm^{-1}, attributed to asymmetric stretching of the sulfate S—O bonds, has practically no counterpart in the Raman spectrum, whereas the corresponding symmetric mode produces a strong Raman band at a frequency (~ 1065 cm^{-1}) at which the IR spectrum displays only a shoulder, the latter being situated on the high-frequency side of the complex group of "carbohydrate" bands between 1050 and 980 cm^{-1} (*193*).

Weak bands in the 900–800 cm^{-1} region, associated with vibrations of the C—O—S group, are present in the IR spectra of sulfated polysaccharides (*196,200,201*). Although frequencies of 825 and 855 cm^{-1} have been reported as typical of sulfate groups attached equatorially and axially, respectively

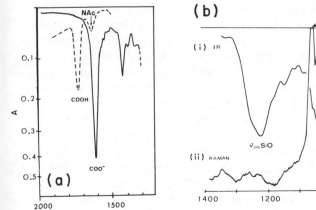
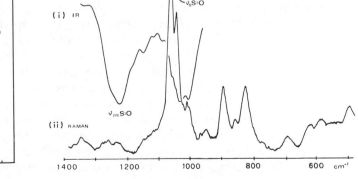

Fig. 14.   IR and Raman spectra of heparin in aqueous solutions. (a) IR "carbonyl" region in D_2O (—) and N DCl (---) (*199*); (b) IR and Raman spectra ("fingerprint" region) in H_2O (*193*). From (*193*), with permission.

(*196*), such assignments cannot be generalized (*202,203*). The same conclusions hold for bands in the 890–800 cm^{-1} region, characteristic of the sulfonyloxy group (*204*). Most likely, vibrations giving rise to bands in the 900–800 cm^{-1} region are strongly coupled with other vibrations, the extent of coupling being also a function of inter- and intramolecular interactions (*193*).

In favorable cases, also, vibrations not strictly localized on specific bonds can be diagnostic of functional groups or stereochemical features. Essentially on the basis of empirical correlations, the weak bands at 891 ± 7 and 844 ± 8 cm^{-1} have been associated with axial and equatorial anomeric C—H bonds, i.e., assuming a $^4C_1$(D) conformation of the pyranose ring, β and α configurations, respectively (*205*). Corresponding bands are also observable in Raman spectra of monosaccharides (*206*), amylose and its oligomers (*207,208*), dextran (*208*), and cellulose (*209*).

Although different positions of glycosidic linkages in homopolysaccharides give rise to different IR patterns (*205*), only in favorable cases are the IR spectra sufficiently reliable for the detection and estimation of specific linkages. Thus, in IR spectra of dextrans, a band at 822 cm^{-1} can be associated with α-D-(1→3) branches; the band is clearly observed in "difference spectra" obtained by subtracting the contribution of the main 6-linked α-D-glucan chain (*210*).

C. Analytical Aspects

Infrared spectra are widely used as "fingerprints" for individual polysaccharides, the limiting factor for this application being the masking effect by strong and broad bands of polar groups on weaker bands associated with more subtle structural features. Since, however, polar groups make weak contributions to Raman spectra, the latter usually provide more specific band patterns than IR spectra. Reference IR spectra are now available either in the specialized literature or in the form of atlases, as for cellulose and cellulose derivative (*211*) and pectic substances (*212*). Raman spectra have been reported for dextran (*207*), amyloses (*208*), celluloses (*195,209*), hyaluronate (*213*), and sulfated glycosaminoglycans (*193*).

Strictly speaking, comparisons of spectra for identification purposes should be restricted to materials in the same physical state and, among solid polysaccharide samples, to those having the same type and degree of crystallinity. That is, the IR and Raman spectra of different polymorphs may vary widely (*214,215*). The effects of crystallinity are even more clearly seen in the spectra of oligomers (*214–216*), especially at low temperatures (*217,218*). "Crystal vibrations" reflecting intermolecular interactions are expected to be especially evident in the far IR region (250–25 cm^{-1}) (*218*).

Crystallinity, as well as scattering effects, usually prevent the reliable use of IR spectra for *quantitative analyses*. Wherever these effects can be minimized, such as in physically homogeneous amorphous films or KBr disks obtained from freeze-dried solutions, the spectra are amenable to quantitative measurements, especially of absorbance ratios that can be ascribed to relative proportions of functional groups (or structural features) in the polysaccharide.

Limitations due to lateral-order and scattering effects are eliminated in solution. Consequently, solutions (in D_2O) have been satisfactorily used for determining the content of uronic acids and esters in pectates (*198*) and of uronates and acetamido groups in glycosaminoglycans (*199*), as well as (in H_2O) that of sulfate in glycosaminoglycans and dextran sulfate (*193*). In principle, any polysaccharide band can be used for quantifying a polysaccharide, although, since the requisite of mutual noninterference is difficult to meet with polysaccharide mixtures, analyses of two or more components are seldom feasible in practice.

Infrared and Raman spectra of solutions in D_2O or H_2O can be used effectively for monitoring enzyme reactions [as illustrated in Fig. 15 for cleavage of chondroitin sulfates by chondroitinase AC (*219*)] or conformational transitions, as reported for xanthan (*215*). Another application of

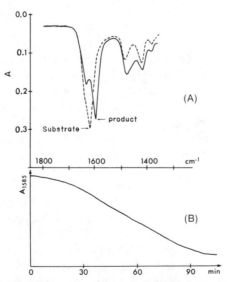

Fig. 15. (A) IR spectra in the carbonyl region of chondroitin sulfates (solid line) and of their unsaturated degradation products with chondroitinase AC, in D_2O; (B) kinetics of the enzyme reaction (monitored from the absorbance at 1585 cm^{-1}) (*219*).

quantitative IR spectroscopy is the estimation of acid dissociation constants of polyuronates from the ratio of the COOH and COO$^-$ band intensities as a function of pH (pD) (*199*). Also, ion binding can be studied from the shifts of carboxylate bands (*220*). It is worth noting that free and bound species show *separate* bands in vibrational spectra, thus providing an "instantaneous picture" of the system, which is in sharp contrast to the "weighted average" usually observed in NMR spectra of rapidly exchanging species.

D. Hydrogen Bonding and Hydration

The molecular interaction most frequently investigated by IR spectroscopy is hydrogen bonding. The stretching frequencies of both donor and acceptor groups are shifted toward lower values when these groups are involved in hydrogen bonding, either inter- or intramolecularly. Deformation (bending) frequencies usually experience smaller shifts, in the opposite direction. Also, increased intensity and band broadening are typically associated with these shifts (*221*).

Infrared studies of hydrogen bonding are easily made when the substance to be examined contains only one or two donor groups (typically OH) per molecule or monomeric residue, and the substance is soluble in nonpolar solvents. Unfortunately, this condition cannot be met with polysaccharides and most of their derivatives. Consequently, much of the work employing nonpolar solvents has been carried out with monosaccharide derivatives and other model compounds. In these solvents (preferably CCl$_4$), free and hydrogen-bonded groups absorb at typically different frequencies. The involvement of a hydroxyl group in *inter*molecular bonding can be studied by measuring the relative intensities of bands associated with the free and bonded (variable frequency) species, as a function of concentration and temperature. *Intra*molecularly bonded hydroxyls can be distinguished by the invariance of their absorption bands with concentration (*222*).

In polar solvents, most OH groups of carbohydrates, polysaccharides, and their derivatives are hydrogen-bonded to the solvent. In DMSO solution, the O—H stretching bands of carbohydrates and polysaccharides are usually broad and centered at 3320–3340 cm^{-1}, with no peaks or shoulders in the region of free OH (~ 3600 cm^{-1}) (*223*). However, *intra*molecularly bonded hydroxyls in dynamic equilibrium with solvent-bonded species can be observed at somewhat lower frequencies (3250–3270 cm^{-1}), as in the case of the O$_2$H $\cdots$ O$_3'$H system of α-(1→4)-linked oligo- and polysaccharides (*223*). The 3-OH of 2,6-di-*O*-methylcyclodextrins is an interesting example of an intramolecularly bonded hydroxyl group, the donor being an OCH$_3$ on a contiguous residue. The corresponding O—H stretching band is at the same frequency (3410 cm^{-1}) in nonpolar solvents, in polar solvents, and in the solid state, and its intrinsic intensity is invariant with concentration over a 100-fold range of concentration in CCl_4 (224).

The hydrogen-bonding patterns of unsubstituted polysaccharides in the solid state produce complex O—H absorptions in the 3600–3200 cm^{-1} region, from which it is usually difficult to assess the contribution of individual hydrogen bonds. Absorption by H_2O (either ordered or disordered) is another complicating factor in this analysis.

One of the most successful approaches to the study of these systems is simplification of the IR pattern by deuterium exchange of relatively loosely bound hydroxyl hydrogens. As illustrated in Fig. 16(1) for cellulose II, exchange with deuterium vapor drastically decreases the intensity of the low-frequency components of the OH band. The two high-frequency bands, barely affected by deuteration, have been attributed to intramolecularly bonded hydroxyls in crystalline regions of cellulose, whereas the easily exchangeable hydroxyls are those from more accessible regions (and from water). Rehydration by exposure to H_2O vapor restores the original pattern, except for the two weak bands on the high-frequency side of the O—D band; this is thought to be a consequence of molecular rearrangement during the treatment rather than of direct penetration of D_2O into the crystalline regions (225). Appropriate intensity measurements of the O—H and O—D peaks provide a method complementary to X-ray diffraction for determining the crystallinity of celluloses.

The use of polarized IR spectra for determining the relative orientation of hydrogen bonds is illustrated in Fig. 16(2). The composite OH band of cellulose obtained under normal experimental conditions (i.e., with unpolarized light) can be resolved into bands due to the intramolecularly bonded OH groups disposed approximately parallel to the cellulose chain and others associated with OH groups intermolecularly bonded between cellulose

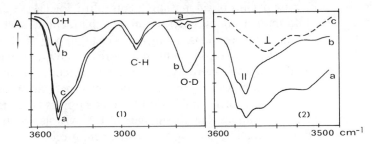

Fig. 16. IR spectra (O—H and C—H stretching regions) of cellulose II. (1) Spectra of a cellulose film before (a) and after (b) exchange with D_2O vapors and of the deuteriated sample after reexposure to H_2O vapors (c). (2) Spectra of an oriented cellulose film obtained with unpolarized light (a) and with the IR beam oriented parallel (b) and perpendicular (c) to the chain axis. Redrawn from Marrinan (225).

chains, which thus lie approximately perpendicular to these chains (225). "Dichroic ratios" obtained by polarized IR spectra have been successfully used, in conjunction with information from X-ray diffraction, for determining the relative orientation of bonds such as O—H, glycosidic C—O—C, and —C(6)$\underset{\text{H}}{\overset{\text{H}}{\diagdown}}$ in a number of oriented crystalline polysaccharides such as cellulose (226,227), β-D-(1→4)- (228) and β-D-(1→3)-xylans (229), and chitin (230). The approximate orientation of the uronate —COO⁻ and acetamido —NH—CO— bonds with respect to the polymer chain has also been determined for the glycosaminoglycans hyaluronic acid, chondroitin 4-sulfate, and dermatan sulfate (231).

The potential of IR and Raman spectroscopy for determining the fine structure of polysaccharides will undoubtedly be exploited more fully when complete vibrational analyses of these systems are available. As yet, analyses of this order of complexity have been attempted only for cellulose (232) and amylose (233). Particularly noteworthy among the findings of these studies has been the rationalization of shifts in IR frequencies observed in the conversion of cellulose I to cellulose II (227) and of the V form of amylose to the B form in terms of different chain conformations and hydrogen-bonding patterns (208).

IV. Comparative Evaluation of Spectroscopic Methods

As noted at the outset of this chapter, electronic, vibrational, and NMR spectra are complementary to each other for studies on polysaccharides. Features characteristic of each type of spectroscopy are summarized here to provide a perspective of the strengths and weaknesses of each.

Aside from their use in chiroptical measurements, ultraviolet (UV) spectra are much less informative about polysaccharides than about such biopolymers as nucleic acids and proteins, owing to the fact that polysaccharides, at most, have only weak electronic chromophores (COO⁻, acetamido groups). However, the scope of electronic spectra can be extended by suitable derivatization or by complexing with aromatic (or other electronically polarizable) chromophores. Their application to solid samples is limited by severe scattering effects. Among the advantages of UV–visible spectra are simplicity, sensitivity, and relatively low cost instrumentation.

Infrared spectroscopy also offers the advantages of simplicity and relatively inexpensive instrumentation. In addition, solid samples can be handled routinely. Infrared spectroscopy is intrinsically less sensitive than UV because of weaker sources, less sensitive detectors, and weaker absorption bands.

However, computer-aided accumulation of spectra can overcome this limitation (at the expense of time and cost). The coupling of a spectrometer with a computer, especially in the FT mode, affords the additional advantage of useful manipulation of spectral data and makes it possible to handle samples requiring compensation for strong solvent or background absorption. Such a capability is expected to boost the range of applications, as with aqueous solutions of biological material, even if turbid or containing other cosolutes.

Infrared spectra usually consist of at least 10 absorption bands with patterns characteristic of each type of polysaccharide. However, strong and broad bands from polar groups often conceal weak but characteristic bands and limit the usefulness of IR spectra as "fingerprints." (It should be realized that the use of more dispersive instruments cannot further improve the resolution of intrinsically broad IR bands.) Some functional groups are easily detectable by IR, which is conveniently utilized to check the derivatization of polysaccharides. A unique advantage of IR, when employed with a polarizer on oriented solid samples, is its use for determining the orientation of functional groups along a polysaccharide chain.

Since polar groups are poor Raman scatterers, Raman spectra usually reveal more details of vibrational spectra and therefore are intrinsically more useful than IR spectra as "fingerprints" (also in aqueous solvents). However, Raman instruments are generally more expensive and more difficult to operate. Moreover, the background fluorescence of samples is often sufficiently strong to prevent the acquisition of useful spectra.

As already noted, NMR spectra are by far the most informative for the structural characterization of polysaccharides. In principle, they provide as many signals as the number of nonequivalent atoms, each signal contributing a wealth of structural information through its chemical shift, coupling, and relaxation characteristics. In practice, however, resolution of all the signals of a polysaccharide can be achieved only in favorable cases, usually through the extra cost of applying a high magnetic field.

At the same strength of magnetic field, ^{13}C spectra are better resolved than ^{1}H spectra, particularly because the latter, in the familiar "proton-coupled" form, usually consist of signals that are dispersed in the form of doublets or multiplets. With the advent of two-dimensional spectroscopy, however, it has become possible to record ^{1}H as well as ^{13}C spectra in the form of simple groups of singlets, without loss of the corresponding spin–spin coupling information. Obviously, this advantage is gained at the expense of longer instrument time and greater cost.

Instrument time and cost are generally the major limiting factors in NMR spectroscopy, especially for ^{13}C and other nuclei of low natural abundance and small magnetic moments. Other limitations, such as interfering resonances due to solvents and the broadness of NMR signals of solid samples, are being minimized by the special approaches described in this chapter. Undoubtedly, solid-state ^{13}C-NMR spectroscopy of polysaccharides will become routine and, with improvements in instrumentation and resolution, is expected to provide even more valuable (although much more expensive) "fingerprints" than IR spectra.

The *quantitative aspects* of these various measurements are of no less importance in evaluating the usefulness of a spectroscopic method for polysaccharide characterization. Whenever an electronic chromophore can be quantitatively attached to a polysaccharide (or a polysaccharide fragment), UV–visible spectra should provide the most sensitive, accurate, and convenient method for quantitation. Infrared spectra can be used directly for quantitative measurements of polysaccharides, the usual requirement being a solubility of at least 1% in a solvent having useful "windows" in the IR region of interest.

Nuclear magnetic resonance spectroscopy has not yet gained the reputation it deserves as a quantitative method, mainly because of sensitivity limitations and problems related to different relaxation characteristics of nuclei, especially in decoupled spectra. However, many of these limitations can be overcome with modern instruments and a proper choice of experimental parameters. Whenever applicable to quantitative problems, both for "internal" analyses—such as the determination of the relative concentrations of functional groups in a molecule—or for analysis of mixtures, NMR offers the great advantages of being specific and of not requiring calibration with reference compounds.

Acknowledgments

The authors express their gratitude to L. Ayotte, P. Dais, G. Gatti, G. K. Hamer, and E. Mushayakarara for spectra included in the figures and to B. Brown and D. S. Lee for assistance in the preparation of the manuscript.

References

1. W. Pigman, ed., "The Carbohydrates. Chemistry, Biochemistry, Physiology," pp. 626, 676. Academic Press, New York, 1957.
2. R. M. McCready and W. Z. Hassid, *J. Am. Chem. Soc.* **65**, 1154–1157 (1943).
3. L. B. Jaques and H. J. Bell, *Methods Biochem. Anal.* **7**, 253–309 (1950).
4. N. Mitchell, N. Shepard, and J. F. Harrod, *Histochemistry* **68**, 245–251 (1980).
5. A. L. Stone, *Methods Carbohydr. Chem.* **7**, 120–138 (1976).
6. E. R. Morris and G. R. Sanderson, "New Techniques in Biophysics and Cell Biology." Wiley, New York, 1973.

7. W. B. Neely, *Adv. Carbohydr. Chem.* **12**, 13–33 (1957).
7a. H. Spedding, *Adv. Carbohydr. Chem.* **19**, 23–49 (1964).
8. F. A. Bovey, "High Resolution NMR of Macromolecules." Academic Press, New York, 1972.
9. R. A. Dwek, "Nuclear Magnetic Resonance in Biochemistry." Oxford Univ. Press (Clarendon), London and New York, 1973.
10. J. L. James, "Nuclear Magnetic Resonance in Biochemistry." Academic Press, New York, 1975.
11. D. W. Jones, "Introduction to the Spectroscopy of Biological Polymers." Academic Press, New York, 1976.
12. L. D. Hall, *Adv. Carbohydr. Chem.* **19**, 51–91 (1964).
13. B. Coxon, *Adv. Carbohydr. Chem. Biochem.* **27**, 7–83 (1972).
14. M. Vincendon, *Bull. Soc. Chim. Fr.* pp. 3501–3511 (1973).
15. G. Kotowycz and R. U. Lemieux, *Chem. Rev.* **73**, 669–698 (1973).
16. I. C. P. Smith, *Acc. Chem. Res.* **8**, 131–145 (1975).
17. A. S. Perlin, *MTP Int. Rev. Sci.: Org. Chem.*, *Ser. Two* **7**, 1–34 (1976).
18. A. S. Perlin, *Methods Carbohydr. Chem.* **7**, 94–100 (1976).
19. A. S. Perlin and G. K. Hamer, *ACS Symp. Ser.* **103**, 123–141 (1979).
20. H. J. Jennings and I. C. P. Smith, *in* "Methods in Enzymology" (V. Ginsburg, ed.), Vol. 50, Part C, pp. 39–50. Academic Press, New York, 1978.
21. H. J. Jennings and I. C. P. Smith, *Methods Carbohydr. Chem.* **8**, 97–105 (1980).
22. L. D. Hall, *in* "The Carbohydrates" (W. Pigman and D. Horton, eds.), Vol. 1B, pp. 1300–1326. Academic Press, New York, 1981.
23. P. A. J. Gorin, *Adv. Carbohydr. Chem. Biochem.* **38**, 13–104 (1981).
24. J. Schaefer and E. O. Stejskal, *Top. Carbon-13 NMR Spectrosc.* **3**, 283–324 (1979).
25. R. R. Ernst, *Adv. Magn. Reson.* **2**, 1–135 (1966).
26. B. D. Sykes and S. L. Patt, *J. Chem. Phys.* **56**, 3182–3184 (1972).
27. A. G. Marshall, T. Marcus, and J. Sallos, *J. Magn. Reson.* **35**, 227–230 (1979).
28. A. G. Redfield, *in* "Methods in Enzymology" (C. H. W. Hirs and S. N. Timosheff, eds.), Vol. 49, Part G, pp. 253–270. Academic Press, New York, 1978.
29. R. R. Ernst and K. Wuetrich, *Biochem. Biophys. Res. Commun.* **96**, 1156–1163 (1980).
30. A. D. Bain, R. A. Bell, J. R. Everett, and D. W. Hughes, *Can. J. Chem.* **58**, 1947–1956 (1980).
31. A. S. Perlin, B. Casu, G. R. Sanderson, and L. F. Johnson, *Can. J. Chem.* **48**, 2260–2268 (1970).
32. Z. Yosizawa, *Biochem. Biophys. Res. Commun.* **16**, 336–341 (1964).
33. S. Inoue and Y. Inoue, *Biochem. Biophys. Res. Commun.* **23**, 513–517 (1966).
34. T. Kotoku, Z. Yosizawa, and F. Yamauchi, *Arch. Biochem. Biophys.* **120**, 553–560 (1967).
35. L. B. Jaques, L. W. Kavanagh, M. Mazurek, and A. S. Perlin, *Biochem. Biophys. Res. Commun.* **24**, 447–451 (1966).
36. A. S. Perlin, M. Mazurek, L. B. Jaques, and L. W. Kavanagh, *Carbohydr. Res.* **7**, 369–379 (1967).
37. A. S. Perlin, *Proc.—IUPAC Symp. Macromol*, July, 1974 (E. B. Mano, ed), pp. 337–348. Elsevier, Amsterdam (1975).
38. S. E. Lasker and M. C. Chin, *Ann. N.Y. Acad. Sci.* **222**, 971–977 (1973).
39. A. S. Perlin, *Fed. Proc., Fed. Am. Soc. Exp. Biol.* **36**, 106–109 (1977).
40. G. Gatti, B. Casu, G. K. Hamer, and A. S. Perlin, *Macromolecules* **12**, 1001–1007 (1979).
41. J. Choay, J. C. Lormeau, M. Petitou, P. Sinay, B. Casu, P. Oreste, G. Torri, and G. Gatti, *Thromb. Res.* **18**, 573–578 (1980).
42. L. Ayotte, E. Mushayakarara, and A. S. Perlin, *Carbohydr. Res.* **87**, 297–301 (1980).

43. I. D. Campbell, C. M. Dobson, R. J. P. Williams, and A. V. Xavier, *J. Magn. Reson.* **11**, 172–181 (1973).

43a. G. Gatti, unpublished results.

44. A. Allerhand, *Pure Appl. Chem.* **41**, 247–273 (1975).

44a. P. Dais and A. S. Perlin, *Carbohydr. Res.* **100**, 103–116 (1982).

45. D. R. Bundle, I. C. P. Smith, and H. J. Jennings, *J. Biol. Chem.* **249**, 2275–2281 (1974).

46. L. Herwats, P. Laszlo, and P. Genard, *Nouv. J. Chim.* **1**, 173–176 (1977).

47. F. Barnoud, D. Gagnaire, L. Odier, and M. Vincendon, *Biopolymers* **10**, 2269–2273 (1971).

48. P. A. J. Gorin, *Carbohydr. Res.* **39**, 3–10 (1975).

49. D. Gagnaire and F. R. Taravel, *J. Am. Chem. Soc.* **101**, 1625–1626 (1979).

50. R. E. London, V. H. Kollman, and N. A. Matwiyoff, *Biochemistry* **14**, 5492–5500 (1975).

51. L. D. Hall and M. Yalpani, *Carbohydr. Res.* **83**, C5–C7 (1980).

52. B. Casu, G. G. Gallo, M. Reggiani, and A. Vigevani, *Staerke* **20**, 387–391 (1968).

53. G. K. Hamer and A. S. Perlin, *Carbohydr. Res.* **49**, 37–48 (1976).

54. R. Freeman and G. A. Morris, *Bull. Magn. Reson.* **1**, 5–26 (1979).

55. L. D. Hall and G. A. Morris, *Carbohydr. Res.* **82**, 175–184 (1980).

56. C. P. Dietrich, H. B. Nader, and A. S. Perlin, *Carbohydr. Res.* **41**, 334–338 (1975).

57. A. S. Perlin and B. Casu, *Tetrahedron Lett.* **34**, 2921–2924 (1969).

58. L. D. Hall and L. F. Johnson, *Chem. Commun.* pp. 509–510 (1969).

59. D. E. Dorman and J. D. Roberts, *J. Am. Chem. Soc.* **92**, 1355–1361 (1970).

60. A. S. Perlin, B. Casu, and H. J. Koch, *Can. J. Chem.* **48**, 2596–2606 (1970).

61. D. E. Dorman and J. D. Roberts, *J. Am. Chem. Soc.* **83**, 4463–4472 (1971).

62. T. Usui, N. Yamaoka, K. Matsuda, K. Tuzimura, H. Sugiyama, and S. Seto, *J. Chem. Soc., Perkin Trans. 1* pp. 2425–2432 (1973).

63. H. J. Koch and A. S. Perlin, *Carbohydr. Res.* **15**, 403–410 (1970).

64. P. A. J. Gorin, *Can. J. Chem.* **51**, 2375–2383 (1973).

65. T. Usui, N. Yamaoka, K. Matsuda, K. Tuzimura, H. Sugiyama, and S. Seto, *Tetrahedron Lett.* pp. 3397–3400 (1973).

66. P. Colson, H. J. Jennings, and I. C. P. Smith, *J. Am. Chem. Soc.* **96**, 8081–8087 (1974).

67. D. K. Dalling and D. M. Grant, *J. Am. Chem. Soc.* **89**, 6612–6622 (1967).

68. A. Bax, R. Freeman, T. A. Frenkiel, and M. H. Levitt, *J. Magn. Reson.* **43**, 478–483 (1981).

69. R. W. Lenz and J.-P. Heeschen, *J. Polym. Sci.* **51**, 247–261 (1961).

70. J. M. van de Ven, *J. Org. Chem.* **28**, 564–566 (1963).

71. A. De Bruyn, M. Anteunis, and G. Verhegge, *Bull. Soc. Chim. Belge* **84**, 721–734 (1975).

72. A. S. Perlin, N. Cyr, H. J. Koch, and B. Korsch, *Ann. N.Y. Acad. Sci.* **222**, 935–942 (1973); R. G. S. Ritchie, N. Cyr, B. Korsch, H. J. Koch, and A. S. Perlin, *Can. J. Chem.* **53**, 1424–1433 (1975).

73. P. A. J. Gorin, E. M. Barretto-Bergter, and F. S. Da Cruz, *Carbohydr. Res.* **88**, 177–188 (1981).

74. J. A. Schwarcz and A. S. Perlin, *Can. J. Chem.* **50**, 3667–3676 (1972).

74a. K. Bock, J. Lundt, and C. Pedersen, *Tetrahedron Lett.* pp. 1037–1040 (1973).

74b. K. Bock and C. Pedersen, *J. Chem. Soc., Perkin Trans. 2* pp. 293–297 (1974).

75. B. Casu, M. Reggiani, G. G. Gallo, and A. Vigevani, *Tetrahedron* **22**, 3061–3083 (1966).

76. W. M. Pasika and L. H. Cragg, *Can. J. Chem.* **41**, 777–782 (1963).

77. D. Gagnaire and M. Vignon, *Makromol. Chem.* **178**, 2321–2333 (1977).

78. P. Dais and A. S. Perlin, unpublished results.

79. D. Gagnaire, R. H. Marchessault, and M. Vincendon, *Tetrahedron Lett.* pp. 3953–3956 (1975).

80. G. G. S. Dutton and T. E. Folkman, *Carbohydr. Res.* **80**, 147–161 (1980).
81. J.-P. Joseleau, M. Lapeyre, M. Vignon, and G. G. S. Dutton, *Carbohydr. Res.* **67**, 197–212 (1978).
82. G. Gatti, B. Casu, G. Torri, and J. R. Vercellotti, *Carbohydr. Res.* **68**, C3–C7 (1979).
83. H. Grasdalen, B. Larsen, and O. Smidsrod, *Carbohydr. Res.* **56**, C11–C15 (1977).
84. F. R. Taravel and P. J. A. Vottero, *Tetrahedron Lett.* pp. 2341–2344 (1975).
85. R. U. Lemieux, R. K. Kullnig, H. J. Bernstein, and W. G. Schneider, *J. Am. Chem. Soc.* **80**, 6098–6105 (1958).
86. A. Parfondry, N. Cyr, and A. S. Perlin, *Carbohydr. Res.* **59**, 299–309 (1977).
87. G. K. Hamer, F. Balza, N. Cyr, and A. S. Perlin, *Can. J. Chem.* **56**, 3109–3116 (1978).
88. K. Bock, private communication.
89. V. S. R. Rao and J. F. Foster, *J. Phys. Chem.* **67**, 951–954 (1963).
90. C. A. Glass, *Can. J. Chem.* **43**, 2652–2659 (1965).
91. M. Falk, D. G. Smith, J. McLachlan, and A. G. McInnes, *Can. J. Chem.* **44**, 2269–2281 (1966).
92. K. Takeo, K. Hirose, and Y. Kuge, *Chem. Lett.* pp. 1233–1236 (1973).
93. H. J. Jennings and I. C. P. Smith, *J. Am. Chem. Soc.* **95**, 606–608 (1973).
94. R. L. Sidebotham, L. Weigel, and W. H. Bowen, *Carbohydr. Res.* **19**, 151–159 (1971).
95. E. J. Bourne, R. L. Sidebotham, and H. Weigel, *Carbohydr. Res.* **22**, 13–22 (1972).
96. T. Usui, M. Yokoyama, N. Yamaoka, K. Matsuda, K. Tuzimura, H. Sugiyama, and S. Seto, *Carbohydr. Res.* **33**, 105–116 (1974).
97. F. R. Seymour, R. D. Knapp, and S. H. Bishop, *Carbohydr. Res.* **74**, 77–92 (1979).
98. H. Friebolin, N. Frank, G. Keilich, and E. Siefert, *Makromol. Chem.* **177**, 845–858 (1976).
99. F. R. Seymour, R. D. Knapp, and S. H. Bishop, *Carbohydr. Res.* **51**, 179–194 (1976).
100. F. R. Seymour, R. D. Knapp, S. H. Bishop, and A. Jeanes, *Carbohydr. Res.* **68**, 123–140 (1979).
101. F. R. Seymour, R. D. Knapp, E. C. M. Chen, A. Jeanes, and S. H. Bishop, *Carbohydr. Res.* **71**, 231–250 (1979).
102. F. R. Seymour, R. D. Knapp, E. C. M. Chen, S. H. Bishop, and A. Jeanes, *Carbohydr. Res.* **74**, 41–62 (1979).
103. F. R. Seymour and R. D. Knapp, *Carbohydr. Res.* **81**, 67–103 (1980).
104. F. R. Seymour, *ACS Symp. Ser.* **103**, 27–51 (1979).
105. P. A. J. Gorin and J. F. T. Spencer, *Can. J. Chem.* **46**, 2299–2304 (1968).
106. P. A. J. Gorin and J. F. T. Spencer, *Can. J. Chem.* **46**, 2305–2310 (1968).
107. P. A. J. Gorin, J. F. T. Spencer, and S. S. Bhattacharjee, *Can. J. Chem.* **47**, 1499–1505 (1969).
108. P. A. J. Gorin, J. F. T. Spencer, and R. J. Magus, *Can. J. Chem.* **47**, 3569–3576 (1969).
109. P. A. J. Gorin, *Can. J. Chem.* **51**, 2105–2109 (1973).
110. P. A. J. Gorin, R. H. Haskins, L. R. Travassos, and L. Mendonca-Previato, *Carbohydr. Res.* **55**, 21–33 (1977).
111. L. Mendonca-Previato, P. A. J. Gorin, and J. O. Previato, *Biochemistry* **18**, 149–154 (1979).
112. P. A. J. Gorin, L. Mendonca-Previato, J. P. Previato, and L. R. Travassos, *J. Protozool.* **26**, 473–483 (1979).
113. E. M. Barreto-Bergter, L. R. Travassos, and P. A. J. Gorin, *Carbohydr. Res.* **86**, 273–285 (1980).
114. H. Grasdalen and T. Painter, *Carbohydr. Res.* **81**, 59–66 (1980).
115. D. Gagnaire and M. Vincendon, *Bull. Soc. Chim. Fr.* pp. 479–482 (1977);
115a. I. Yokota. S. Shibata, and H. Saito, *Carbohydr. Res.* **69**, 252–258 (1979).

116. A. S. Perlin, N. M. K. Ng Ying Kin, S. S. Bhattacharjee, and L. F. Johnson, *Can. J. Chem.* **50**, 2437–2441 (1972).
117. A. I. Usov, S. V. Yarotsky, and A. S. Shashkov, *Biopolymers* **19**, 977–990 (1980).
118. S. S. Bhattacharjee, W. Yaphe, and G. K. Hamer, *Carbohydr. Res.* **60**, C1–C3 (1978).
119. Y. M. Choy and G. G. S. Dutton, *Can. J. Chem.* **51**, 3021–3026 (1973).
120. J. M. Berry, G. G. S. Dutton, L. D. Hall, and K. L. Mackie, *Carbohydr. Res.* **53**, C8–C10 (1977).
121. G. G. S. Dutton and A. V. Savage, *Carbohydr. Res.* **83**, 351–362 (1980).
122. C.-C. Cheng, S.-L. Wong, and Y.-M. Choy, *Carbohydr. Res.* **73**, 169–174 (1979).
123. G. G. S. Dutton, K. L. Mackie, A. V. Savage, D. Rieger-Hug, and S. Stirm, *Carbohydr. Res.* **84**, 161–170 (1980).
124. K. Okutani and G. G. S. Dutton, *Carbohydr. Res.* **86**, 259–271 (1980).
125. I. R. Poxton, E. Tarelli, and J. Baddiley, *Biochem. J.* **175**, 1033–1042 (1978).
126. W. Egan, *Magn. Reson. Biol.* **1**, 197–258 (1980).
127. D. Lambertus, J. Haverkemp, J. F. G. Vliegenthart, G. Spik, B. Fournet, and J. Montreuil, *Eur. J. Biochem.* **100**, 569–574 (1979).
128. J. F. G. Vliegenthart, *Adv. Exp. Med. Biol.* **125**, 77–91 (1980).
129. J. F. G. Vliegenthart, H. Van Halbeek, and L. Dorland, *Proc. Int. Congr. Pure Appl. Chem.* **27**, 253–262 (1980).
129a. P. H. Atkinson, A. Grey, J. P. Carver, J. Hakimi, and C. Ceccarini, *Biochemistry* **20**, 3979–3986 (1981).
130. A. S. Perlin, D. M. Mackie, and C. P. Dietrich, *Carbohydr. Res.* **18**, 185–194 (1971).
131. B. Casu, G. Gatti, N. Cyr, and A. S. Perlin, *Carbohydr. Res.* **41**, C1–C8 (1975).
132. A. Penman and G. R. Sanderson, *Carbohydr. Res.* **25**, 273–282 (1972).
133. D. R. Kearns and R. G. Shulman, *Acc. Chem. Res.* **7**, 33–39 (1974).
134. M. St.-Jacques, P. R. Sundararajan, K. J. Taylor, and R. M. Marchessault, *J. Am. Chem. Soc.* **98**, 4386–4391 (1976).
135. K. K. Ghosh and R. D. Gilbert, *Text. Res. J.* 326–330 (1971).
136. L. Kenne, B. Lindberg, and J. K. Madden, *Carbohydr. Res.* **73**, 175–182 (1976).
137. J. P. McTague, V. Toss, and J. H. Gibbs, *Biopolymers* **2**, 163–172 (1964).
138. G. Govil and I. C. P. Smith, *Biopolymers* **12**, 2589–2598 (1973).
139. A. Darke, E. G. Finer, R. Moorhouse, and D. A. Rees, *J. Mol. Biol.* **99**, 477–486 (1975).
140. H. Saito, T. Ohki, Y. Yoshioka, and F. Fukuoka, *FEBS Lett.* **68**, 15–18 (1976).
141. H. Saito, T. Ohki, N. Takasuka, and T. Sasaki, *Carbohydr. Res.* **58**, 293–305 (1977).
142. H. Saito, T. Ohki, and T. Sasaki, *Biochemistry* **16**, 908–914 (1977).
143. H. Saito, E. Miyata, and T. Sasaki, *Macromolecules* **11**, 1244–1250 (1978).
144. H. Saito, R. Tabeta, and T. Harada, *Chem. Lett.* pp. 571–574 (1981).
145. H. Saito and R. Tabeta, *Chem. Lett.* pp. 713–716 (1981).
146. O. Smidsrod, I.-L. Andresen, H. Grasdalen, B. Larsen, and T. Painter, *Carbohydr. Res.* **80**, C11–C16 (1980).
147. P. A. J. Gorin and M. Mazurek, *Carbohydr. Res.* **72**, C1–C5 (1979).
148. K. Dill and A. Allerhand, *J. Biol. Chem.* **254**, 4524–4531 (1979).
149. A. Allerhand, D. Doddrell, and R. Komoroski, *J. Chem. Phys.* **55**, 189–198 (1971).
150. R. H. Atalla, J. C. Gast, D. W. Sindorf, V. J. Bartuska, and G. E. Maciel, *J. Am. Chem. Soc.* **102**, 3249–3251 (1980).
151. W. L. Earl and D. L. Vander Hart, *J. Am. Chem. Soc.* **102**, 3251–3252 (1980).
152. R. E. Dehl, *J. Chem. Phys.* **48**, 831–835 (1968).
153. R. E. Dehl and C. A. J. Hoeve, *J. Chem. Phys.* **50**, 3245–3251 (1969).
154. W. W. Fleming, R. E. Fornes, and J. D. Memory, *J. Polym. Sci., Polym. Phys. Ed.* **17**, 199–211 (1979).

155. D. Gagnaire and M. Vincendon, *Bull. Soc. Chim. Fr.* pp. 472–474 (1965).
156. D. Gagnaire and M. Vincendon, *Bull. Soc. Chim. Fr.* pp. 3413–3414 (1968).
157. D. Gagnaire, L. Odier, and M. Vincendon, *J. Polym. Sci., Part C* **28**, 27–43 (1969).
158. H. Friebolin, G. Keilich, and E. Siefert, *Angew. Chem., Int. Ed. Engl.* **8**, 766–767 (1969).
159. V. W. Goodlett, J. T. Dougherty, and H. W. Patton, *J. Polym. Sci., Polym. Chem. Ed.* **9**, 155–171 (1971).
160. P. Månsson and L. Westfelt, *Cellulose Chem. Tech.* **14**, 13–17 (1980).
161. D. Gagnaire, N. Heran, R. Le Fur, L. Pouit, and M. Vincendon, *Bull. Soc. Chim. Fr.* pp. 4326–4330 (1970).
162. A. Parfondry and A. S. Perlin, *Carbohydr. Res.* **57**, 39–49 (1977).
163. F. F.-L. Ho and D. W. Klosiewicz, *Anal. Chem.* **52**, 913–916 (1980).
164. J. R. De Member, L. D. Taylor, S. Trummer, L. E. Rubin, and C. K. Chiklis, *J. Appl. Polym. Sci.* **21**, 621–627 (1977).
165. F. F.-L. Ho, R. R. Kohler, and G. A. Ward, *Anal. Chem.* **44**, 178–181 (1972).
166. D. S. Lee and A. S. Perlin, *Carbohydr. Res.* **91**, C-5–C-8 (1981); *ibid*, **106**, 1–19 (1982).
167. T. K. Wu, *Macromolecules* **13**, 74–79 (1980).
168. B. Casu, M. Reggiani, G. G. Gallo, and A. Vigevani, *Tetrahedron* **24**, 803–821 (1968).
169. G. Keilich, E. Siefert, and H. Friebolin, *Org. Magn. Reson.* **3**, 31–36 (1971).
170. D. Gagnaire, D. Mancier, and M. Vincendon, *Org. Magn. Reson.* **11**, 344–349 (1978).
171. J. N. C. Whyte and J. R. Englar, *Can. J. Chem.* **49**, 1302–1305 (1971).
172. J. N. C. Whyte and G. A. Strasdine, *Carbohydr. Res.* **25**, 435–441 (1972).
173. L.-Å. Fransson, A. Malmström, I. Sjöberg, and T. N. Huckerby, *Carbohydr. Res.* **80**, 131–145 (1980).
174. M. Vincendon and A. S. Perlin, cited in Perlin (*39*).
175. D. E. Eveleigh and A. S. Perlin, *Carbohydr. Res.* **10**, 87–95 (1969).
176. I. V. Vikha, V. G. Sakharovsky, V. F. Bystrov, and A. Ya. Khorlin, *Carbohydr. Res.* **25**, 143–152 (1973).
177. S. L. Patt, D. Dolphin, and B. D. Sykes, *Ann. N.Y. Acad. Sci.* **222**, 211–219 (1973).
178. B. Casu and A. S. Perlin, cited in Perlin (*39*).
179. J. Boyd, F. B. Williamson, and P. Gettins, *J. Mol. Biol.* **137**, 175–190 (1980).
180. P. A. J. Gorin and M. Mazurek, *Can. J. Chem.* **51**, 3277–3286 (1973).
181. A. D. Bain, D. R. Eaton, R. A. Hux, and J. P. K. Tong, *Carbohydr. Res.* **84**, 1–12 (1980).
182. T. J. Baker, L. R. Schroeder, and D. C. Johnson, *Carbohydr. Res.* **67**, C4–C7 (1978).
183. D. Gagnaire, D. Mancier, and M. Vincendon, *J. Polym. Sci.* **18**, 13–25 (1980).
184. N. B. Colthup, L. H. Daly, and S. E. Wiberley, "Introduction to Infrared and Raman Spectroscopy." Academic Press, New York, 1964.
185. R. N. Jones, and C. Sandorfy, *in* "Chemical Applications of Spectroscopy" (W. West, ed.), pp. 271–290. Wiley (Interscience), New York, 1956.
186. D. N. Kendall, ed., "Applied Infrared Spectroscopy." Van Nostrand-Reinhold, Princeton, New Jersey, 1966.
187. J. R. Ferraro and L. J. Basile, eds., "Fourier-Transform Infrared Spectroscopy. Applications to Chemical Systems," Vol. 1. Academic Press, New York, 1978.
188. R. S. Tipson and F. S. Parker, *in* "The Carbohydrates" (W. Pigman and D. Horton, eds.), Vol. 1B, pp. 1394–1436. Academic Press, New York, 1981.
189. A. J. Michell, *Appita* **26**, 25–29 (1972).
190. J. D. S. Goulden, *in* "Laboratory Methods in IR Spectroscopy" (F. A., Miller, ed.), p. 106. Heyden, London, 1965.
191. B. Casu and M. Reggiani, *Stärke* **18**, 218–229 (1966).
192. H. A. Szymanski, ed., "Raman Spectroscopy. Theory and Practice," Vol. 1. Plenum, New York, 1967.

193. F. Cabassi, B. Casu, and A. S. Perlin, *Carbohydr. Res.* **63**, 1–11 (1978).
194. P. D. Vasko, J. Blackwell, and J. L. Koenig, *Carbohydr. Res.* **23**, 407–416 (1972).
195. A. G. Walton and J. Blackwell, "Biopolymers," p. 204. Academic Press, New York, 1973.
196. S. F. D. Orr, *Biochim. Biophys. Acta* **14**, 173–181 (1954).
197. J. D. S. Goulden and J. E. Scott, *Nature* (*London*) **220**, 698–699 (1968).
198. S. M. Bociek and D. Welti, *Carbohydr. Res.* **42**, 217–226 (1975).
199. B. Casu, G. Scovenna, A. J. Cifonelli, and A. S. Perlin, *Carbohydr. Res.* **63**, 13–27 (1978).
200. A. G. Lloyd, K. S. Dogson, R. G. Price, and F. A. Rose, *Biochim. Biophys. Acta* **46**, 108–115 (1961).
201. D. A. Rees, *J. Chem. Soc.* pp. 1821–1832 (1963).
202. N. S. Anderson, T. C. S. Dolan, A. Penman, D. A. Rees, G. P. Mueller, D. J. Stancioff, and N. F. Stanley, *J. Chem. Soc. C* pp. 602–606 (1968).
203. M. J. Harris and J. R. Turvey, *Carbohydr. Res.* **15**, 51–56 (1970).
204. R. C. Chalk, M. E. Evans, F. W. Parrish, and J. A. Sousa, *Carbohydr. Res.* **61**, 549–552 (1978).
205. S. A. Barker, E. J. Bourne, and D. H. Whiffen, *Methods Biochem. Anal.* **3**, 213 (1956).
206. C. Y. She, N. D. Dinch, and A. T. Tu, *Biochim. Biophys. Acta* **372**, 345–357 (1974).
207. P. D. Vasko, J. Blackwell, and J. L. Koenig, *Carbohydr. Res.* **19**, 297–310 (1971).
208. J. J. Cael, J. L. Koenig, and J. Blackwell, *Carbohydr. Res.* **29**, 123–134 (1973).
209. R. H. Atalla and B. E. Dimick, *Carbohydr. Res.* **39**, C1–C3 (1975).
210. F. R. Seymour, R. L. Julian, A. Jeanes, and B. L. Lamberts, *Carbohydr. Res.* **86**, 227–246 (1980).
211. R. G. Zbankov, in "Infrared Spectra of Cellulose and its Derivatives" (B. I. Stepanov, ed.)., Consultants Bureau, New York, 1966.
212. M. P. Filippov, in "Infrared Spectra of Pectic Substances" (G. V. Lazur'evsky, ed.), Shtiinza Publ., Kishinev, URSS, 1978.
213. A. T. Tu, N. D. Dinh, C. Y. She, and J. Maxwell, *Stud. Biophys.* **63**, 115–131 (1977).
214. H. G. Higgins, C. M. Stewart, and K. J. Harrington, *J. Polym. Sci., Polym. Symp.* **51**, 59–84 (1961).
215. J. Blackwell, *ACS Symp. Ser.* **45**, 103 (1977).
216. B. Casu and M. Reggiani, *J. Polym. Sci. C* **7**, 171–185 (1964).
217. A. J. Michell, *Aust. J. Chem.* **21**, 1257–1266 (1968).
218. V. M. Tul'chinski, S. E. Zurabyan, K. A. Asankozhoev, G. A. Kogan, and A. Y. Khorlin, *Carbohydr. Res.* **51**, 1–8 (1976).
219. S. V. Vercellotti, unpublished.
220. B. Casu and A. S. Perlin, unpublished.
221. G. C. Pimentel and A. L. McClellan, "The Hydrogen Bond," p. 67. Freeman, San Francisco, California, 1960.
222. A. J. Michell and H. G. Higgins, *Tetrahedron* **21**, 1109–1120 (1965).
223. B. Casu, M. Reggiani, G. G. Gallo, and A. Vigevani, *Tetrahedron* **22**, 3061–3086 (1966).
224. B. Casu, M. Reggiani, G. G. Gallo, and A. Vigevani, *Tetrahedron* **24**, 803–821 (1968).
225. H. J. Marrinan, in "Recent Advances in the Chemistry of Cellulose and Starch" (J. Honeyman, ed.), p. 147. Heywood, London, 1959.
226. C. Y. Liang and R. H. Marchessault, *J. Polym. Sci.* **39**, 269–279 (1959); **37**, 385–395 (1959); R. H. Marchessault and C. Y. Liang, *ibid.* **43**, 71–84 (1960).
227. J. Blackwell, P. D. Vasko, and J. L. Koenig, *J. Appl. Phys.* **41**, 4365–4379 (1970).
228. E. D. T. Atkins, K. D. Parker, and R. D. Preston, *Proc. R. Soc. London, Ser. B* **173**, 209–221 (1969).
229. R. H. Marchessault and C. Y. Liang, *J. Polym. Sci.* **59**, 357–378 (1962).

230. J. Blackwell, Ph.D. Thesis, University of Leeds, England, 1967; see also Walton and Blackwell (*195*).
231. J. J. Cael, D. H. Isaac, J. Blackwell, J. L. Koenig, E. D. T. Atkins, and J. K. Sheehan, *Carbohydr. Res.* **50**, 169–179 (1976).
232. J. J. Cael, K. H. Gardner, J. L. Koenig, and J. Blackwell, *J. Chem. Phys.* **62**, 1145–1153 (1975).
233. J. J. Cael, J. L. Koenig, and J. Blackwell, *Biopolymers* **14**, 1885–1903 (1975).

5

Shapes and Interactions of Carbohydrate Chains

DAVID A. REES, EDWIN R. MORRIS,
DAVID THOM, AND JOHN K. MADDEN

I. Conformational Principles 196
  A. Monosaccharide Building Blocks 196
  B. Interresidue Linkages 197
  C. Disaccharides 200
  D. Carbohydrate Chains 203

II. Experimental Characterization 204
  A. X-ray Fiber Diffraction 204
  B. Chiroptical Techniques 206
  C. Nuclear Magnetic Resonance 214
  D. Disorder–Order Transition Kinetics 216
  E. Measurement of Molecular Weight and Particle Size 217
  F. Hydrodynamic and Rheological Techniques 219
  G. Stoichiometry of Cation Binding 222
  H. Differential Scanning Calorimetry 223

III. Homopolysaccharides 224
  A. Conformational Types 224
  B. Ribbon Sequences in Plant Cell Walls 226
  C. Ribbon Sequences in Crustaceans, Insects, and Bacteria . . . 227
  D. Ribbon Sequences That Bind Cations 228
  E. Hollow Helices with Energy Reserve Functions 235
  F. Hollow Helices with Structural Functions 237
  G. Flexible Coils and Linkages 237

THE POLYSACCHARIDES, VOL. 1

IV. Regular Copolysaccharides   238
   A. Introduction   238
   B. Algal Polysaccharides of the Agar–Carrageenan Family . . . 240
   C. Bacterial Copolymers   245
   D. Glycosaminoglycans of the Hyaluronate–Chondroitin
      Family   248
   E. Heparin and Heparan Sulfate   254

V. Disordered Chains in Solution   255
   A. Dilute Solution   255
   B. Concentrated Solution   259

VI. Hydrated Networks   263
   A. Influence of Interruptions   263
   B. Higher-Order Interactions   269

VII. Mixed Interactions   276
   A. Polysaccharide–Protein Interactions   276
   B. Inclusion Complexes   278

   References   281

I. Conformational Principles

A. Monosaccharide Building Blocks

Exploration of the overall shapes of carbohydrate chains usually starts from knowledge of or assumptions about the conformations of the component sugar rings (1–3). Since six-membered (pyranose) rings are by far the most common in naturally occurring polysaccharides and their individual ring conformations are more stable and well defined, most of the successful work has been done with those systems. The only major exceptions are in the field of polynucleotides, for which the interconversion between furanoside conformations plays an important role in the determination of the polymer form.

The pyranose ring conformations that are important in polysaccharides are the two chair conformations designated $^4C_1$ and $^1C_4$ (Fig. 1) to indicate the disposition of atoms above and below the plane of the ring. In older notation the same conformations were denoted by C1 and 1C, respectively. Boat conformations, two of which are illustrated in Fig. 1, probably have fleeting existence at low proportions in disordered (random coil) polysaccharide chains.

The approaches available for predicting the relative stabilities and hence equilibrium populations of alternative conformations have been thoroughly reviewed elsewhere (1,2), and for free sugars this can, of course, extend to comparison of the relative stabilities of alternative isomers, such as five-

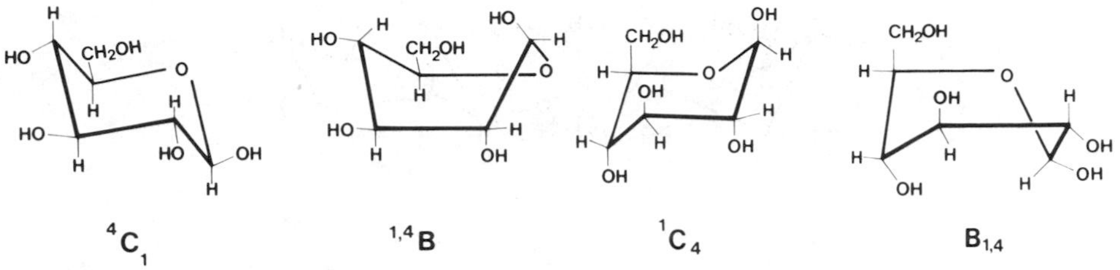

Fig. 1. Pyranose ring conformations, illustrated for β-D-glucose.

membered (furanose) rings, or open chains. They begin from two complementary, but quite different approaches. (a) Measured equilibrium constants
are used empirically to derive the corresponding free energies and then to
make use of trends and regularities in these values (4,5). (b) In a more general
approach, attempts are made to build up a complete picture of all the actual
attractions and repulsions between atoms in terms of van der Waals forces,
polar interactions, hydrogen bonding, and torsional contributions (6,7).
The calculations are now refined to a point at which an accurate prediction
can usually be made of the shapes and configurations that will predominate
in sugar solutions, as well as the proportions of each form. They may even
indicate minor distortions and deviations from "ideal" chair geometry,
although they make no attempt to consider terms involving solvent. It
appears that these terms usually cancel when differences in free energy are
considered, but exceptions are known in which additional stabilization of
particular conformations by water is indicated. In general, however, the
shape of naturally occurring pyranose rings in carbohydrate chains may be
regarded as fixed in that chair conformation in which C-6 is equatorial ($^4C_1$
for D-sugars and $^1C_4$ in the L series), with the single exception of L-iduronate
residues in certain glycosaminoglycan states (Section IV,E). Overall chain
geometry is therefore determined predominantly by the relative orientations
of adjacent residues.

B. Interresidue Linkages

For glycosidic bonds in which linkage is through an oxygen atom attached
to a ring carbon, the relative orientations of the participating residues can
be defined completely by the two dihedral angles ϕ and ψ shown in Fig. 2.
When the connecting linkage is between C-1 of one residue and C-6 of its
neighbor, there is an extra covalent bond and torsion angle (ω), giving these
units markedly increased freedom to adopt a wide variety of orientations
relative to each other. The term *linkage conformation* (8) is used to define
a distinct set of values for these angles (ϕ,ψ) or (ϕ,ψ,ω).

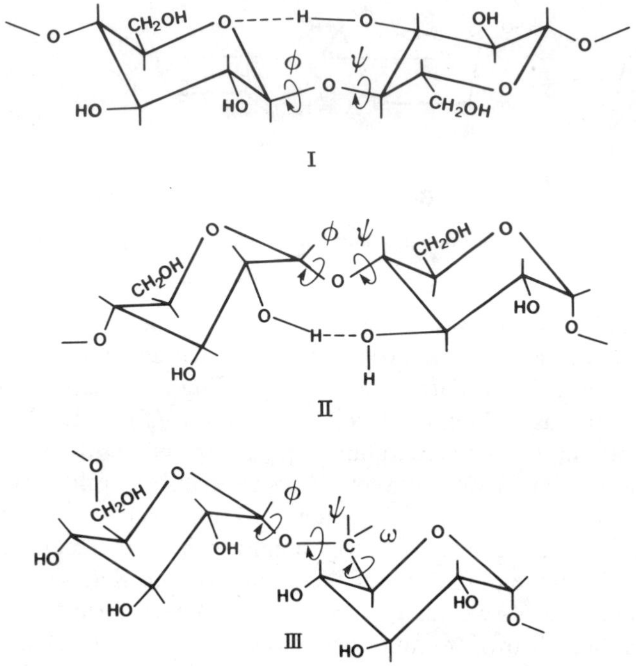

Fig. 2.   Interresidue linkages in carbohydrate chains, illustrated for the glucose homopolymers cellulose (I), amylose (II), and dextran (III). Linkage conformation is defined by the dihedral angles (ϕ,ψ) or (ϕ,ψ,ω). The intramolecular hydrogen bonds discussed in the text are also indicated.

In considering the overall conformations of carbohydrate chains it is useful to start by distinguishing between (a) ordered conformations in which the values of the torsion angles are fixed by cooperative interactions between residues and (b) disordered conformations (random coils) in which continuous fluctuation occurs, with the possibility of any individual state being determined by its relative energy according to the Boltzmann principle. In either case, any attempt to predict the likely conformation requires an approach to the estimation of interaction energies. It is not possible to deal with linkage geometry in polysaccharides by simple extension of the classical methods for conformational analysis of small molecules, since we would first have to identify a limited number of discrete conformations that correspond to the important low-energy forms in equilibrium. The alternative approach is to calculate conformational energies directly from energy functions by listing the stereochemical constraints that have been established from structural studies, e.g., bond lengths, bond angles, and particular values for van der Waals radii, and then systematically varying the remaining parameters with energy calculations at each stage to derive an "ideal" or "average" conformation as appropriate. The computer methods now invariably used to do this have been reviewed extensively elsewhere (*8–13*) and are not discussed in detail here. The methods can be classified into four broad types:

1. In the "hard-sphere" approach, the atoms are regarded as solid spheres with radii corresponding to the van der Waals radii. Conformational angles (ϕ and ψ) are "stepped" systematically, and for each conformation thus generated the distance between each atom on one residue and all atoms on the other residue is calculated. As soon as any distance is found to be less than the sum of the radii of the atoms involved, the conformation is rejected. If none is, the conformation is accepted as a possibility. Slightly more sophisticated methods allow for some distortion of the molecule to alleviate minor clashes, leading to distinctions between "fully allowed" and "marginally allowed" conformations. This approach has been used successfully, particularly in conjunction with X-ray diffraction studies (*14–16*) (Section II,A).

2. In the so-called empirical energy function approach, it is assumed that molecular attractions and repulsions can be treated according to the laws of classical mechanics. The total energy is calculated as a sum of terms representing van der Waals attractions and repulsions, polar interactions, bond torsion, and hydrogen bonding. Each term is evaluated from a simple function of interatomic distance, which has the form to be expected from theory but may include constants with empirical values. Bond angles as well as torsion angles can usefully be varied, and another equation used to calculate angle strain. The validity of the potential functions and the strengths and limitations of the methods have been extensively reviewed (*8–13*). Further progress with polysaccharides will depend on careful checking of assumptions and more extensive experimental verification of calculated energies. Even in their present form, however, such energy calculations have been useful for practical problems. It appears that van der Waals repulsion has a dominant influence on many polysaccharide conformations, more so, for example, than in polypeptides and polynucleotides. This accounts for much of the success of both the hard-sphere and potential function approaches, and in fact most of the achievements of polysaccharide conformational analyses have occurred because it is relatively easy to rank conformations in order of increasing or decreasing van der Waals repulsion.

3. An important development from the previous approach is the method of empirical force field calculations, the principle being to use mathematical minimization methods to locate energy minima rather than to perform systematic mapping of energies over all conformations. The advantage is greater economy of computer time so that more detailed analysis is possible,

with inclusion of additional variables such as energy terms associated with bond angle deformations and, in principle, interactions with solvent molecules. This method does not give such a ready overview of internal energy changes with the relevant bond rotations, although the energetics of selected conformational interconversions can be explored. The results (17) for maltose, cellobiose, and gentiobiose are in broad agreement with other approaches.

4. The alternative to empirical energy functions is the use of quantum mechanics. In principle this is more theoretically sound and more genuinely predictive and should therefore be more satisfactory. Calculations have been attempted for several biopolymers and biopolymer fragments, including disaccharides (18–20). The requirements in computer time, however, are impracticably high without drastic simplifications of rigorous quantum mechanical principles. Results obtained using these approximate treatments must be treated with some reservation until sufficient calculations have been checked against experiment, although for disaccharides there is reasonable agreement with calculations using empirical functions.

C. Disaccharides

One source of unambiguous information on the rotational angles between adjacent sugar residues consists of X-ray diffraction studies of crystalline disaccharides and their derivatives. These may then be compared with values predicted from calculation. Irrespective of the method of conformational analysis employed, the strategy is to develop a map showing the variation of internal energy with rotation around the glycosidic angles ϕ and ψ (Fig. 2). To achieve this the conformational energy is first calculated for a model in which ϕ and ψ have fixed, but arbitrary values. One bond to the linkage oxygen (say ϕ) is then fixed while the model is rotated incrementally through $360°$ about the other (ψ). The atomic coordinates are recalculated after each increment and used to derive each new conformational energy. The process is then repeated about the other bond, to sample all possible permutations of ϕ and ψ. The result is a contour map of conformational energy in which the lowest troughs are predicted to be the most stable conformations. Except for ($1\rightarrow6$) linkages all such calculations show that more than 90% of the conformations on the map are energetically disallowed.

For cellobiose and maltose and derivatives, a sufficient number of crystal structures have been determined to allow useful comparisons to be made (8,21). As shown on the map for cellobiose (Fig. 3a) different forms do not have exactly the same conformation at the glycosidic linkage, but lie close together within the predicted boundary and very close to the overall minimum of the energy map ($\phi = 31°$, $\psi = -26°$). Presumably, the slight adjust-

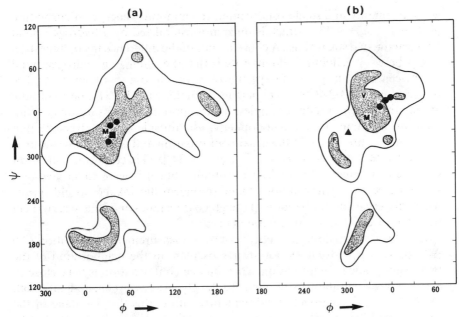

Fig. 3. Conformational energy maps for (a) cellobiose and (b) maltose. Shaded areas are within 5 kcal/mol of the potential energy minimum (M). The outer contour lines indicate conformations within 10 kcal/mol of M. Linkage conformations are shown for crystalline disaccharide derivatives (●), the cellulose bent chain conformation (■), the sixfold V-amylose helix (V), the proposed folded structure of maltose (F), and the crystal structure of 6′-iodophenyl α-maltoside (▲), which is unusual in lying toward the subsidiary minimum of the maltose energy map. For definitions of ϕ and ψ see Section II,B. From (21), with permission.

ments from one derivative to another occur to facilitate packing in a different crystal environment. Although the cellobiose map shows two troughs of favored conformational energy, only one of these is populated. This zone is larger in area and hence more "probable" than the smaller area; it also contains zones of somewhat lower energy states and allows the formation of a hydrogen bond between O-3 on one residue and O-5 on its neighbor (Fig. 2). Interpretations of optical rotation studies (22) in dimethyl sulfoxide and water in terms of contributions to the overall optical rotation from the linkage conformation (Section II,B) also show good agreement between the observed values and those calculated from the solid-state conformation. This agreement does not imply that the disaccharide in solution is locked in the crystal conformation, but it does suggest a preference for ϕ and ψ values in the neighborhood of the minimum energy conformation.

Analogous calculations (8,21) for maltose (Fig. 3b) show that the crystal conformations characterized by X-ray diffraction are more scattered but again lie mainly in a single zone around the minimum energy conformation ($\phi = -32°$, $\psi = -13°$), which is once more stabilized by a hydrogen bond between the two sugar rings. A consequence of the difference in configuration at C-1 between cellobiose and maltose is that the stabilizing hydrogen bond is between different pairs of oxygen atoms. In maltose it is between O-3 of one residue and O-2 of its neighbor (Fig. 2). In this case there is one exception, among the dozen or more examples, which approaches the subsidiary minimum on the map. This is 6'-iodophenyl α-maltoside (23), a glycoside that apparently cannot pack in the usual conformation and is therefore captured near the alternative zone ($\phi = -52°$, $\psi = -34°$). This conformation is not stabilized by interresidue hydrogen bonding, but if it were to survive into solution (see below) the rotation around the glycosidic linkages would appear to fold together the apolar faces of the glucose residues and thus screen them from the external aqueous environment.

In contrast to cellobiose, optical rotation measurements in solution (22) indicate that maltose does not always oscillate in the neigborhood of the hydrogen-bonded conformation. This linkage conformation is very close to eclipsed positions about both bonds to the glycosidic oxygen, and, although the O-2 · · · O-3' hydrogen bond evidently offsets this disadvantage in the solid state (8,21), this is not possible in strongly hydrogen-bonded solvents. In dioxan and dimethyl sulfoxide, both optical rotation (22) and proton nuclear magnetic resonance (^{1}H-NMR) measurements (24–26) suggest that, although the time-averaged conformation is significantly displaced from the crystal structure, a large proportion of molecules exist in the major zone in which the hydrogen bond is possible. In aqueous solution, by contrast, the evidence suggests that the molecule must spend a much larger portion of its time in the area of the subsidiary minimum. Carbon-13 NMR observations (27) confirm that in water steric or proximity effects occur between the sugar residues, as would be expected for the alternative "folded" structure. It is also possible that the arrangement of sugar hydroxyl groups in the alternative zone matches more closely the organization in transiently ordered "clusters" of water molecules to provide the extra stability that the conformational calculations fail to predict (22).

The solvent dependence of conformation seen in maltose, however, appears to be a relatively uncommon effect. More generally, conformational predictions based on hard-sphere (see above) constraints alone, give calculated optical rotation values (Section II,B) that correlate well with experiment, thus confirming that these influences dominate carbohydrate linkage geometry (28,29). For some linkages at least, it is possible to detect a substantial influence on the conformation about the glycosidic bond from the exo-anomeric effect; this evidence is from [13]C–[1]H coupling constants (30,31) and optical rotation (29).

For diequatorial linkages calculations show that the numerically averaged values of ϕ and ψ can be predicted independently, ϕ being determined by the size of R_1 (Fig. 4) and ψ by R_2 and R_3. On this basis, groups of glucosyl oligomers, for example, could then be assigned the value of ϕ known for cellobiose from both solution and crystal studies, whereas that of lactose could be applied to galactosyl homologues, and the measured optical rotations then analyzed to calculate the values of ψ (29). For a large number of compounds, the results demonstrated with remarkable consistency that (a) each group of oligosaccharides having a similar substitution around the aglycone bond show only very small variations in ψ and (b) replacement of a substituent with a less bulky one causes ψ to shift as expected toward the extra space that is created.

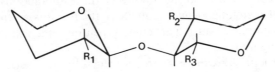

Fig. 4. Restriction of conformational mobility by equatorial substituents adjacent to the interresidue linkage. From (29), with permission.

The factors affecting ϕ and ψ for α linkages are more interdependent, but a similar general picture emerges if it is assumed that, as in the β series, ϕ is the same for all related compounds. Conformational analysis also shows that increasing the number of axial linkages to the glycosidic oxygen decreases the number of accessible conformations.

D. Carbohydrate Chains

Even though the majority (>90–95%) of linkage conformations for disaccharides are shown by the conformational energy maps to be forbidden by steric considerations alone, the remaining conformational space allows significant freedom of oscillation about the bonds to the glycosidic oxygen. In solution the molecules are not normally constrained in a unique state but oscillate around it because of collisions and thermal energy. A carbohydrate chain typically contains a large number of such linkages, and the overall shape is the result of these independent oscillations. At any instant, such a chain would show a spectrum of linkage conformations, most of them in favorable low-energy orientations, although thermal energy would also induce a few linkages to adopt much higher energy forms and a few monomer units to adopt twist, boat, and alternative chair conformations. The greater the internal freedom at each linkage, the greater the number of conformations available to each individual segment and the less likely it will be for the chain to adopt a unique ordered shape in which each linkage is fixed close to the minimum energy form. Chain flexibility thus provides a strong entropic drive, which generally overcomes energy considerations and induces the chain to adopt disordered or random coil states in solution (*3,8*). This influence is known as the *conformational entropy*.

Under particular circumstances, however, favorable nonbonded energy terms (hydrogen bonding, dipolar and ionic interactions, and solvent effects) can combine to fix macromolecules in ordered shapes (*3,8,32*). For carbohydrate chains the interactions between individual pairs of monomers is insufficient to do this, and it occurs by the *cooperative interaction* of energy terms along extended sequences of the chain, which reinforce each other to outweigh the conformational entropy. Since these interactions almost invariably occur between long, *regular* sequences, the result is a helix, because any interactions favored within a particular repeating unit will also be favored for neighboring units.

Ordered structures are more favored for the solid state because cooperative interactions can then operate between chains as well as within them. However, as we shall see, they can also occur in solution, and we can then investigate the transition between ordered and disordered states to gain information about the kinetics, thermodynamics, and molecular mechanisms.

II. Experimental Characterization

Spectroscopic techniques such as IR and NMR are now well established as an essential element in the experimental repetoire of the carbohydrate chemist. Their application to analysis of polysaccharide primary structure, and more recent elegant extension to the problems of conformational analysis, are reviewed comprehensively in the previous chapter. We confine ourselves here to less familiar physical and physicochemical approaches, aimed specifically at characterisation of secondary and higher levels of structural organisation.

A. X-ray Fiber Diffraction

At present the only technique for characterizing polysaccharide ordered structures at a level approaching atomic resolution is X-ray diffraction from oriented specimens. Although it cannot, of course, be assumed that structures in the condensed phase will necessarily persist under hydrated conditions (in solutions, gels, or biological tissues), they do represent a reliable and precise reference state for other experimental approaches such as are outlined later. In contrast to single-crystal studies, it is seldom possible to establish atomic coordinates in polymer fibers from diffraction intensities alone. However, by intelligent use of computer model building (Section I) to select sterically reasonable models for detailed comparison with X-ray data, the details of the ordered structures have been derived for many polysaccharides (3,8,33).

Mathematically, any ordered chain conformation may be considered a helix defined by two parameters: n, the number of residues per turn, and h, the projected length of each unit on the axis. By convention a negative value for n defines a left-handed helix. A complementary means of describing helix geometry that is commonly used in diffraction studies is to give the helix repeat distance, the number of primary structural units per repeat (i.e., helix symmetry), and the number of turns of the helix per repeat. For example, fivefold helix symmetry can be achieved with relative orientations of adjacent units of 72°, 144°, 216°, or 288°, and these are known as 5_1, 5_2, 5_3, and 5_4 helices, respectively. In the 5_1 form the helix repeats after one turn, whereas two turns are required for 5_2 geometry, although five primary structural units are involved in each case ($n = 5$ and 2.5, respectively). Since a rotation of 288° in one direction is exactly equivalent to one of 72° in the opposite direction, the 5_4 helix is a left-handed version of 5_1 ($n = -5$ for 5_4), and 5_2 and 5_3 are similarly related ($n = -2.5$ for 5_3).

Mathematical relationships for converting particular values of the glycosidic angles ϕ and ψ (Fig. 2) to helix parameters are well established (8), given that the geometry of the monomer units can be considered fixed. Conformational energy maps (e.g., Fig. 3) can therefore be used to define the "allowed" range of n and h for homopolymeric sequences, and the approach can be extended to regular copolymers by using the appropriate energy map for each type of interresidue linkage.

The level of structural information that can be deduced from diffraction evidence is dependent on the level of organization within the fiber. When the ordered chains are aligned, but without regular packing arrangement, analysis is limited initially to calculation of the helix repeat distance from layer line spacings, and helix symmetry from the pattern of meridional reflections. Further refinement is achieved by calculation of the cylindrically averaged Fourier transform for trial models for comparison with the experimental data followed by adjustment of the model to obtain the best fit.

Ordered lateral packing of adjacent molecules into polycrystalline bundles adds an extra dimension of structural information by giving rise to Bragg diffractions, from which unit cell dimensions and sometimes the space group can be derived. Computer modeling can then define not only the possible conformations of individual chains, but also likely packing arrangements, which can be tested against the observed intensity distribution. Information about the position of bound cations or water molecules incorporated within the ordered structure can sometimes be obtained from differences between the observed and calculated intensities. If the residual diffraction pattern is consistent with a physically reasonable structure (e.g., cations located in likely relationships to charged groups on anionic polysaccharides, or water molecules in suitable positions for hydrogen bonding), this in itself may give further confidence in the conformation of the polymer chains and their location within the unit cell.

Unfortunately, it is not uncommon for more than one sterically feasible model to be reasonably consistent with the diffraction data. Distinction is then normally made by quantitative comparison of the standard of agreement between observed diffraction intensities and those calculated for the various contending structures. Common problems include distinguishing between parallel and antiparallel arrangements of chains within a unit cell, between left-handed and right-handed helix sense, and between coaxial double helices and noncoaxial packing of single helices. In the two latter cases, evidence from other techniques may also be of assistance in reaching a final decision. For example, the dynamics of conformational ordering in solution (Section II,D) may sometimes distinguish between coaxial helices and nested single helices. It may also be possible to distinguish between left and right helix sense by using established correlations between optical rotation and the dihedral angles between adjacent residues in the polymer chain (see, for example, Section VII,B).

B. Chiroptical Techniques

Application of such familiar techniques as ultraviolet (UV), visible, and infrared (IR) spectroscopy and the related Raman approach to carbohydrate systems is at present confined largely to characterization of primary structure, although the IR technique may provide evidence of the involvement and orientation (through dichroism measurements) of specific groups in, for example, hydrogen-bonding schemes in ordered structures (16). The use of circularly polarized rather than natural light, however, offers a direct route to conformational information, since the interaction with dissymetric molecules of circularly polarized radiation of opposite handedness is in general different and critically sensitive to stereochemical organization (34,35). Two closely related spectroscopic methods are based on this principle: circular dichroism (CD), which measures the differential absorption of left and right circularly polarized light, and optical rotatory dispersion (ORD), which measures differential refraction. The fundamental band form for these two techniques is shown in Fig. 5. For a single electronic transition both CD and ORD at any wavelength are determined completely by the same three

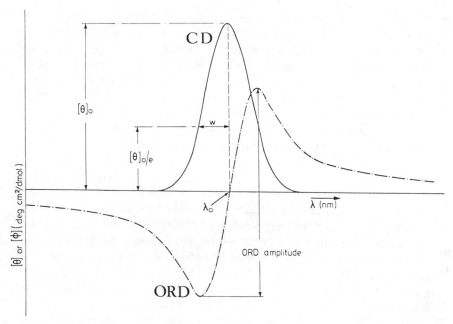

Fig. 5. Fundamental spectral band form of CD and ORD. For a single optically active transition, both CD molar ellipticity $[\theta]$ and ORD molar rotation $[\phi]$ at any wavelength are determined completely [Eqs. (1) and (2) respectively] by the same three parameters: position λ_0, intensity $[\theta]_0$, and width w (= half-width at $1/e$ of maximum intensity). ORD amplitude between peak and trough is approximately equal to $1.22[\theta]_0$.

parameters: position λ_0, intensity $[\theta]_0$, and width w by the gaussian and Kronig–Kramers equations [Eqs. (1) and (2), respectively] (35).

$$[\theta]_\lambda = [\theta]_0 e^{-(\lambda - \lambda_0)^2 / w^2} \tag{1}$$

$$[\phi]_\lambda = \frac{2[\theta]_0}{\sqrt{\pi}} \left(e^{-(\lambda - \lambda_0)^2 / w^2} \int_0^{(\lambda - \lambda_0)/w} e^{x^2}\, dx - \frac{w}{2(\lambda + \lambda_0)} \right) \tag{2}$$

In principle, the two methods therefore contain the same structural and conformational information, but in practice each has particular advantages and disadvantages. The simpler CD band form facilitates spectral resolution and assignment, but since CD drops off rapidly to zero on moving away from the band center its use is confined to chromophores that absorb light in an accessible spectral region. For proteins and polynucleotides the electronic transitions of the polymer backbone lie within the wavelength range of current commercial CD instruments and are used extensively to characterize secondary structure (36). The corresponding conformation-sensitive backbone transitions of carbohydrate chains, however, are centered in the vacuum UV, below the lower wavelength limit of current commerical CD equipment (~ 190 nm) (37). These bands were first observed directly for monosaccharides and oligosaccharides by Nelson and Johnson (38) and for polysaccharides by Stevens and co-workers (39). Two optically active transitions of opposite sign centered at ~ 150 and 170 nm have now been reported (39–43) for a number of polysaccharides, the lower-wavelength band being the more intense and determining the sign of optical rotation at longer wavelengths (e.g., the sodium D line). In some cases, notably homopolymers of glucose, a third, smaller band is also observed at higher wavelength (~ 185 nm) (44,45).

Quantitative comparison of CD and ORD shows that for at least one family of polysaccharides (the galactomannans) (41) these two bands account completely for the full ORD envelope at all accessible wavelengths and vary systematically in intensity with composition (i.e., relative content of galactose and mannose residues) without altering bandwidth or position. Variation in vacuum ultraviolet circular dichroism (VUCD) intensity with conformation has been reported (40) for the sol–gel transition of agarose (Fig. 6) and follows the same temperature course as optical rotation at higher wavelength. Finally, the VUCD of alginate (42) shows systematic changes with ionic environment that parallel the changes in carboxyl CD on cation (Ca^{2+}) binding, as discussed below. Thus, VUCD appears to offer a direct and simple route to primary, secondary, and tertiary structural information that has previously been obtainable only indirectly through optical rotation. A small number of research groups (currently five) have constructed CD instruments that in favorable cases can penetrate to ~ 140 nm, but VUCD is still a specialist technique rather than a generally available method.

Since ORD remains finite at wavelengths remote from the band center (Fig. 5) it can be used (46) to monitor the same transitions indirectly. One of the earliest approaches to systematic correlation of optical activity to carbohydrate stereochemistry was that of Hudson (47), who developed a system of empirical rules by treating the sugar ring as a series of "asymmetric centers" each contributing separately to the overall optical rotation. In many cases these rules gave good agreement with observed optical rotation values at a single wavelength (the sodium D line at 589 nm) but broke down for a number of specific classes of carbohydrates, including monosaccharides in which O-2 is axial and sterically crowded disaccharides. Since later developments in chiroptical theory (48) have shown that rotational strength is determined by interactions between groups rather than additive contributions from each, Hudson's rules must be regarded as fundamentally empirical and subject to possible exceptions.

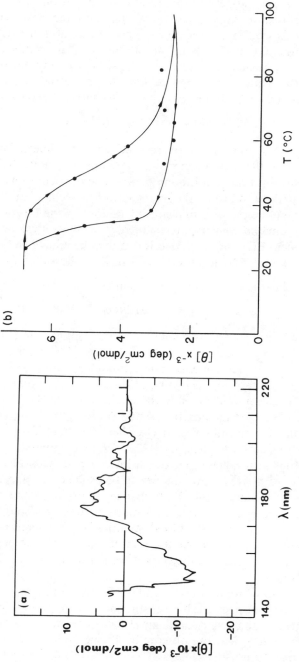

Fig. 6. VUCD of agarose. (a) Solid film; (b) changes in intensity of the higher-wavelength CD band that accompany the sol–gel transition. From (40), with permission.

A number of "second-generation" treatments based on this new under-standing have been developed, first by Whiffen (49) and Brewster (50) and later by Lemieux (51,52), to systematize and predict optical rotation from the relative orientation of substituents about covalent bonds within the sugar ring. This approach was extended to the relative orientation of adjacent residues in carbohydrate chains by Rees (28,53), who introduced the concept of "linkage rotation" $[\Lambda]$. In the case of a disaccharide this is defined by

$$[\Lambda] = [M]_{NR} - ([M]_R + [M]_{MeN}) \tag{3}$$

where $[M]_{NR}$, $[M]_R$, and $[M]_{MeN}$ are, respectively, the D-line molar rotations of the disaccharide, the component reducing residue, and the methyl glycoside of the nonreducing residue in the same anomeric configuration. The para-meter $[\Lambda]$ is understood to be the increment of molecular rotation arising from all the new interactions that are introduced in the formation of the disaccharide from the constituent residues. By a theoretical development from the Whiffen–Brewster approach it can then be related to the dihedral angles ϕ and ψ between adjacent residues (Fig. 2), as shown in Eqs. (4) and (5).

$$\alpha \text{ linkages:} \qquad [\Lambda] = -120(\sin \phi + \sin \psi) - 105 \tag{4}$$

$$\beta \text{ linkages:} \qquad [\Lambda] = -120(\sin \phi + \sin \psi) + 105 \tag{5}$$

Zero values of both ϕ and ψ are defined from the conformation in which the relevant C—O and C—H bonds are eclipsed. Viewing from the nonreducing end of the molecule in this reference state, any conformation can be generated by *clockwise* rotation of the *remote* bond through the appropriate positive angle. This convention is identical to that defined by Rees (28) (although for simplicity we have used the notation ϕ and ψ rather than $\Delta\phi$ and $\Delta\psi$) but differs from other conventions used elsewhere (16,21,54,55). Equations (4) and (5) have been verified extensively (22,28,29) by comparison of observed linkage rotations with values calculated from di- and oligosaccharide crystal structures. The same treatment can clearly be extended to polysaccharides to rationalize and quantify the sharp changes in optical rotation (56) that normally accompany cooperative transitions between ordered and disor-dered chain conformations (46). Good quantitative agreement between ob-served and calculated optical rotation values has been demonstrated for agarose (57) and ι-carrageenan (53) (Section IV) in the double helical and random coil states.

This use of optical rotation as an index of polysaccharide chain confor-mation may be complicated by substituent chromophores, particularly carboxylic acids, salts, esters, and amides, which absorb (44) at higher wavelengths and may therefore obscure backbone optical activity. In favor-able cases it may be possible (58,59) to overcome this difficulty by the combined use of CD and ORD. Thus, the ORD contribution of accessible chromophores can be calculated from their CD behavior by the Kronig–Kramers transform [Eq. (2)] and subtracted from overall ORD to unmask the optical activity of deeper-lying transitions.

In some cases, particularly polyuronates, CD of substituent chromophores may be of direct value in characterizing different levels of structural organization. As shown in Fig. 7, the $(1\rightarrow4)$-linked linear homopolymeric sequences from alginate, poly(α-L-guluronate) and poly(β-D-mannuronate), and poly(α-D-galacturonate) from pectin give CD spectra (predominantly from the carboxyl $n \rightarrow \pi^*$ transition) that are similar in magnitude and spectral form to those of the corresponding monomeric methyl glycosides (60). These spectra illustrate three general features of uronate CD: (a) that in polyuronates contributions from each residue are almost additive, (b) that residues in which O-4 is axial (galacturonate, guluronate, and iduronate)

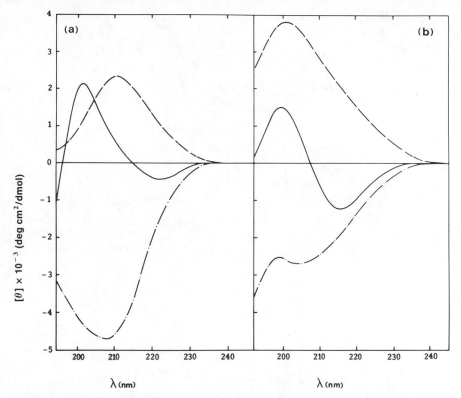

Fig. 7. Uronate CD. (a) β-Methyl D-mannuronoside (——); α-methyl L-guluronoside (—·—·—); α-methyl D-galacturonoside (————) (60). (b) Alginate poly(D-mannuronate) (——), and poly(L-guluronate) (—·—·—) sequences (61); poly(L-galacturonate) (————) from pectin (62).

show a single n → π* transition, whereas those with O-4 equatorial (gluc-uronate and mannuronate) give a second band of opposite sign at higher wavelength [attributed (60,63) to rotational isomerism about the C-5—C-6 bond], and (c) that the lower-wavelength (or sole) n → π* band is positive for D residues and negative for L residues. Typical alginate spectra show a peak at ~200 nm and a trough at ~212 nm, the relative magnitudes of which vary systematically with composition, so that the ratio of peak height to trough depth provides an index of the relative proportions of mannuronate and guluronate present (60,61). As shown in Fig. 8, the CD behavior of alginate mixed sequences, in which L-guluronate and D-mannuronate occur

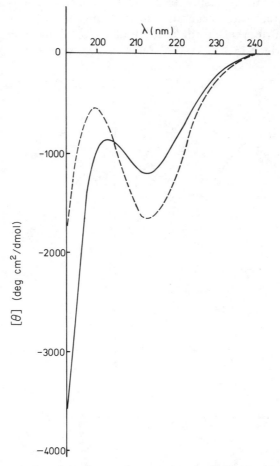

Fig. 8. The sensitivity of polyuronate CD to residue sequence is evident from a comparison of the observed CD of alginate heteropolymeric mixed sequences (———) with a spectrum (———) synthesized by linear combination of 50% of the spectra for homopolymeric poly(D-man-nuronate) and poly(L-guluronate) chain sequences. [From (61), with permission.]

in an approximately alternating structure, is not identical to that of an equimolar mixture of homopolymeric blocks of each type, thus demonstrating a limited sensitivity of uronate CD to adjacent residues (61). This sensitivity to primary sequence can be used (64) to estimate the relative proportions of each block type present, by matching observed alginate CD by means of linear combination of different proportions of the CD spectra for the three-component sequences using an iterative least-squares computer technique.

Gelation of polyuronates with calcium or related divalent metal ions is accompanied by large spectral changes (Section III,D), which characterize and quantify interchain association and cation chelation (62,65–69). Finally, the geometry of hydrated interchain junctions can be compared directly (62,67) with that in the solid state by comparison of the CD behavior of gels and of solid films cast from them (see Section III,D). Acetamido sugars also show n → π* CD bands of comparable magnitude over the same spectral range (70–75) and may be even more sensitive to the presence of neighboring residues; this is illustrated in Fig. 9, in which the CD of the methyl glycoside

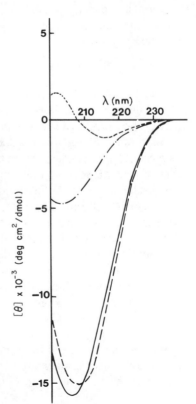

Fig. 9. Acetamido CD in monosaccharides and polysaccharides. From the observed CD behavior of sodium hyaluronate (———) and sodium β-methyl D-glucuronate (– – – –), the probable spectral contribution from the 3-linked N-acetyl-β-D-glucosamine residues in the polymer chain can be calculated by difference (——) and compared with the observed CD (–·–·–) of the methyl glycoside of 3-O-methyl-N-acetyl-β-D-glucosamine. The marked CD enhancement on incorporation in the chain contrasts with the CD behavior of uronic acid glycosides and polymers (Fig. 8). [From (71,74), with permission.]

of 3-O-methyl-2-acetamido-2-deoxy-β-D-glucose is compared with that of hyaluronate (74). The marked enhancement of optical activity on incorporation in the polymer chain is attributed to restriction of the conformational mobility of the chromophore by adjacent residues.

C. Nuclear Magnetic Resonance

Although CD and ORD (Section II,B) are extremely powerful methods for characterizing conformational transitions, they are of less use in detecting ordered structures that remain stable under all possible experimental conditions. A particularly successful technique for dealing with this problem is NMR relaxation (67,76) using either ^{1}H or ^{13}C nuclei. Of particular interest is the dissipation of the energy of the excited state to adjacent nuclei, as characterized by the spin–spin relaxation time T_2. Since thermal motions interfere with this process, T_2 values for small molecules or for polymers with rapid segmental motions are very much larger than those for solids or conformationally rigid chain sequences in solutions or gels. Relaxation behavior can be measured directly as the time constant for exponential decay of magnetization or indirectly from high-resolution NMR linewidth ($\Delta v_{1/2}$), from the relationship

$$T_2 = 1/\pi \Delta v_{1/2} \tag{6}$$

Linewidths for ordered polysaccharides are normally so great that the entire high-resolution spectrum is effectively flattened into the baseline (77,78). The process of spectral collapse provides a convenient index of the adoption of conformational order (Fig. 10), whereas "permanent" order can be detected by direct measurement of T_2, which typically has a value of the order of tens of microseconds for polysaccharides in rigid conformations, compared with tens of milliseconds in the random coil state. New techniques for measuring relaxation parameters, such as the use of nuclear Overhauser effects and two-dimensional spectroscopy have great promise here (30,31).

The conventional use of high-resolution NMR to determine the ring conformations of monosaccharide residues is applicable to polymers (79,80) as well as to derivatives of low molecular weight, although the poorer spectral resolution often gives rise to practical difficulties. Carbon-13 coupling constants can in some cases be used to give information about linkage conformations (81,82). These topics are discussed in more detail in Chapter 4.

In dimethyl sulfoxide solutions, hydroxyl proton resonances may be shifted by interresidue hydrogen bonding, and this allows NMR measurements to be used for assessing the extent to which various hydrogen-bonded linkage conformations are populated (24,25,83). Several possibilities for the future

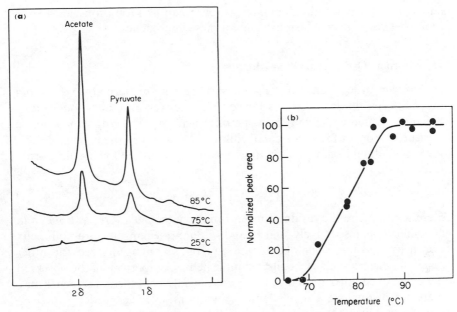

Fig. 10. High-resolution NMR linewidth as an index of polysaccharide chain flexibility, illustrated for the disorder–order transition of xanthan. (a) Adoption of the ordered chain conformation on cooling is accompanied by extreme broadening of the well-resolved ^{1}H resonances from acetate and pyruvate substituents. (b) The temperature course of the transition can be monitored by the reduction in measurable peak area. With permission from (77) *J. Mol. Biol.* **110**, 1 (1977). Copyright by Academic Press, Inc. (London) Ltd.

extension of NMR methods have been powerfully presented in a recent study of the linkage conformations in a series of di- and oligosaccharides that correspond to fragments of blood group substances (30). Trial conformations were set up by potential energy calculations, which took into account contributions from the exoanomeric effect as well as van der Waals interactions and which were based on coordinates from crystal structure determinations. The predictions were then tested by the following nmr criteria. (a) Rate of ^{1}H relaxation ($1/T_1$) should be directly proportional to the sum (calculated from the trial model) of the sixth powers of all interproton distances for the appropriate nucleus; (b) ^{1}H chemical shifts should provide a fingerprint for conformational near identity between pairs of compounds; (c) ^{13}C–^{1}H coupling constants across the glycosidic linkage should be consistent with the predicted dihedral angles; (d) measurements of nuclear Overhauser enhancements to the accuracy possible with modern instruments should be related to individual interproton distances. Good consistency was found between the conclusions from all these approaches.

D. Disorder–Order Transition Kinetics

Disorder–order transitions in polysaccharides are normally very rapid (84–86) (typically less than 1 s for complete conversion), but modern fast-reaction techniques (87,88) now permit such transition dynamics to be measured. For an intramolecular coil–helix transition, the extent of helix formation should change with time according to the following first-order rate equation:

$$\ln[x_e/(x_e - x)] = (k_1 + k_{-1})t \qquad (7)$$

Here, x and x_e are the residue concentrations in the helix form at time t and at equilibrium, respectively, and k_1 and k_{-1} are rate constants for coil–helix and helix–coil (89). Analogous rate equations can be derived for other reaction schemes such as 2 coil $\rightleftharpoons$ double helix. Comparison of the observed dynamics of conformational change with the different reaction schemes may therefore distinguish between inter- and intramolecular order (which is sometimes difficult to resolve by fiber diffraction evidence), while variation of rate constants with temperature or solvent environment is diagnostic of the nature and strength of interaction.

The choice of perturbation must take account of both the nature of the system under investigation and practical difficulties. For thermally induced transitions (e.g., Fig. 6) a rapid change in temperature ("T jump") seems an obvious choice. Conventional temperature control (e.g., using a jacketed cell), however, is suitable only for relatively slow transitions with a time course of several seconds. A very rapid temperature increase is possible if an electric discharge is used but for charged polymers this may cause partial molecular alignment, with complications both in monitoring the subsequent transition and in interpreting the kinetics. Polysaccharide ordered structures are frequently disrupted by alkali and reformed on neutralization. The use of pH changes for kinetic studies, however, is complicated by associated temperature changes on neutralization.

For charged polysaccharides a particularly convenient perturbation is a rapid increase in ionic strength ("salt jump") to promote structural order. Solutions of the polysaccharide in the low-salt disordered state and of an appropriate salt solution are held in separate syringes and are discharged simultaneously through a high-speed mixer into a flow-through cell. When the flow of the mixed solutions is arrested, the rate of conformational change is monitored, usually at present by optical rotation (polarimetric stopped-flow, (87), as illustrated in Fig. 11. For certain systems, other techniques (88) such as CD or light scattering may be more appropriate.

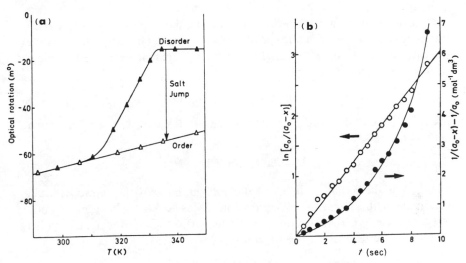

Fig. 11. Characterization of the dynamics of conformational ordering in polysaccharide systems by stopped-flow polarimetry, illustrated for the salt-induced disorder–order transition of xanthan. (a) The ordered conformation is stabilized to progressively higher temperatures with increasing ionic strength and can therefore be induced isothermally by the rapid addition of salt solution ("salt jump"). (b) Analysis of the dynamics of conformational change in terms of first-order (left-hand axis) and second-order (right-hand axis) kinetic schemes indicates an intramolecular rather than an intermolecular process. (From (86), with permission.

E. Measurement of Molecular Weight and Particle Size

When the adoption of intermolecular structure is accompanied by a change in average mass of the isolated species in solution, the stoichiometry of the ordered state may be characterized by a number of physical and physicochemical techniques sensitive to molecular weight (90). Reducing end group analysis (91,92) measures chain length and is unaffected by non-covalent association, and it therefore provides a baseline for conformational studies. Colligative properties, by contrast, reflect the number of independently moving particles, a multistranded molecular assembly making the same contribution as a single chain. Since carbohydrate polymers are often highly polydisperse, no single measurement can fully characterize the distribution of chain length or particle size. Comparison of the changes in number-average and weight-average molecular weights—for example, from osmometry and light scattering—with conformational change (*93*) can give more stoichiometric information than either alone and in principle can help in the study of possible mismatching of chain length in intermolecular structures.

In conventional light scattering (*90*), the angular dependence of scattering intensity is measured over a range of concentrations. The experimental index of scattering intensity at an angle of θ to the incident beam, the Rayleigh ratio R_θ, is related to $\bar{M}_w$ by

$$K(1 + \cos^2 \theta)c/R_\theta P(\theta) = (1 + 2Bc + \cdots)/\bar{M}_w \qquad (8)$$

where c is concentration, B is the second virial coefficient, and K can be defined by

$$K = 2\pi^2 n_0 (dn/dc)^2/\lambda^4 N_0 \qquad (9)$$

where n_0 is the refractive index of the solvent, dn/dc is the refractive index increment with increasing concentration, N_0 is Avogadro's number, and λ is the wavelength of the light used. The particle scattering factor $P(\theta)$ is related to the root-mean-square radius of gyration r by

$$P(\theta) = 1 - \{16\pi^2 r^2/3\lambda^2[\sin^2(\theta/2) + \cdots]\} \qquad (10)$$

Thus, on extrapolation to zero angle and zero concentration,

$$Kc/R_\theta = 1/\bar{M}_w \qquad (11)$$

Normally, both extrapolations are carried out simultaneously by construction of a Zimm plot (*90*), in which Kc/R_θ is plotted against $\sin^2(\theta/2) + kc$, where k is an arbitrary constant (usually 100 or 1000) chosen to separate lines for solutions of different concentration.

In addition to providing molecular weight information, the slope of the zero-angle line gives the second virial coefficient [Eq. (8)], and the slope of the zero-concentration line gives the radius of gyration of the polymer [Eq. (10)]. The latter value, however, is in general reliable only for molecular weights in excess of $\sim 10^5$. Information about the shape of ordered molecular structures or the hydrodynamic volume of disordered polymer coils may also, in favorable cases, be obtained from the related technique of quasi-elastic light scattering, whereby rates of molecular diffusion are determined from changes in frequency of the incident laser light beam on scattering (*94*).

For large particles scattering is biased heavily toward low angles, and the presence of a relatively small amount of aggregated material can lead to a sharp downturn in Zimm plots toward low angle. If such aggregates are present as an interfering contaminant (e.g., undissolved solid) in the solution under investigation, it is best to ignore departures from linearity at low angles and use scattering at higher angles for molecular weight determination. For studies of higher levels of organization, however, very large molecular assemblies may be of direct interest, and measurements at angles lower than those normally possible on conventional scattering photometers (95) become essential. Direct measurement of scattering very close to the forward direction of the incident beam is now possible on modern low-angle laser instruments and can be used to characterize supermolecular organization.

F. Hydrodynamic and Rheological Techniques

Higher levels of organization, and in particular intermolecular networks, can also be probed by rheological studies of solutions or gels. These fall into two categories: small-deformation measurements, in which structure is retained, and large-deformation, in which structural breakdown is measured. A powerful, general small-deformation technique is mechanical spectroscopy (96), whereby an oscillatory distortion is applied to the sample and the resistance to deformation is measured. For true solids the greatest deformation (strain) occurs when applied stress is at a maximum (i.e., at either extreme of the oscillatory cycle), whereas for a perfect liquid the greatest resistance to flow (stress) occurs when the rate of deformation is greatest, i.e., in the middle of the oscillatory cycle, when net deviation from the null position (strain) is zero. Thus, stress and strain are in phase for perfect solids and exactly out of phase for perfect liquids. Polysaccharide solutions, gels, and biological assemblies show intermediate behavior (97,98), and resolution of induced stress into in-phase and out-of-phase components gives a measure of the degree of solidlike and liquidlike nature of the material, characterized by the rigidity modulus G' and viscosity modulus G'', respectively. For permanent networks (e.g., true gels) G' is much larger than G'', and both show little frequency dependence.

At high frequencies of oscillation concentrated solutions of disordered polymers behave similarly to gels, whereas at lower frequencies G'' becomes predominant, as illustrated in Fig. 12. The origin of this behavior lies in the interpenetration, or "entanglement," of individual coils to develop a transient network structure. The principal response to slowly applied deformation is network rearrangement to accommodate the strain (i.e., flow). When the time scale of deformation is rapid relative to that of molecular rearrangement, however, the predominant effect is distortion of the network, with substantial recovery of the original structure on release (i.e., elasticity). At even higher frequencies the main effect is distortion (segmental oscillation) of individual chains, and G' and G'' become nearly equal. The magnitude of

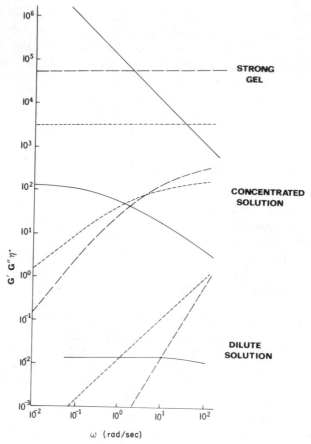

Fig. 12. Typical mechanical spectra of polysaccharide solutions and gels. Solid-like and liquid-like behavior are characterized by the storage (———) and loss (-----) moduli G' and G'', respectively, and the overall response to applied deformation can be monitored by dynamic viscosity η^* (———). The samples shown are 2% w/v agar, 5% w/v λ-carrageenan, and 5% w/v dextran.

G' over this "plateau region" of gel-like character (Fig. 12) gives a measure of cross-link density. Thus, mechanical spectroscopy measurements reflect both the number and time scale of transient intermolecular associations.

The ratio of total stress to frequency of oscillation ω is known as the dynamic viscosity η^* and is related to ω, G', and G'' as shown in Eq. (12).

$$\eta^* = (G'^2 + G''^2)^{1/2}/\omega \tag{12}$$

At low frequencies, where the time scale of imposed movement is slow relative to that of network rearrangement, η^* remains constant. With increasing frequency, however, a progressively smaller fraction of individual chains have sufficient time to reentangle within the period of oscillation, and η^* drops. Closely analogous behavior (*99*) is observed in large-deformation studies of the shear rate ($\dot{\gamma}$) dependence of viscosity η and is discussed in greater detail in Section V. For many polymer solutions the frequency dependence of η^* and shear rate dependence of η are closely superimposable when the same numerical values of ω and $\dot{\gamma}$ are compared. This empirical correlation (the Cox–Merz rule) (*100*) is often obeyed in polysaccharide solutions.

Viscosity measurements at concentrations below those required for coil entanglement can be used to characterize the hydrodynamic behavior of individual molecules. A particularly useful parameter is intrinsic viscosity,

$$[\eta] = \lim_{c \to 0} \left[(\eta_{\text{solution}} - \eta_{\text{solvent}})/\eta_{\text{solvent}}c \right] \tag{13}$$

which has the units of reciprocal concentrations. For each polymer/solvent system,

$$[\eta] = KM^{\alpha} \tag{14}$$

where K and α are constants, and thus intrinsic viscosity, once calibrated, can be used as a convenient, practical method for molecular weight determination. More fundamentally, intrinsic viscosity provides an index of random coil dimensions and hence of chain conformation (Section V).

A useful, practical large-deformation measurement for gels is yield stress, the force required to fracture gel samples of given fixed geometry (typically cylindrical plugs of ~ 1 cm height and diameter) on compression between parallel plates. Although in no way a fundamental molecular property, yield stress is none the less of considerable value in comparative studies of network properties in related systems. One such approach (*101*) is competitive inhibition of network structure by chain segments of sufficient length to form one stable association with the parent molecule but not two or more, thus blocking associations between intact chains, without contributing to crosslinking (Fig. 13). Polysaccharide gels incorporating short segments that are chemically identical to the junction-forming sequences of the gelling polymer show yield stress values that may be an order of magnitude lower than those of gels formed from the intact polymer alone, whereas control experiments using comparable (or greater) levels of segments that are not incorporated into the network show little if any change. Competitive inhibition of network structure therefore offers a simple method of investigating primary structural requirements for the formation of stable interchain junctions in polysaccharide gels. It may also be extended to specific interchain association in

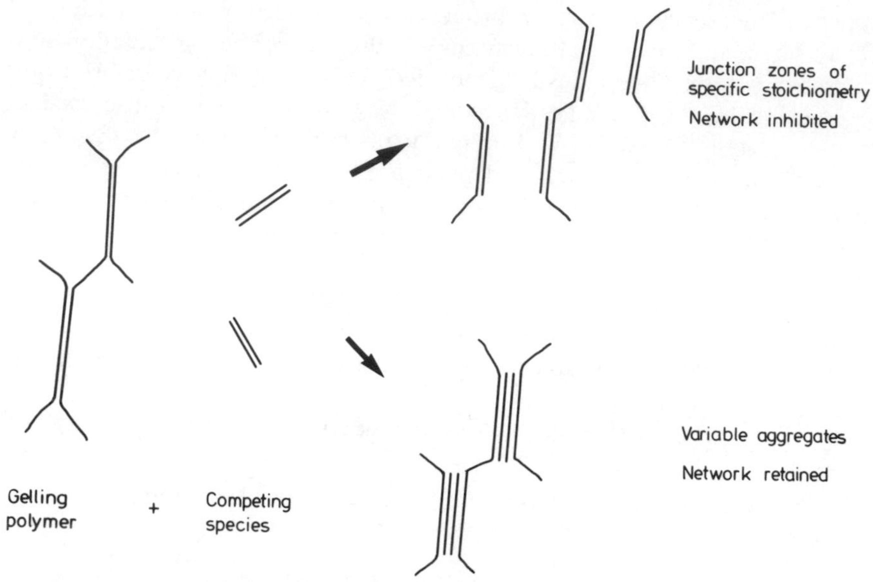

Fig. 13. Competitive inhibition of polysaccharide gel structure by short chain segments, which form stable associations with binding sequences on intact chains without contributing to network formation. With permission from (*101*) *J. Mol. Biol.* **138**, 363 (1980). Copyright by Academic Press, Inc. (London) Ltd.

solution, using mechanical spectroscopy (Fig. 12) to characterize the resultant changes in network properties (*102*).

G. Stoichiometry of Cation Binding

Another physicochemical technique for characterizing ordered interchain junctions is equilibrium dialysis, which has proved to be particularly valuable in studies of Ca^{2+}-induced association of polyuronates (*62,68*). As shown schematically in Fig. 14 the stoichiometry of cation chelation between chains is sensitive to both chain symmetry and the number of chains involved. Determination of the level of site-bound cations is complicated, however, by nonspecific polyelectrolyte binding. To overcome this problem, solutions of the polysaccharide under investigation are dialyzed to equilibrium against mixed solutions of the ion implicated in junction formation and much higher concentrations of another counterion that shows no specific binding behavior but may satisfy polyelectrolyte requirements. The stoichiometry of cation binding can then be determined from the amount of bound cation

Twofold dimer
50% Ca^{2+} binding

Large aggregate
→ 100% Ca^{2+} binding

● = Ca^{2+}

Viewed along
chain axis

Threefold dimer
~ 33% Ca^{2+} binding

Fig. 14. Stoichiometry of cation binding within ordered interchain junctions. With permission from (*102a*) *J. Mol. Biol.* **155**, 507 (1982). Copyright by Academic Press, Inc. (London) Ltd.

that is not available for equilibration across the dialysis membrane, several different concentrations of the competing counterion being used to verify that nonspecific charge–charge interactions have, in fact, been suppressed.

H. Differential Scanning Calorimetry

One advantage of NMR (Section II,C) is that it can be applied to turbid or opaque systems. The same is true of differential scanning calorimetry (DSC), in which thermally induced changes in conformation or packing are detected by the absorption or release of heat. Differential scanning calorimetry transitions normally give traces that approximate a gaussian band form, and this may be of particular value (*103*) in detecting sequential processes that are not widely separated in their temperature course, as illustrated in Fig. 15.

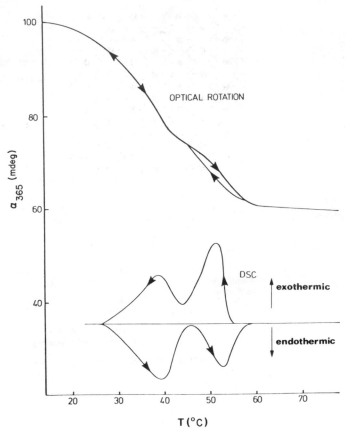

Fig. 15. Use of DSC to monitor changes in carbohydrate chain conformation. Thermal changes (lower curves) give clearer evidence than optical rotation (upper curves) of two discrete transitions in the order–disorder behavior of ι-carrageenan (K^+ salt form). From (*163*), with permission.

III. Homopolysaccharides

A. Conformational Types

The allowed regular chain conformations of homopolysaccharide sequences (e.g., Fig. 16) can be categorized into four distinct types (*8,29*), each with a characteristic range of *n* and *h* values (Section II). Physical and computer model-building studies (*29,104,105*) show that the principal distinction does not arise so much from different interactions between adjacent residues (since these can be very similar for the different classes) as from the relative orientation of the glycosidic bonds to and from each sugar ring. The con-

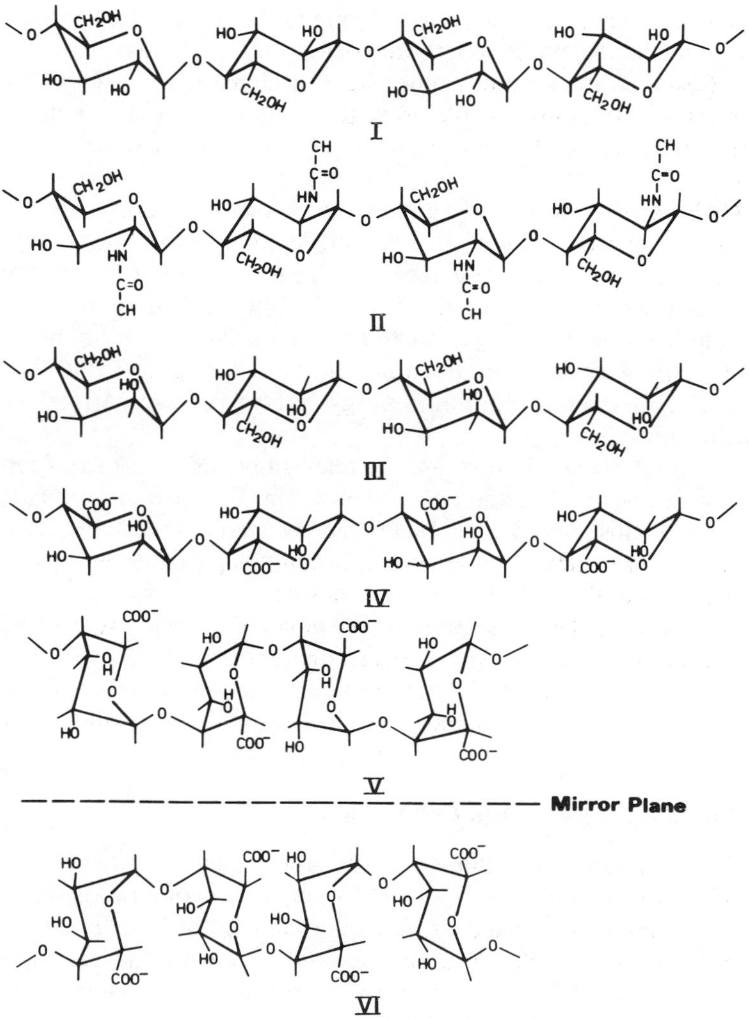

Fig. 16. Stereochemical comparison of homopolysaccharide primary structures. I, cellulose; II, chitin; III, mannan; IV, poly(D-mannuronate); V, poly(L-guluronate); VI, poly(D-galacturonate).

straints of tetrahedral bonding then limit the range of possible orientations to the following classes:

1. *Extended ribbons.* For these structures the bonds connecting each sugar residue to its two glycosidic oxygens are almost parallel or subtend an obtuse angle (for example, the O—C-4 and C-1—O bonds across a β-D-glucose residue in cellulose). For this family of polysaccharides, computer model

building shows that allowed values of n are in the range $\pm 2-4$ and h values are close to the maximum length for a sugar residue. This means that the plane of each monomer unit cannot lie at a large angle to the axis of the helix but must be almost parallel to it, thus leading to extended, drawn-out shapes, which, as we shall see, seem to correlate with a structural role in biology.

2. *Coiled springs.* In these structures the bonds from each sugar unit to its glycosidic oxygens subtend an acute angle (for example, the O—C-4 and C-1—O bonds across an α-D-glucose residue in amylose). For this family of polysaccharides, computer model building shows that allowed values of n cover a much wider range, $\pm 2-10$ and, more significantly, h can be close to zero. The structure therefore resembles a coiled spring in various states of extension, and several important energy reserve polysaccharides seem to belong to this type.

3. *Crumpled ribbons.* In this case the relevant bonds are far from coplanar for example, the O—C-2 and C-1—O bonds in $(1 \rightarrow 2)\beta$-D-glucan. Extended sequences of this type are characterized by steric clashes between non-adjacent residues, making ordered conformations particularly unfavorable. Few examples of this class are found in nature.

4. *Flexible coils.* The most common examples of this type are $(1 \rightarrow 6)$-linked polysaccharide chains. At each linkage the extra bond between the sugar rings leads to diminished steric contacts and allows additional bond rotation. The adoption of ordered shapes is therefore opposed by increased conformational entropy.

B. Ribbon Sequences in Plant Cell Walls

The known forms of cellulose in the solid state are extended ribbons (*106–108*), usually with $n = 2$, $h = 0.515$ nm, and a conformational energy close to the minimum predicted by conformational analysis and stabilized by hydrogen bonding of each successive residue to its two nearest neighbors through O-3 and O-5 (Fig. 2). The great strength, fibrous character, insolubility, and inertness that are characteristic of cellulose and so important to its skeletal function in plant cell walls (*3,109*) arise from ordered packing of these chains to form compact and tightly bonded aggregates.

Such structures can in principle have chains packed in either parallel (i.e., with all the nonreducing ends occurring at one end of a packed bundle and all the reducing ends at the other) or antiparallel arrangements. In the natural organization in plant tissue, the extended ribbons of cellulose probably occur in parallel rather than antiparallel arrays (*108,110*). The flattened sheets of the chains lie side by side and are joined by hydrogen bonds. These sheets are laid on top of each other in a way that staggers the chains, just as bricks are staggered to give strength and stability to a wall.

In industrial processing cellulose often undergoes severe treatment with alkali, as, for example, in making cotton easier to dye by the so-called mercerization process or in the regeneration process used in the manufacture of rayon or cellophane. This may cause the carbohydrate chains to rearrange to a different mode of packing (108,111) known as cellulose II, in contrast to the native state (cellulose I) (110). At present there is no known way of reconverting cellulose II to the natural organization. The chain conformation is essentially similar in both forms, the significant differences being that in cellulose II the sheets of extended ribbons are packed in antiparallel arrays and laid directly on top of one another rather like planks in a timber yard as opposed to bricks in a wall (3,108,110,111).

The ribbon conformation of β-(1→4)-xylan is twisted with a left-handed, threefold sense, displaced (14) from the cellulose (and chitin) conformation by roughly 30°–40° in both ϕ and ψ. Such twisted ribbons are less firmly bound into the plant cell wall, as judged by the criteria of the ease with which they can be extracted from the tissue and a lower extent of crystallinity in the native state. The conformation and packing are conserved (112) in natural derivatives that carry dense substitution at C-2 and C-3 of the xylose backbone and are apparently accommodated in the lattice by displacement of water molecules, which otherwise form continuous columns parallel with the xylan chains and closely adjacent to them. Displacement can be partial to accommodate irregular branching (113). Certain arabinoxylans, after extraction, show temperature-dependent changes in optical rotation, viscosity, and other properties, which appear to be due to the reversible renaturation of this native structure (114). Acetylation of the xylose backbone, by contrast (115), converts the chain from the threefold shape to a twofold cellulose-like conformation.

β-(1→4)-Mannan (Fig. 16,III) adopts a twofold helical structure in the solid state (116,117), which, in contrast to xylan, converts to a threefold helix when the polymer is acetylated (118).

C. Ribbon Sequences in Crustaceans, Insects, and Bacteria

The polysaccharide chitin is a fundamental skeletal material whose role in crustaceans, insects, and spiders is somewhat analogous to that of cellulose in plants, in that it contributes to the structure of the exoskeleton, the lining of the gut, the tendons, the wing coverings, and the internal skeleton. This polysaccharide is also present in the cell walls of fungi.

The structure of chitin (Fig. 16, II) is very closely related to that of cellulose and, indeed, is identical except that the —OH group on each C-2 is formally replaced by —NHCOCH$_3$. The way in which the chains pack side by side in a crystalline, strongly hydrogen-bonded manner is also very similar. As with cellulose, different forms (α-, β-, and γ-chitin) are known (119); all of these are built up from sheets of parallel chains. Again the sheets may be either parallel (β-chitin) or antiparallel (α-chitin) to each other. In γ-chitin it seems that the arrangement is more complex, possibly involving pairs of parallel sheets separated by single antiparallel sheets. As in the case of cellulose, the antiparallel form is always obtained in the laboratory when chains come together from a swollen or solution state (i.e., cellulose II and α-chitin, respectively, are always obtained) (108). One difference between the two polysaccharides is that cellulose occurs in nature in one form only (cellulose I), whereas all three forms of chitin are found to exist, sometimes in the same organism.

A rather rigid cell wall structure provides the chief protection against mechanical and osmotic damage to bacterial cells and is the chief determinant of cell shape. The form and strength of this structure derive from the peptidoglycan component, which can be regarded as a form of chitin in which up to about half the sugar units are further modified by attachment of a peptide chain at O-3 through a linkage involving the formation of an ester of lactic acid. The polysaccharide chain lengths are variable and generally shorter than in other ribbon-type polymers; the peptide "tails" vary somewhat in their composition from one bacterial genus to another and are themselves cross-linked to build up a three-dimensional network that has been called a "bag-shaped macromolecule." This acts as a framework to which other cell wall polymers are anchored.

There is little definitive evidence about chain conformations in peptidoglycans or whether the interactions of ordered shapes have any importance in determining properties of the wall. However, a number of possible models have been proposed and discussed in the literature (120), and there is general agreement on some likely features. When the stiff, ribbonlike carbohydrate chains are aligned in sheets like those in cellulose and chitin and the peptide tails are put into a likely conformation, it is found that each tail can drape around the ribbon to project away at right angles. A gridlike structure is then formed when they connect. In principle, hydrogen bonding and other noncovalent forces could exist between carbohydrate chains in this arrangement, similar to those in the cellulose and chitin sheets. However, X-ray diffraction of isolated wall preparations shows no evidence of ordered packing over long ranges (121). Although other explanations are possible, perhaps the wall is best regarded as consisting of disordered and flexible sheets of carbohydrate chains, reinforced by peptide cross-links to compensate for the loss of strength that must be a consequence of disorder.

D. Ribbon Sequences That Bind Cations

β-D-Mannopyranosyl uronate and α-L-gulopyranosyl uronate occur as regular $(1\rightarrow4)$-linked sequences in alginate (122,123), the major structural polysaccharide of marine brown algae (Phaeophyceae). Both homopolymeric sequences are almost invariably found together in all alginate molecules (124–126), albeit to widely differing extents, and heterotypic mixed sequences containing both monomer units are usually also present (126). The implications of structural interruptions are considered in Section VI; here we concentrate on the properties of the two types of homopolymeric sequences.

From their structures and ring conformations (Fig. 16) we might predict that poly(β-D-mannuronate) would show some similarities to cellulose, although the change from an equatorial to an axial hydroxyl at C-2 will increase chain flexibility, and cation binding by the charged polyanion might also influence any ordered conformation. In the solid state a twofold cellulose-like form does in fact exist in fibers of the free acid (127), but in the lithium, potassium, sodium, and calcium salts this converts to a threefold helix, presumably to accommodate the bound cations (128).

For poly(α-L-guluronate) (Fig. 16,V), the axial–axial configuration of the glycosidic linkage leads to a distinctly buckled ribbon with limited flexibility. In the solid state both the free acid and its salts pack as twofold chains (128,129). Examination of molecular models shows that large interstices exist between chains packed in this conformation. Cooperative interactions between such buckled ribbons would therefore be strong only if the interstices could be filled effectively by water molecules or cations.

Evidence that polyguluronate does indeed have a strong tendency to incorporate appropriate cations in this way has been obtained by an elegant study of ion-binding behavior. When calcium ion activity is measured as a function of chain length in dilute solutions of oligouronates (Fig. 17a), the mannuronate series shows the expected polyelectrolyte behavior (130,131). Ion activity decreases continuously with chain length and levels off at high degrees of polymerization. A similar effect is observed for oligoguluronate sequences, but between chain lengths of 20–28 residues there is a stepwise change in ion activity (130,131), indicating a cooperative mechanism.

Support for this conclusion is derived from the binding of Ca^{2+} to alginates in the presence of increasing concentrations of monovalent cations (Fig. 17b). A proportion of Ca^{2+} is strongly resistant to displacement (68), corresponding to half the stoichiometric equivalent of poly(L-guluronate) sequences only, thus showing that these segments have a strong, preferential binding mechanism that is not available to the other types of sequence and indicating that this primarily involves the association of pairs of polyguluronate chains rather than large aggregates. A similar interpretation can be made from earlier ion-exchange studies (132) and from competitive inhibition of alginate gelation by isolated blocks (Section II,F) (101).

Circular dichroism studies provide a sensitive probe of the local environment of the carboxyl chromophores in uronate residues (60,63,133) (Section II,B) and give spectacular support for specific site binding by

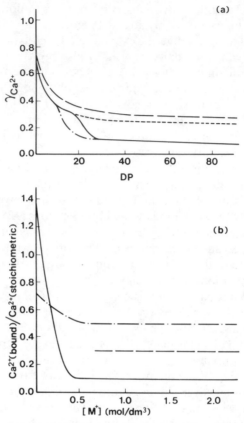

Fig. 17. Cooperative cation binding to polyuronate chains. (a) Variation of calcium ion activity coefficient ($\gamma_{Ca^{2+}}$) as a function of the degree of polymerization (DP) in solutions of the calcium salt form of poly(D-mannuronate) (——), poly(L-guluronate) (——), and poly(D-galacturonate) (–·–·–). Theoretical values of $\gamma_{Ca^{2+}}$ for polyguluronate and polygalacturonate, based on simple polyelectrolyte behavior, are also shown (---). From (*131*), with permission. (b) Equilibrium dialysis investigation of the level of bound Ca^{2+} resistant to displacement by swamping concentrations of univalent cation (M^+) for poly(D-galacturonate) (–·–·–) and alginate samples with ∼60% (——) and ∼20% (——) of poly(L-guluronate) chain sequences. In all cases the concentration of tightly bound calcium ions is equivalent to half the stoichiometric requirement of the polyguluronate or polygalacturonate present, consistent (Fig. 14) with binding within dimers of 2_1 chain symmetry. From (*62,68*), with permission.

polyguluronate segments (*62,65–69*). Gelation of alginate solutions by controlled addition of divalent cations is accompanied by dramatic changes in CD in which the negative trough arising predominantly from guluronate residues is swamped by a new positive transition. Comparison of the magnitude of difference spectra (gel CD minus solution CD) for a series of alginates of different composition (69,134) suggests specific involvement of polyguluro-nate. The behavior of isolated blocks of each structural type confirms this view (68). As shown in Fig. 18, large CD changes are observed when Ca^{2+} ions are added to polyguluronate chain segments, but this only applies if their chain length is above the critical threshold for cooperative binding. The available evidence suggests that the spectral changes arise through perturbation of both n and π orbitals of polyguluronate carboxylate chro-mophores in coordination of divalent cations, although almost certainly to different extents. Polymannuronate blocks show no measurable change in CD at any chain length when they are treated similarly (68).

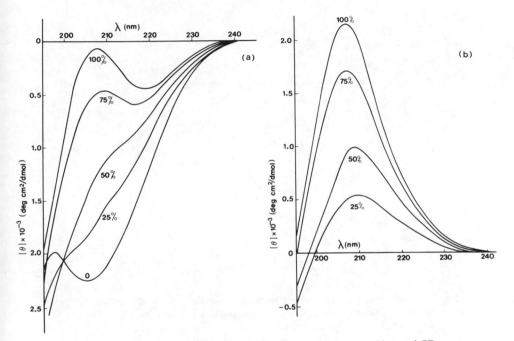

Fig. 18. Ca^{2+}-Induced changes in the CD of poly(L-guluronate). (a) Observed CD spectra in the presence of increasing concentrations of Ca^{2+} ions, expressed as a fraction of the total stoichiometric requirement of the polyuronate. (b) Corresponding difference spectra (CD in the presence of Ca^{2+} minus CD in the absence of Ca^{2+}). From (68), with permission.

Comparison of the environment of carboxyl chromophores in alginate gels and solid films by CD (Fig. 19) suggests very similar chain geometry and packing between the hydrated and solid states (67). Physical and math-ematical model building (66) indicates that the buckled ribbons of poly(α-L-guluronate) could pack together, with the cations strongly coordinated in

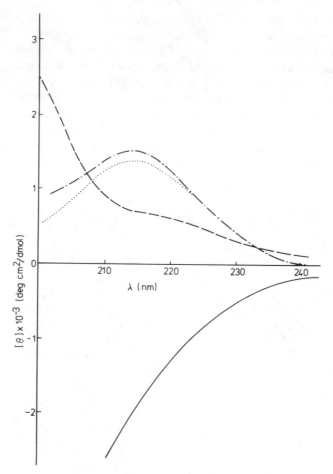

Fig. 19. Comparison of CD behavior for polyuronate Ca^{2+} gels and solid films prepared from them. Alginate shows little additional spectral change between the gel ($\cdots\cdots$) and film ($-\cdot-\cdot-$), whereas pectate (polygalacturonate) gels ($----$) show a massive change in CD on drying to the solid state ($\longrightarrow$). From (*62,67*), with permission.

oxygen-lined cavities between the chains like eggs in an egg box (Fig. 20). For each chain the likely coordination site involves a carboxylate oxygen and O-5 from one residue, with the glycosidic oxygen and O-2 and O-3 of the next residue toward the nonreducing chain terminus.

Homopolymeric sequences of $(1\rightarrow4)$-linked α-D-galactopyranosyl uronate are stereochemically analogous to the polyguluronate sequences discussed above in that the chains are exact mirror images (Fig. 16) except at C-3. They occur naturally as the partial methyl ester in the backbone segments

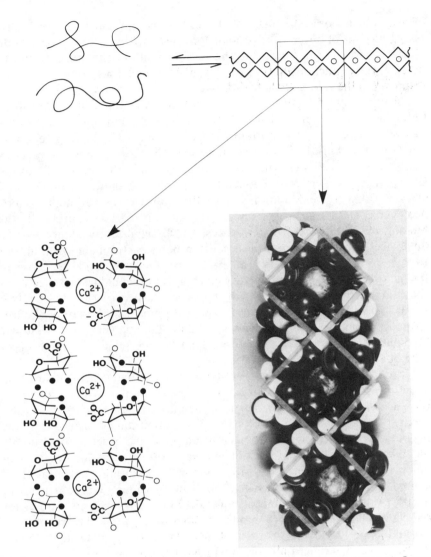

Fig. 20. "Egg box" model for Ca²⁺-induced dimerization of poly(L-guluronate). Oxygen atoms involved in cation chelation are shown (bottom left) as filled circles. Interchain packing of the array of bound ions is illustrated in the CPK space-filling model (bottom right), and chain contours are traced to show the regular, buckled, twofold conformation. A closely analogous structure has been proposed (*62*) for calcium poly(D-galacturonate). From (*69*), with permission.

of pectin, a major constituent of the cell walls and soft tissue of higher plants. Pectins of lower ester content form firm gels with Ca^{2+} ions, as do alginates. However, in contrast to the twofold conformation for polyguluronates, X-ray diffraction analysis of polygalacturonate fibers shows threefold geometry in the solid state (135–137).

Calcium-induced gelation of pectins of widely differing degrees of substitution provides strong evidence that galacturonate and not methyl galacturonate sequences are important for Ca^{2+} binding (138). Gel strength, a function of network formation by intermolecular associations, increases as the content of methyl esterified units decreases, and the magnitude of the CD change associated with cooperative ion binding shows a similar trend (69,138). Competitive inhibition studies also show clearly that methyl esterified segments have no effect on gel strength, in marked contrast to the binding of fully deesterified sequences, which significantly reduce it. The differences between the spectra of sodium polygalacturonate solution and its calcium gel (62) are very similar in form and magnitude to those of polyguluronate but are of opposite sign (Fig. 21), as would be expected from the near mirror image relationship between the two species. It therefore seems likely that the local dissymmetric environment of the carboxyl chromophores, and hence the geometry of the ion-binding sites, is very similar in both and that pectin gelation can also be interpreted in terms of an "egg box" mechanism (66,102a).

Cation-binding studies (Fig. 17a) confirm that oligogalacturonates also show the sharp, cooperative increase in Ca^{2+} binding (131), although in this case the critical sequence length is significantly less than for polyguluronate (~ 15 residues). Support for this observation can be derived from the magnitude of CD change on Ca^{2+}-induced gelation of a series of randomly methyl esterified pectins (138). For polygalacturonates, as for polyguluronates, the level of bound Ca^{2+} decreases with increasing concentration of competing monovalent ions (Fig. 17b), again reaching a constant value at $\sim 50\%$ of the total stoichiometric equivalent (62). This suggests that, as for polyguluronate, the primary mode of association is by dimerization of twofold buckled ribbons, consistent with early conclusions from X-ray diffraction studies of partially oriented pectin gels (139).

As illustrated in Fig. 19, calcium pectate gels, in contrast to those of alginate, show massive changes in CD, including a reversal of sign, when dried to solid films (62,102a). The onset of spectral change is observed at hydration levels of $\sim 50\%$ by weight and has been attributed to a conversion from twofold to threefold chain symmetry in response to packing constraints and more favorable intermolecular hydrogen-bonding possibilities in the condensed phase. It therefore appears that, in contrast to the majority of polysaccharide systems, the stable ordered conformations of polygalacturonate are different in the hydrated and solid states.

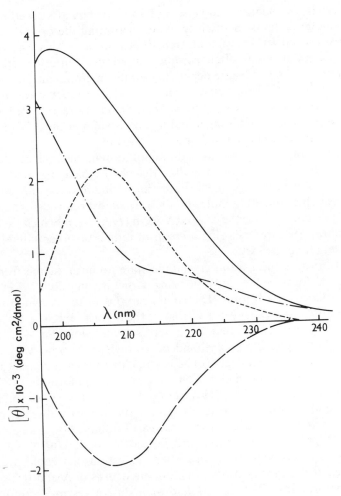

Fig. 21. Ca^{2+}-Induced changes in the CD of poly(D-galacturonate). Spectral differences (———) between the Na$^+$ solution (———) and Ca^{2+} gel (–·–·–) are similar in form and magnitude to those observed for poly(L-guluronate) (––––) but of opposite sign, consistent with closely analogous geometry of cation binding to near mirror image chains (Fig. 16). From (*62*), with permission.

E. Hollow Helices with Energy Reserve Functions

A common function of homopolymers that preferentially adopt "coiled spring" conformations is not in generating tissue structure, but in serving as energy reserves in biological systems. Starch and its counterpart in animal tissues, glycogen, are by far the most common, occurring almost as widely as cellulose, although in smaller quantities. The two component polysaccharides of starch, amylose and amylopectin, are stored together in crystalline granules, structured to avoid disrupting the osmotic balance of plant tissues and yet accessible and reactive to enzymes. Amylose, an α-$(1\rightarrow4)$-linked glucan, is essentially linear and can adopt a variety of ordered conformations in the solid state depending on its environment. Its best known ordered states have the spring very compressed, its geometry varying between 6, 7 and 8 residues per helical turn (140–144), but more extended forms also occur, the most extreme being found in a potassium bromide complex (145), which is believed to have $n = 4$ and $h = 0.40$ nm.

Such structures do not occur in dilute solution in the absence of cosolutes, where there is ample evidence that the chains have the hydrodynamic properties of a random coil (146) but can be induced by such complexing agents as iodine and n-butanol (140–142). The most rigorously characterized example (Section VII) is the six-fold V form (143,144), in which the helix has six residues per turn enclosing a chain of iodine or other suitable "guest" molecules along the axis.

The B form is the natural crystalline state found in starch from tubers, and the A form is the corresponding structure in cereals. Retrograded amylose adopts the B form. The conformations of the A and B forms are considerably more extended than that of the V form. Initially, the structure of the B form was interpreted (54) in terms of insertion of a water molecule into the O-6—O-2 hydrogen bond of V-amylose (Section VII) since this could produce the exact degree of helix stretching that is required to account for the observed differences between these ordered conformations. However, the ready formation of the B form from other crystalline states and from the random coil in solution did imply a stability that was always difficult to reconcile with hollow helices in the absence of stabilization from an included species (8). There is now convincing evidence that both structures are double helices (147), thus confirming an earlier suggestion that could not be substantiated at the time (148,149). The double helices of A and B forms seem to be conformationally identical, and their differences in X-ray diffraction patterns arise from their crystal packing and water content, B-amylose having a much more openly packed structure with additional water molecules located in channels between the hexagonally packed double helices.

Other examples of coiled springs with energy reserve functions are the β-$(1\rightarrow3)$-glucans (laminaran and paramylon), which occur as reserve materials in brown algae and some unicellular algae, and a branched $(1\rightarrow3)$-linked polymer of β-D-galactopyranose, which takes the place of glycogen as the reserve polysaccharide in snails (3).

In most examples of hollow helix polysaccharides, the chains are quite short. In amylopectin they are about 20–25 units in length, in glycogen about 10–12 units, in laminaran about 20–30, and in snail galactan about

10. These chains are typically joined by linkages of the loosely jointed type in a highly branched overall pattern. Amylose is exceptional in being longer (1000–2000 units), but even this usually contains some degree of branching.

F. Hollow Helices with Structural Functions

Although the carbohydrate chains that function as extracellular framework materials are usually members of the ribbon family, there are a few examples of hollow helices sufficiently extended to twist around each other to build up a strong network that performs the function in a different way. One example is the (1→3)-linked polymer of β-D-glucose, which seems to serve as the structural component of the cell walls of certain yeasts and occurs as callose in certain higher plant tissues, especially sieve tubes and pollen. When isolated, these chains can be crystallized as triple helices (*150–152*). The geometry is such that O-2 of each unit projects into the center of the helix to hydrogen bond with corresponding atoms on the other two chains. Another example is the β-D-(1→3)-linked β-D-xylan chains in certain green algae, which exist as a similar triple helix that was characterized earlier (*153*). Such twisting together of polysaccharide molecules cross-links the chains in units of three, rather than in the larger assemblies formed by packing of ribbonlike polysaccharides. This alternative mechanism for building biological cohesion might perhaps create a more flexible and open framework, analogous in this respect to the bacterial cell wall.

G. Flexible Coils and Linkages

Although regular sequences containing linkages of the flexible type do occur in nature, they are not widespread. The best known example is the α-D-(1→6)-linked bacterial polysaccharide dextran. As previously indicated, the increased flexibility of the extra glycosidic bond imparts substantially greater conformational entropy, which militates against ordered states in solution. No ordered conformations of such chains have yet been characterized, although microcrystalline arrays have been induced in unoriented films (*154*).

Isolated linkages of the (1→6) type are widespread in nature (*155*), especially in branched polymers, and indeed there are few examples of multichain polysaccharides that do not have (1→6) linkages in their structure. The characteristic freedom of rotation about these linkages could conceivably provide a brushlike molecule with flexible joints to facilitate biological interactions such as access of enzymes to the interior. Examples are amylopectin, glycogen, and various plant exudate gums. In animal systems such multichain structures may have the carbohydrate chains attached to a polypeptide backbone through linkages that are stereochemically analogous in that they also are separated by three covalent bonds.

IV. Regular Copolysaccharides

A. Introduction

Repeating units in heterotypic polysaccharides (e.g., Fig. 22) may range from the relatively simple A-B type to the complex periodic sequences found

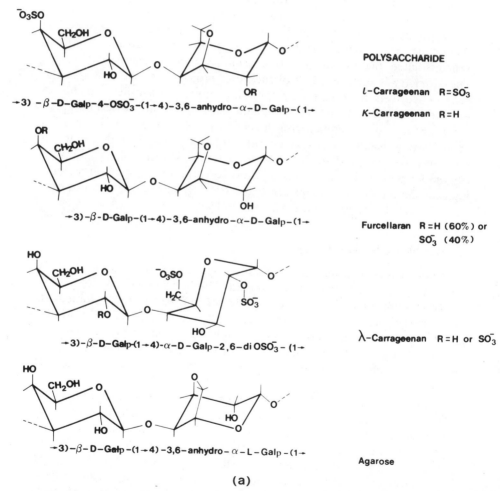

POLYSACCHARIDE

ι-Carrageenan   R=SO$_3^-$

κ-Carrageenan   R=H

Furcellaran   R=H (60%) or SO$_3^-$ (40%)

λ-Carrageenan   R=H or SO$_3^-$

Agarose

(a)

Fig. 22. Primary structures of heterotypic polysaccharide chains. (a) Alternating copolymers of algal origin; (b) bacterial copolysaccharides.

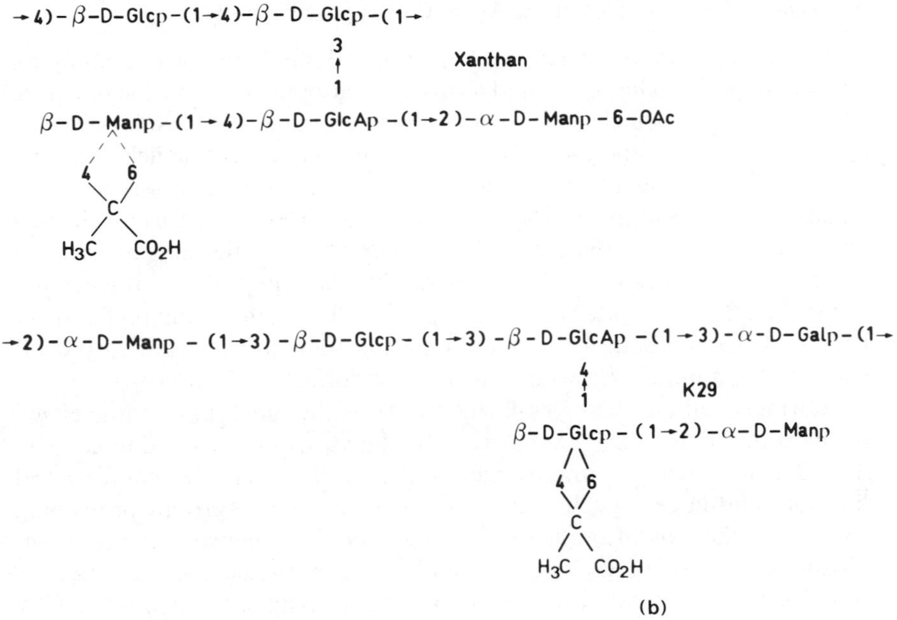

→ 4)–β–D–Glcp–(1→4)–β–D–Glcp–(1→

           3
           ↑   Xanthan
           1

β–D–Manp–(1→4)–β–D–GlcAp–(1→2)–α–D–Manp–6–OAc

     4   6
     \\   /
       C
     /   \\
   H_3C CO_2H

→2)–α–D–Manp–(1→3)–β–D–Glcp–(1→3)–β–D–GlcAp–(1→3)–α–D–Galp–(1→

                                4
                                ↑   K29
                                1

              β–D–Glcp–(1→2)–α–D–Manp
                      / \\
                      4 6
                      \\ /
                       C
                     / \\
                H_3C CO_2H

                      (b)

Fig. 22. (*Continued*)

in the cell surface bacterial polysaccharides. The simple A-B types include the carrageenans, agarose, and some of the glycosaminoglycans of animal connective tissue in which there is an alternating arrangement of (1→3) and (1→4) diequatorially linked residues. If present as the sole constituent of a homopolysaccharide, the 4-linked residue would give rise to an extended ribbon, whereas the 3-linked unit would generate a hollow helix (Section III,A). The possible conformations predicted for this family by computer model-building calculations range from undulating ribbons to extended hollow helices (*8*), and, experimentally, X-ray diffraction studies indicate that all are variants of an extended hollow helix differing in the degree of extension. In the more complex periodic sequences chain branching can have an important influence on the conformation of the backbone. For example, xanthan, an extracellular bacterial polysaccharide with a five-sugar repeating unit (Fig. 22b), has a (1→4)-β-D-glucan backbone, which does not, however, adopt the twofold ribbon conformation of cellulose; instead, it crystallizes as an extended fivefold helix probably as a result of interactions with the trisaccharide side chain present on every alternate residue of the backbone.

B. Algal Polysaccharides of the Agar–Carrageenan Family

As described in more detail in Chapter 4, Volume II the carrageenans are a family of polysaccharides found as matrix components of marine red algae (Rhodophyceae), which contain derivatized D-galactose residues linked alternately α-(1→3) and β-(1→4), as shown in Fig. 22a. In the helix-forming members, the basic repeating unit contains (1→3)-linked β-D-galactose 4-sulfate and (1→4)-linked 3,6-anhydro-α-D-galactose, with various degrees of 2-sulfation of the latter (*156,157*). The repeating sequence is interrupted or "masked" by partial replacement of the 3,6-anhydride by D-galactose sulfate or disulfate residues. In *ι*-carrageenan, for example, the replacement level is typically about 10%. These masking units are relevant to the capacity to form infinite three-dimensional networks (see Section VI).

ι-Carrageenan (see Chapter 4, Volume II) is the most fully characterized in the solid state (*16,158*) and has been shown to exist as a double helix (Fig. 23) in which the strands are parallel (rather than antiparallel) and threefold and have a pitch of 2.66 nm, with the sulfate groups protruding away from the axis of the helix. Hydrogen bonding between the only unsubstituted hydroxyl groups, O-2 and O-6, of D-galactose residues on complementary strands is repeated by helical symmetry to form a fully hydrogen-bonded structure.

Oriented fibers of κ-carrageenan, a related but less highly sulfated polysaccharide from other species of red algae (Chapter 4, Volume II) gives diffraction patterns with some similarities to those of *ι*-carrageenan, but the interchain packing is more disordered and therefore the patterns are more difficult to interpret (*16*). The provisional interpretation is in terms of an analogous double helix of somewhat shorter pitch (2.46 nm) and with the relative positions of the chains shifted.

Furcellaran, an additional member of the carrageenan family, which is even less highly sulfated (*159*), gives still poorer diffraction data. There is some evidence from solution properties to suggest a similarity with *ι*- and κ-carrageenan.

λ-Carrageenan has a different backbone structure, being composed predominantly of (1→3)-linked β-D-galactose 2-sulfate and (1→4)-linked α-D-galactose 2,6-disulfate and containing few anhydride residues (*160,161*). A very preliminary interpretation (*55*) of early diffraction data (*162*) suggests a rather flat, extended ribbon rather than a coaxial double helix.

The agarose backbone (*163*) is diastereoisomeric with *ι*- and κ-carrageenans in that it contains (1→4)-linked 3,6-anhydro-α-L-galactose rather than the D-enantiomer (see Chapter 4, Volume II). As in the carrageenan series, but perhaps to an even greater extent, the agar backbone is substituted to various extents with neutral and charged groups (*164–166*). In this context agarose itself is one idealized extreme of a range of naturally occurring structures.

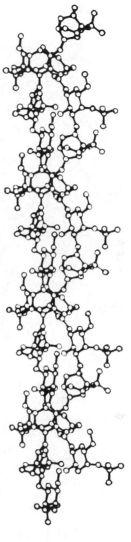

Fig. 23. *ι*-Carrageenan double helix; the structure is threefold, parallel, and right-handed and has a pitch of 2.66 nm. With permission from (*158*), *J. Mol. Biol.* **90**, 253 (1974). Copyright by Academic Press, Inc. (London) Ltd.

ι-Carrageenan

The X-ray diffraction evidence (*57*) from a study of agarose and derivatives supports a double helix model (Fig. 24), again with parallel threefold chains but now with a left-handed sense and a pitch of 1.90 nm. The reversal of the screw sense is probably related to the enantiomorphic change in the anhydride residue. With a pitch of 1.90 nm, the agarose helix is less extended than either *ι*-carrageenan (2.66 nm) or *κ*-carrageenan (2.46 nm), following a

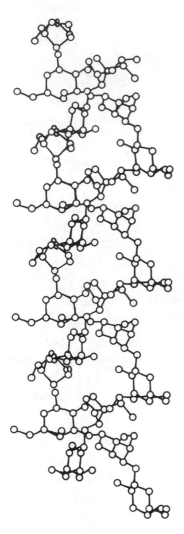

Fig. 24. Agarose double helix; as in the case of *ι*-carrageenan (Fig. 23) the structure is threefold and parallel but left-handed and with a shorter helix pitch (1.90 nm). With permission from (*57*), *J. Mol. Biol.* **90**, 269 (1974). Copyright by Academic Press, Inc. (London) Ltd.

Agarose

trend with changing sulfate content. Whereas the carrageenan helix changes in its extension and perhaps the relative positioning of strands with degree of sulfation, the geometry of the agarose helix apparently remains unaltered by changed substitution in the derivatives studied. The double helical conformations of agar and carrageenan polysaccharides can persist under hydrated conditions, in solutions and gels, and an important consequence of alterations in the nature and extent of substitution is then a change in the stability of the ordered structure and its aggregation states (Section VI).

In the first instance on the basis of the X-ray structure, Anderson *et al.* (*16*) proposed that the formation of gels when solutions of ι- and κ-carrageenan were cooled involves the association of chains through double helices to develop a three-dimensional network structure. The first direct evidence (*167*) in support of this proposition came from the optical rotation changes that accompany the sol–gel transition and show a sigmoidal form consistent with an order–disorder transition.

To confirm that these effects are not caused simply by strain birefringence in the gel as it forms, additional studies were conducted with short, structurally regular chain segments that do not form gels (Section VI). The small proportion of the $(1\rightarrow4)$-linked residues that occur in both ι- and κ-carrageenan as galactose 6-sulfate are susceptible to periodate oxidation. Subsequent reduction with sodium borohydride and mild hydrolysis with acid (*168*), followed by elimination of residual 6-sulfate with alkaline borohydride (*169*), produced segments of more regular structure. Solutions of ι-carrageenan segments prepared in this way showed large changes in optical rotation with temperature (Fig. 25), which were attributed to a random coil ⇌ double helix transition (*56*). The sign and magnitude of the optical rotation shift agreed closely with values calculated from the double helix geometry in the condensed state using the semiempirical method for correlation of optical activity with dihedral angles between adjacent residues in the polysaccharide backbone (*28*) (Section II,B). Thermodynamic measurements (*170*) and the temperature course of optical rotation change (*171*) were consistent with a "two-state, all-or-none" dimerization process.

The temperature course and nature of the transition have also been investigated by NMR (Section II,C). The high-resolution ^{13}C- and ^{1}H-NMR spectra of ι-carrageenan segments (*67,172*) give sharp, well-defined signals for the random coil at 80°C, but on cooling to 15°C the peaks collapse (Fig. 25). This can be understood in terms of adoption of the conformationally rigid helical form. At intermediate temperatures, neither ^{13}C- nor ^{1}H-NMR signals are broadened or shifted, again consistent with a two-state, all-or-none transition. Pulse NMR studies (*172*) show a characteristic ^{1}H spin–spin relaxation time T_2 of approximately 30 μs for ι-carrageenan segments in the ordered form compared with approximately 5 ms for the mobile random coil. These results are fully consistent (Section II,C) with the observed collapse of high-resolution signal due to extreme line broadening.

Molecular weight determinations by light scattering and membrane osmometry (Section II,E) show an approximate doubling of both $\bar{M}_w$ and $\bar{M}_n$ for ι-carrageenan segments on cooling (*93*), as would be expected for double helix formation. [Note, however, that under certain other conditions

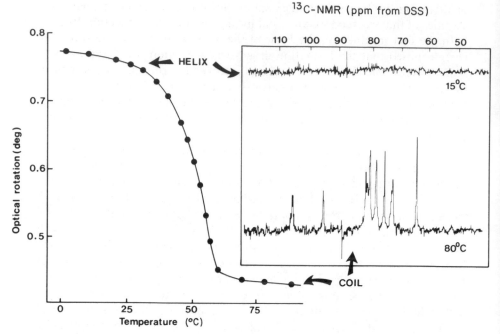

Fig. 25.   Coil–helix transition in segmented *ı*-carrageenan, as monitored by optical rotation (546 nm) and line broadening in high-resolution ^{13}C-NMR. From (*56,67*), with permission.

it has been claimed (*173*) that a conformation transition can occur without a change in $\bar{M}_n$.] The double helix proposal has received further conformation from an investigation of the dynamics of salt-induced conformational ordering using stopped-flow polarimetry (*84*) (Section II,D). The transition obeys second-order kinetics, again indicative of a dimerization process.

As mentioned above, the solid-state evidence (*16*) for the existence of κ-carrageenan as a double helix is less unambiguous than in the case of *ı*-carrageenan, and alternative models such as nested single helices cannot be entirely discarded from this evidence alone. However, the dynamics of the salt-induced conformational transition of κ-carrageenan have been studied (*85*) and shown to obey second-order kinetics, pointing to a dimerization step that is consistent with double helical entities. Both native κ-carrageenan (*167*) and the more regular segments (*174*) prepared as for *ı*-carrageenan show hysteresis with heating and cooling, which is believed to be associated with aggregation of the ordered helices. Because of this aggregation many of the techniques used to characterize the *ı*-carrageenan conformation in solution cannot be applied to κ-carrageenan.

Segmented agarose chains show an extent of aggregation that is so marked that precipitation occurs with conversion to the ordered state. However, optical rotation measurements are possible for the sol–gel transition of the intact polymer (*174*), probably because aggregation is limited by constraints within the gel network (see Section VI). The temperature-dependent changes exhibit cooperativity and hysteresis, and the sign and magnitude are consistent with the double helix model. Agarose gelation and liquefaction have also been monitored (*40*) by VUCD (Section II,B). Temperature-dependent changes in CD intensity were observed, which again showed both cooperativity and hysteresis (Fig. 7b).

C. Bacterial Copolymers

Xanthan, the extracellular polysaccharide from the gram-negative bacterium *Xanthomonas campestris*, has a branched pentasaccharide repeating unit (*175,176*) based on a cellulosic [(1→4)-β-D-glucan] backbone with alternate glucose residues substituted by charged trisaccharide side chains, each containing one D-glucuronic acid and two D-mannose residues. Different xanthan samples are substituted to various degrees by acetylation at O-6 of inner mannose residues and 4,6-pyruvate ketal on terminal mannose (*177, 178*). Aqueous solutions are highly viscous at relatively low concentration and show the unusual properties of insensitivity to large changes in temperature or ionic strength but high sensitivity to shear rate (*179*). These properties are consistent with the adoption of an ordered conformation in solution (*180*).

Single-wavelength optical rotation measurements (*77,180–182*) show a sigmoidal temperature dependence similar to that accompanying the order–disorder transition in other polysaccharide systems. Circular dichroism studies (*77,180*) showed that the observed changes in optical rotation did not arise entirely from changes in the optical activity of the acetate, pyruvate, and uronate chromophores, but from conformation-sensitive electronic transitions of the polymer backbone centered in the vacuum UV spectral region (Section II,B).

The mobility of the polymer has been studied by means of NMR relaxation (*77,180*) (Section II,C). At elevated temperatures, salt-free xanthan solutions show T_2 values in the region of milliseconds, as expected for disordered conformations, which convert to the microsecond range on cooling, typical of rigid ordered structures. High-resolution ^{1}H-NMR shows (Fig. 10a) sharp signals for the methyl protons of the pyruvate and acetate substituents at high temperature, which broaden dramatically on cooling, again indicating coversion to a rigid ordered conformation. Furthermore, the plot of NMR peak area versus temperature (Fig. 10b) is sigmoidal, consistent with a cooperative order–disorder process.

The spectroscopic evidence thus points to the existence of an ordered structure in solution. The absence of any detectable concentration dependence in the transition suggests the possibility of a single-stranded structure (77,180,182) stabilized by intramolecular bonding. The X-ray diffraction analysis (183) of oriented xanthan specimens points to a fivefold helix with an axial periodicity of 4.70 nm, consistent with a rise per backbone disaccharide of 0.94 nm (Fig. 26). Evidently, therefore, the consequence of chain branching and the removal of one potential cellulose O-3—O-5 hydrogen bond is to prevent the adoption of the extended twofold helical conformation of cellulose. Due to the poor crystallinity of the oriented fibers the X-ray evidence has not yet distinguished conclusively between single helix and double helix models.

Although the evidence presented above from solution studies supports an intramolecular structure the simplest interpretation of which would be a single-stranded model, other evidence from electron microscopy (184) and hydrodynamic studies (185) has been interpreted in terms of a double helix. A recent kinetic study (86) of the salt-induced disorder–order transition of xanthan has shown first-order rather than second-order behavior, consistent with an intramolecular rather than an intermolecular reaction. Molecular weight measurements by low-angle light scattering (186) showed no increase in $\bar{M}_w$ over the temperature course of the optical rotation transition, which again would preclude intramolecular double helix formation. Although an intramolecular double helix model could be reconciled with the experimental evidence for solutions, it seems unlikely on stereochemical grounds because it would require chain folding. It seems more likely that the ordered structure of xanthan in solution is a single-stranded helix stabilized by noncovalent interactions between side chain and backbone. The unusual rheological properties are discussed in terms of the ordered conformation and its further associations in Section VI.

The capsular polysaccharides (187) from Escherichia coli serotype K29 and from its two mutants, M13 and M41, gave identical diffraction patterns from oriented fibers (188), showing a similarity of primary structure and conformation. The data for M41 were interpreted in detail to derive a twofold helix model that persists under all conditions of water content and replacement of sodium by calcium. In M13, Bayer and Thurow (189) found that the capsular polysaccharide is attached to the exterior of the cell in the form of radial strands or coiled structures, and it is likely (188) that the in vivo conformation of these strands is similar to the highly hydrated fibers characterized by X-ray diffraction. The K29 capsular polysaccharide is the receptor for phage 29, which is highly specific for this capsule and the two mutants. Since the penetration of the capsule by the phage occurs through cleavage of the polymer chain, some particular conformational feature of

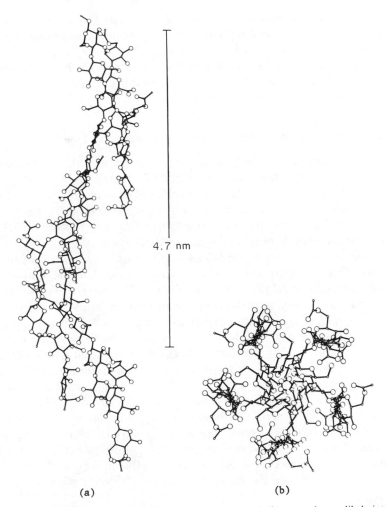

(a)   (b)

Fig. 26. Xanthan fivefold helix. The computer-generated diagrams show a likely interpretation of X-ray fiber diffraction evidence, projected (a) perpendicular to and (b) along the helix axis. An alternative, sterically feasible double helical structure gives calculated diffraction intensities that agree equally well with current X-ray data, but this model is less consistent with evidence from solution studies discussed in the text. Reproduced with permission from (*183*), *Am. Chem. Soc. Symp. Ser.* **45**, 90. Copyright (1977) American Chemical Society.

the polysaccharide may be relevant to the targeting of the phage to a specific site on the cell surface.

An analogous role for capsular polysaccharides has been postulated from a recent fiber diffraction study (*190*) on type III pneumococcal polysaccharide, which, being an alternating copolymer of $(1 \rightarrow 3)$-linked β-D-glucuronic acid and $(1 \rightarrow 4)$-linked β-D-glucose, has a linkage geometry that is similar to that of many of the glycosaminoglycans. The crystalline conformation is an extended threefold helix with $h = 0.92$ nm.

A number of *Klebsiella* capsular polysaccharides have been studied by X-ray fiber diffraction (*191–193*) and shown to adopt extended conformations of twofold or threefold chain symmetry.

D. Glycosaminoglycans of the Hyaluronate–Chondroitin Family

Glycosaminoglycans are polysaccharide components of animal connective tissues and are found mainly in the extracellular matrix. With the apparent exception of hyaluronate, they are covalently linked to protein in the natural state and in this form are known as proteoglycans (*194–197*). Since their chemistry and biochemistry are discussed in Chapter 5, Volume III we confine our attention here to conformational behavior.

The hyaluronate–chondroitin family of glycosaminoglycans has primary structures (Fig. 27) that are strikingly similar to each other and to the agarose–carrageenan family (Fig. 22a). Glycosidic linkages are alternately $(1 \rightarrow 3)$ and $(1 \rightarrow 4)$, and, with the component residues in likely ring conformations, all bonds to glycosidic oxygens are equatorial. The chain conformations in both families can therefore be regarded as a single class of somewhat extended helices (Fig. 28), which may be either left- or right-handed (*55*).

The substitution pattern varies widely on this common backbone (*55*), and this evidently influences the degree of chain extension. As discussed in Section I,C, the chain contour generally, although not invariably, becomes more extended with increasing equatorial substitution by bulky groups adjacent to glycosidic oxygens and with electrostatic charge. The more contracted forms (agarose, carrageenan, and one form of hyaluronate), but not the more extended chains, are found as double helices; such forms are probably impossible for the latter because of interchain clashes near the helix axis.

Hyaluronate has been the subject of detailed X-ray diffraction analysis during the past decade, and much information is now available on its conformation in the solid state (*33*). There is considerable versatility in chain conformation and packing, although all forms appear to crystallize with left-handed screw symmetry and are packed so that nearest-neighbor chains are antiparallel. The conformations can be classified in broad terms into compressed and extended types, on the basis of the axial repeat per disaccharide (h), which is around 0.85 nm for the compressed forms and around 0.95 nm for the extended forms. Hyaluronate appears to be the only glycosaminoglycan that adopts compressed conformations and does so with monovalent cations in the absence of calcium to form 4_3 helices.

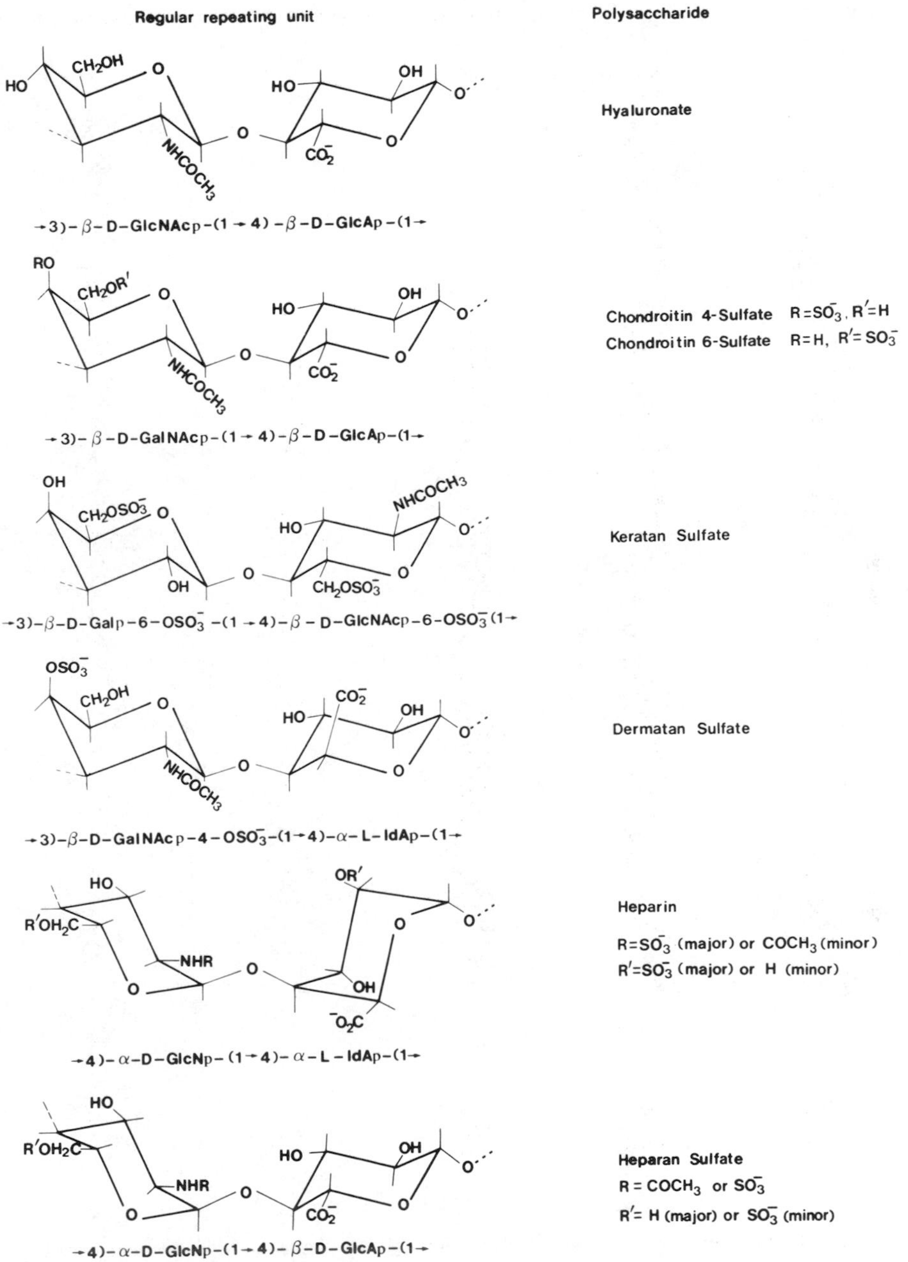

Fig. 27. Primary structures of glycosaminoglycans. Considerable structural variability occurs in many of these materials, and the structures shown are idealized repeating units. Dermatan sulfate can contain substantial quantities of D-glucuronate, and in heparin it may represent 15–40% of the total uronate content; conversely, heparan sulfate may have up to 50% of the uronate residues present as L-iduronate. Iduronate and glucuronate residues are not distributed randomly but are grouped in blocks. The $^4C_1$ conformation shown for L-iduronate in dermatan sulfate is that found in the solid state; solution studies indicate adoption of the alternative $^1C_4$ chair form. L-Iduronate residues in heparin are believed to adopt only the $^1C_4$ conformation shown.

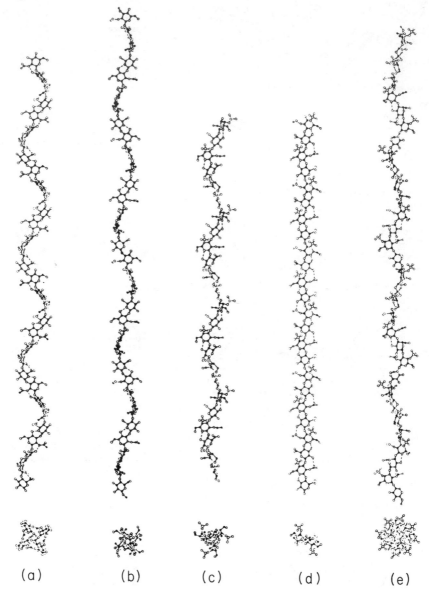

(a)   (b)   (c)   (d)   (e)

Fig. 28. Ordered conformations of glycosaminoglycan chains (*33*) viewed along (lower structures) and perpendicular to the helix axis. (a) Compressed 4_3 helix of hyaluronate ($h = 0.82$–0.85 nm). (b) Extended 4_3 hyaluronate conformation ($h = 0.94$ nm). (c) Chondroitin 4-sulfate 3_2 helix; similar structures ($h = 0.94$–0.95 nm) also occur for chondroitin 6-sulfate, dermatan sulfate, and hyaluronate. (d) The 2_1 conformation of chondroitin 4-sulfate; analogous 2_1 structures ($h = 0.93$–0.98 nm) can be adopted by all glycosaminoglycans. (e) Dermatan sulfate 8_3 helix; this allomorphic form ($h = 0.92$–0.98 nm) is also found for chondroitin 6-sulfate.

For the sodium salt (*198*), the detailed conformation and packing depend on the water content of the oriented specimen. When thoroughly dried, the chain is a left-handed fourfold helix that crystallizes in a tetragonal unit cell with the antiparallel chains packed in highly interdigitated fashion. In specimens prepared at higher humidities, the symmetry is relaxed so that the unit cell is orthorhombic and the polysaccharide helix, although retaining four disaccharide residues in the same helix rise of 3.3 nm, is now formally twofold because alternate disaccharide residues are conformationally non-identical. All disaccharide residues in both structures appear to be stabilized by interresidue hydrogen bonding between (a) the glucuronate carboxylate and the amide nitrogen of the acetamido residue across the $(1{\to}4)$ linkage and (b) O-4 of the acetamido residue and the ring oxygen of glucuronate across the $(1{\to}3)$ linkage. Only in the tetragonal form, however, do both hydrogen bond lengths suggest that stabilization is strong. In the orthorhombic form, (a) and (b) are alternately lengthened along the chain to about 0.29 nm; it would appear that water is accommodated by the orthorhombic lattice at some expense to interchain hydrogen bonding.

The contracted 4_3 conformation has been observed (*199*) with spectacularly altered interchain relationships when hyaluronate is crystallized in the presence of K^+ at low pH. This new form is an antiparallel double helix. The carboxyl groups of the two strands are brought close together in this structure with a countercation being bound within the cavity between them, and it is believed that partial protonation aids in stabilizing the structure by further reducing electrostatic repulsions and leading to interchain hydrogen bonding. The double helix is formed in the presence of K^+, NH_4^+, Rb^+, and Cs^+ but not Na^+; steric hindrance and ionic repulsion apparently prevent sufficient contraction to accommodate the smaller Na^+ ion.

Several more extended forms of hyaluronate (0.92 nm $< h <$ 0.98 nm) have also been found; an extended 4_3 helix is observed in fibers of the potassium salt after dehydration and rehydration (*200*); a 2_1 form crystallizes at low pH (*201*); and a 3_2 helix can be obtained in either of two crystal forms depending on whether it is in fibers of calcium hyaluronate (*202*) or sodium hyaluronate contaminated with Ca^{2+} ions (*203*). Structure refinements of the 3_2 forms (in which the helix pitch is 2.84 nm) show intramolecular hydrogen bonding from O-4 of *N*-acetylglucosamine to the uronate ring oxygen across each $(1{\to}3)$ linkage and from O-3 of glucuronate to O-5 of the acetamido residue across each $(1{\to}4)$ linkage. In comparison with the fourfold forms (see above), therefore, the former hydrogen bond is conserved, whereas the latter replaces the acetamido–carboxylate bond across the $(1{\to}4)$ glycosidic linkage. These structures also suggest how Ca^{2+}, even in trace amounts, can induce formation of the extended threefold conformation. This ion is site bound between carboxy groups on adjacent antiparallel chains in a manner that apparently cannot be achieved for the compressed fourfold forms.

Viscosity measurements (*204,205*), light scattering (*206*), and other investigations have shown that in dilute solution at neutral pH hyaluronate shows the overall properties typical of a random coil. Nevertheless, spectroscopic data (*76,207*) and the resistance of hyaluronate to periodate oxidation (*208*) suggest the possibility of a degree of local conformational ordering and consequent chain stiffening, even under conditions of extensive hydration. The capacity (*102*) of hyaluronate chain segments (~60 disaccharide units) to inhibit viscoelasticity in solutions, acid "putties," and gels (*209*) of the intact polymer also indicates some contribution to network formation by the specific association of stiffened chain sequences in addition to nonspecific polymer entanglement (see Sections II,F and V,B).

One possible contribution to chain stiffening could be the existence of interresidue hydrogen bonding, as indeed has been proposed (*208,210*) to account for the relative lack of reactivity of certain glycosaminoglycans to periodate oxidation. Reexamination of the published atomic coordinates of hyaluronate structures derived by fiber diffraction analysis and molecular model building has suggested (*211*) that two hydrogen bonds might be simultaneously possible at each glycosidic linkage to generate a "super-hydrogen-bonded" conformation of hyaluronate (Fig. 29). This is consistent with the results of recent NMR studies (*212,213*) of hyaluronate and the chondroitin sulfates and with an examination of the twofold hyaluronate conformation formed at acidic pH (*214*).

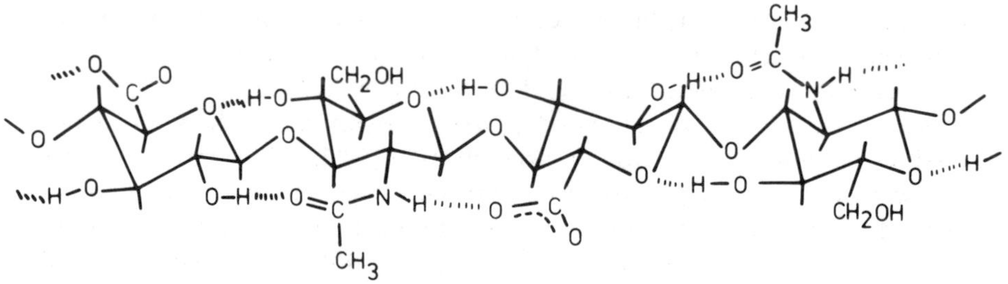

Fig. 29. Possible "super-hydrogen-bonded" conformation of hyaluronate, as discussed in the text. With permission from (*211*), *J. Mol. Biol.* **138**, 383 (1980). Copyright by Academic Press, Inc. (London) Ltd.

As in the case of hyaluronate, the chondroitin and dermatan sulfates can adopt several alternative chain conformations in the condensed phase, depending on the experimental conditions (*33*). One of the forms that has been subjected to detailed refinement (*215*) is sodium chondroitin 4-sulfate, which crystallizes as a 3_2 helix having a projected residue rise per disaccharide of

$h = 0.96$ nm. The structure incorporates an O-3 $\cdots$ O-5 intramolecular hydrogen bond across the (1→4) linkage similar to that found in the corresponding 3_2 form of hyaluronate. However, the ester sulfate and inversion of configuration at C-4 of the acetamido residue preclude an O-4 $\cdots$ O-5 hydrogen bond across the (1→3) linkage as in hyaluronate and it appears that no intramolecular bonds stabilize the conformation at this linkage. This chondroitin 4-sulfate structure differs from the hyaluronate form in its intermolecular as well as intramolecular hydrogen bonding, in that O-2 of each glucuronate is bonded to the acetamido carbonyl on an adjacent chain.

As for hyaluronate (above), the nature of the counterion is an important conformational determinant for chondroitin 4-sulfate. The Ca^{2+} salt (216) crystallizes as a 2_1 form in which there is interresidue hydrogen bonding across the (1→4) linkage, involving the ring oxygen of the galactosamine residue and O-3 of glucuronate. The chains are arranged in sheets within which neighbors are antiparallel, and the Ca^{2+} ions and water molecules are located mainly between the sheets. The chondroitin 4-sulfate chains within the intact proteoglycan show (216) analogous sensitivity to the nature of the countercation, since they also convert between threefold (Na^+) and twofold (Ca^{2+}) forms.

For chondroitin 6-sulfate (216), an 8_3 chain conformation (33) is accessible in addition to the 3_2 and 2_1 states corresponding to the 4-sulfate. Again, these are associated with different salt forms: Na^+, Ca^{2+}, and H^+, respectively. Owing to the greater charge density of chondroitin 4- and 6-sulfates compared with that of hyaluronate, the counterion in each calcium salt form is chelated across the (1→3) linkage by the sulfate ester and carboxyl within a single chain; this contrasts with calcium hyaluronate, in which the cation is bound between chains.

Dermatan sulfate likewise can adopt three distinct conformations in the condensed phase (217–219), which are a threefold ($h = 0.95$ nm), a twofold ($h = 0.94$ nm), and an eightfold helix ($h = 0.92$ nm), now shown (89) to have right-handed (8_3) rather than left-handed (8_5) helix sense. The influence of the nature of the counterions is less clear than for the other glycosaminoglycans, since all these forms have been obtained with the sodium salt. To account for the extended pitch of all of these helices, it is necessary to postulate that the iduronate residues adopt the $^4C_1$ conformation, in which the carboxyl is axial rather than equatorial to the ring and both glycosidic linkages are equatorial. This is somewhat contrary to expectation from the known conformational equilibria of model idopyranosides (2) and from NMR studies (220–222) of the polymer itself in solution, and it therefore seems necessary to invoke a conformational change between the solid state and solution. The susceptibility of dermatan sulfate to periodate oxidation (223), which has been considered to provide evidence for a dominant $^4C_1$

ring geometry in which the hydroxyl groups at C-2 and C-3 are trans–diequatorial rather than trans–diaxial as in the $^1C_4$ chair form, might then be explained (220) in terms of a degree of ring flexibility or distortion in solution that is favorable for complexing with periodate ions.

Keratan sulfate differs from the other glycosaminoglycans in containing no uronic acid residues. It adopts a twofold helix in the solid state (224) with an axial rise per disaccharide residue of 0.95 nm, similar to the conformation adopted by hyaluronic acid and the chondroitin sulfates at low pH. Thus, when the charge on the uronic acid residues of these glycosaminoglycans is suppressed, the chemical similarity between them and the uronate-free keratan sulfate becomes apparent, and a similar conformation results.

X-ray fiber diffraction studies (216) of intact proteoglycan indicate that in the condensed phase the "bottle brush" structure (225–227) adopts an approximately planar "double-edged comb" conformation, with glycosaminoglycan side chains lying at right angles to the protein core. Adjacent polysaccharide chains within the fiber are antiparallel, thus implying interdigitation of chains from different proteoglycan molecules. Interactions of this type, perhaps stabilized by interchain hydrogen bonding or Ca^{2+} chelation, as discussed above for specific glycosaminoglycans, might, if stable in an aqueous environment, provide a mechanism for network formation in vivo. Closely similar X-ray diffraction patterns have been observed (216) for a proteoglycan in which condroitin 4-sulfate predominates and for the free polyanion, indicating that the polysaccharide chain conformation is unaffected by covalent attachment to the protein core.

E. Heparin and Heparan Sulfate

These alternating copolymers of derivatized glucosamine and hexuronate residues do not share the $(1\rightarrow3)$, $(1\rightarrow4)$ pattern common to the other glycosaminoglycans and the carrageenan–agar family but instead are linked entirely $(1\rightarrow4)$. Within this alternating sequence the uronate residues may occur as β-D-glucuronate or α-L-iduronate, and the α-D-glucosamine residues as 2-acetamido-2-deoxy or 2-sulfamino-2-deoxy derivatives; there is much variation in relative proportions within both the uronate and the amino residue types and in the pattern of further sulfation (see Chapter 5, Volume III). Heparin contains mostly iduronate 2-sulfate and sulfaminodeoxyglucose 6-sulfate residues, in contrast to heparan sulfate, which has more acetamidodeoxyglucose and glucuronate. However, polymers isolated from natural sources always seem to contain all these features in some proportion. Owing to the complexity of these primary structures and the (as yet) moderate quality of the X-ray fiber diffraction photographs, no chain conformations have been uniquely established.

For heparin, the diffraction evidence is interpreted (228) in terms of a twofold helix with an axial rise per disaccharide residue (h) that varies as a function of relative humidity from 0.80 to 0.87 nm. Sterically reasonable models can be built to fit the observed diffraction distribution with L-iduronate residues in the $^1C_4$ rather than the $^4C_1$ chair form derived for dermatan sulfate in oriented fibers (see Section IV,D) and with interresidue hydrogen bonding between O-3 of the amino sugar and the ring oxygen of iduronate. In solution, NMR studies (221,222,229–232), including a complete spectral analysis (229), are also consistent with the $^1C_4$ ring form of L-iduronate in heparin but with some skewing of the chair, which appears to become more pronounced at elevated temperatures.

When L-iduronate adopts the $^1C_4$ conformation, its carboxyl group is equatorial and both glycosidic linkages are axial as in the calcium-binding sequences of alginate and pectin [poly(L-guluronate) and poly(D-galacturonate), respectively; Section III]. Evidence that heparin is also analogous in specific site binding has come from CD and optical rotation (233,234), which show Ca^{2+}-induced changes closely similar to those for polyguluronate and polygalacturonate and from changes in NMR spectra (233–235) attributed to a conformational transition induced by calcium–carboxyl interactions. However, there is no evidence that such binding causes the chain to adopt a rigidly ordered overall conformation as it does for polyguluronate and polygalacturonate. Equilibrium dialysis measurements (Section II) show that binding can occur to the extent of one calcium ion per tetrasaccharide unit of heparin (234).

X-ray diffraction studies (228,236,237) of heparan sulfate in the sodium salt form indicate a twofold helix with $h = 0.93$ nm, which is a conformation close to the maximum of chain extension. Sterically reasonable models can be built in which each of the two different glycosidic linkages is stabilized by hydrogen bonding between O-3 of the acetamido residue and the ring oxygen of glucuronate and from the acetamido group itself to O-3 of glucuronate. When converted to the calcium salt (228), the diffraction pattern changes ($h = 0.84$ nm) to become very similar to that of heparin (Ca^{2+}), suggesting that heparin-like sequences (238) in heparan sulfate now dominate the diffraction patterns.

V. Disordered Chains in Solution

A. Dilute Solution

One of the most familiar properties of carbohydrate polymers is their capacity to generate high solution viscosities at relatively low concentrations, which is of importance both for their biological function in, for example, interstitial fluids and for technological exploitation. The origin of this behavior can be traced to restricted rotation about the glycosidic bonds between adjacent residues (Fig. 2), which in turn restricts chain flexibility, thus leading to large overall coil dimensions for the disordered polymer.

The degree of coil expansion can be quantified by "random flight" theory (9). For a freely jointed chain of n residues each of length l, the average separation of the ends of the chain (root-mean-square end-to-end distance L) varies with chain length, as shown in Eq. (15). A chain of restricted

$$L^2 = nl^2 \tag{15}$$

flexibility can be regarded as a freely jointed chain with a smaller number (n') of segments of greater length (b) but the same overall average dimensions. The characteristic ratio (C_∞) of statistical segment length (b) to actual residue length (l) then gives a measure of the degree of conformational restriction [Eq. (16)].

$$C_\infty = b/l \tag{16}$$

This treatment is valid only if the number of statistical segments per chain is sufficiently large ($n' \gtrsim 10$) for overall random coil behavior. An alternative and more general (if more complex) approach is the wormlike chain model (9), in which the degree of directional correlation between adjacent residues is characterized by a persistence length a. For random coils persistence length and statistical segment length are directly related [Eq. (17)]. In both theories,

$$2a = b \tag{17}$$

however, the fundamental index of chain stiffness is the characteristic ratio (C_∞).

From conformational energy calculations (Section I) the relative probability of occurrence of different relative orientations of adjacent residues in the polymer chain can be calculated from the Boltzmann distribution [Eq. (18)],

$$p(\phi,\psi) = ke^{-V(\phi,\psi)/RT} \tag{18}$$

where ϕ and ψ are the dihedral angles defining the conformation of the linkage (Fig. 2), $V(\phi,\psi)$ is the conformational potential energy, $p(\phi,\psi)$ is the probability of occurrence, R is the gas constant, T is absolute temperature, and k is the constant of proportionality such that $\sum p = 1$ when summed over all combinations of ϕ and ψ. An elegant general method has been developed (9) for subsequent calculation of C_∞ and has been widely applied to polysaccharide chains (239–243).

The dependence of overall chain conformation on local linkage geometry, which has been demonstrated for ordered structures (Section III), is also evident in the disordered state (8). Thus, chains that pack as extended ribbons in the condensed phase give characteristic ratios in the range 24–358 (typically 100), whereas for those that form hollow helices calculated C_∞ values are normally less than 10. Within each class, variations in chain flexibility can be correlated with local primary structure. In general, axial linkages allow less conformational mobility than do equatorial linkages, and bulky equatorial substituents (OH or larger) adjacent to the glycosidic or aglycone bond introduce further constraints. As a simple empirical guide, restrictions on freedom of rotation about a given linkage can be characterized by a "steric crowding factor," defined as the number of axial bonds in the linkage plus the number of equatorial substituents immediately adjacent to it (*29*). Steric crowding factors show reasonable correlation both with calculated values of C_∞ and with those obtained experimentally.

For a true random flight coil, experimental measurement of C_∞ is straightforward. Root-mean-square end-to-end distance L is directly related [Eq. (19)] to radius of gyration r, which can be experimentally determined by, for

$$L^2 = 6r^2 \tag{19}$$

example, light scattering. Hence, knowing the residue length l from the primary structure and the number of residues n from molecular weight, one can calculate C_∞ [Eqs. (15) and (16)] as the ratio of the experimental value of L^2 to the theoretical value for a freely jointed chain nl^2.

For real chains the overall coil dimensions are not determined only by local interactions between adjacent residues, but may be "perturbed" by interactions between residues that are remote from each other in the primary sequence (e.g., by steric clashes, electrostatic repulsion, van der Waals interactions, etc.). Under these conditions L^2 is no longer proportional to n [Eq. (15)] and C_∞ becomes meaningless.

By adjusting solvent conditions (e.g. changing temperature or adding a limited amount of a poor solvent for the polymer) it may be possible to balance segment–segment and segment–solvent interactions and mimic random flight behavior experimentally ("theta" conditions) (*9,244,245*). For most carbohydrate polymers no such conditions are experimentally possible. It may still be feasible, however, to characterize unperturbed coil dimensions, and hence C_∞, by studying a range of polymer fractions of different average chain length and extrapolating results to zero molecular weight, since nonnearest-neighbor interactions become relatively less important with decreasing chain length.

A particularly convenient index of polymer coil dimensions in solution (*246*) is intrinsic viscosity $[\eta]$, the fractional increase in viscosity per unit concentration c for isolated chains (i.e., $c \to 0$). For both perturbed and unperturbed chains intrinsic viscosity increases with coil dimensions according to the Flory–Fox relationship [Eq. (20)] and with molecular weight according to the Mark–Houwink equation [Eq. (21)]. It follows directly (for

$$[\eta] = \Phi L^3/M_r \tag{20}$$

$$[\eta] = KM_r^\alpha \tag{21}$$

a detailed derivation see, for example, 98) that under theta conditions [i.e., in which Eqs. (15) and (16) are obeyed] $\alpha = \frac{1}{2}$ and K is related to C_∞, as shown in Eq. (22),

$$K_\theta = \Phi l^3 (C_\infty/m_r)^{3/2} \tag{22}$$

where m_r is residue molecular weight and Φ is a constant ($\approx 2.5 \times 10^{26}$ kg^{-1}). The best known extrapolation from perturbed to unperturbed coil dimensions, the Stockmayer–Fixman method, uses a relationship of the form

$$[\eta]/M_r^{1/2} = K_\theta (1 + C' M_r^{1/2}) \tag{23}$$

where C' is a constant. Thus, the intercept of a plot of $[\eta]/M_r^{1/2}$ against $M_r^{1/2}$ gives K_θ and hence C_∞.

The Stockmayer–Fixman approach takes account of perturbation of coil dimensions by steric clashes. For polyelectrolytes an additional source of coil expansion is intramolecular electrostatic repulsion, and in this case it is also necessary to extrapolate to infinite ionic strength I, where such effects are suppressed. As illustrated in Fig. 30, polyelectrolyte intrinsic viscosity (247) increases almost linearly with $I^{-1/2}$, providing a convenient method

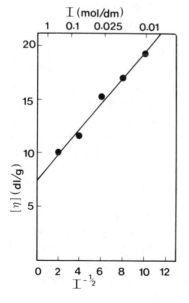

Fig. 30. Variation in polyelectrolyte intrinsic viscosity $[\eta]$ with ionic strength I, illustrated for λ-carrageenan. From (241), with permission.

for extrapolation. The sensitivity to salt (i.e., slope of plots of the type shown in Fig. 30) decreases with increasing chain stiffness (i.e., with C_∞). Smidsrød and Haug (248) have made direct use of this behavior to develop a convenient empirical index of the flexibility of polyelectrolytes, which correlates well with measured persistence lengths. A major advantage of this approach is that results can be obtained for a single sample of unknown molecular weight.

The measured viscosity η of polymer solutions is also dependent on the rate of shear $\dot{\gamma}$ applied. In general, η decreases with increasing $\dot{\gamma}$ (shear thinning). The most probable overall shape for a random coil is not spherical, but somewhat elongated, with typical dimensions in the approximate ratio 12:3:1 along mutually perpendicular axes (249). For isolated coils in dilute solution, shear thinning can be traced to partial alignment of the long axes in the direction of flow, to present a narrower profile to the solvent. At low shear rates this effect becomes negligible, and η tends toward a maximum constant value η_0, which should be used in all calculations of intrinsic viscosity for correlation with primary structure. Whereas shear thinning in dilute polysaccharide solutions is a relatively minor effect, far greater changes are observed in concentrated solution and, as outlined below, have an entirely different molecular origin.

B. Concentrated Solution

Since the volume v occupied by a polymer coil must vary as the cube of its linear dimensions (e.g., as L^3), it follows from Eq. (20) that $v \propto [\eta]M_r$, whereas the number of molecules is proportional to c/M_r. The total volume occupied by all the chains present can therefore be characterized by the dimensionless parameter $c[\eta]$, irrespective of chain length and polymer type. With increasing concentration a point is reached at which the occupied volume equals the total volume of the solution and additional chains can be accommodated only by interpenetration ("entanglement") of the polymer domains (250). The concentration at which this occurs is known as c^*.

As shown in Fig. 31 the transition from dilute to concentrated solution conditions is accompanied by a dramatic change in the concentration dependence of η_0. For aqueous solutions of disordered polysaccharides the following striking generalities are observed (99). (a) The onset of coil entanglement invariably occurs at a value of $c^* \approx 4/[\eta]$. (b) At this point $\eta_0 \approx$ 10 mPa·s (in older units 10 cP). (c) Specific viscosity (fractional increase in viscosity due to the polymer; η_{sp}) varies as $\sim c^{1.4}$ for dilute solutions and $\sim c^{3.3}$ for concentrated solutions. The only exceptions that have been observed are the plant galactomannans locust bean gum and guar gum (99,251,252) and hyaluronate at low pH and high ionic strength (211), which

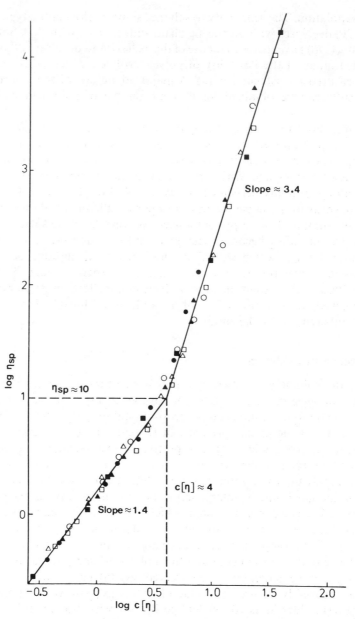

Fig. 31. Concentration dependence (*99*) of zero-shear specific viscosity (η_{sp}) for dextran (○), carboxymethylamylose (●), alginate of high mannuronate content (△), alginate of high guluronate content (▲), λ-carrageenan (□), and hyaluronate (■). From (*99*), with permission.

show a somewhat lower value of c^* and higher concentration dependence above c^*. This behavior is attributed to an increase in effective molecular weight by specific intermolecular associations analogous to those characterized in the condensed phase (Sections III and IV) and in gels (Section VI) but may in principle be of shorter lifetime.

Another striking generality of behavior is evident in the shear rate ($\dot{\gamma}$) dependence of viscosity (η) in concentrated random coil polysaccharide solutions (99). The degree of shear thinning in different solutions can be compared directly by expressing measured viscosities as a fraction of η_0. Similarly, shear rate dependencies can be compared (250) by referring the applied shear rates to the shear rate at which viscosity has decreased to some fixed fraction of η_0. A convenient shear-thinning parameter (99) is $\dot{\gamma}_{0.1}$, the shear rate at which $\eta = \eta_0/10$. When the observed flow curves for disordered polysaccharides of widely differing primary structure and molecular weight are compared in this way (Fig. 32), it is immediately evident that the form of shear thinning is essentially invariant, and thus differences in flow properties can be adequately described by the two parameters η_0 and $\dot{\gamma}_{0.1}$. The response to oscillatory deformation (Section II,F) is closely similar, and for most concentrated random coil polysaccharide solutions (99) the frequency (ω) dependence of dynamic viscosity (η^*) is virtually identical to the shear rate dependence of η.

Interpenetration of polymer coils in concentrated solution gives rise to a dynamic "entangled" network structure (250). At low rates of shear, those entanglements that are disrupted by the imposed deformation are replaced by new interactions between different partners, with no net change in the extent of entanglement and hence no reduction in viscosity. This situation corresponds to the horizontal plateau region in Fig. 32. The onset of shear thinning occurs when the rate of externally imposed movement becomes greater than the rate of formation of new entanglements; thus, the cross-link density of the network is depleted, and viscosity is reduced.

The time t required for reentanglement increases [Eq. (24)] with molecular weight M_r and chain stiffness (250), which for a specific molecular weight is characterized by the magnitude of η_0 at a particular concentration c.

$$t \propto M_r \eta_0 / c \tag{24}$$

Thus, $\dot{\gamma}_{0.1}$, which decreases with increasing t, is proportional to $c/M_r\eta_0$. A useful consequence of this behavior (98) is that measured values of η_0 and $\dot{\gamma}_{0.1}$ provide a rapid estimate of molecular weight, as shown in Eq. (25),

$$\log M \approx \log c - \log \eta_0 - \log \dot{\gamma}_{0.1} + 12 \tag{25}$$

where η_0 is in millipascal seconds, c is in percent weight per volume, and $\dot{\gamma}_{0.1}$ is in reciprocal seconds.

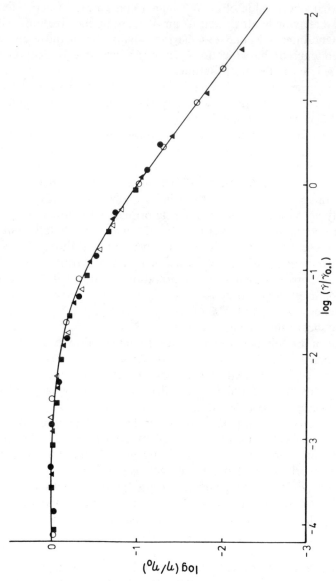

Fig. 32. Shear rate ($\dot{\gamma}$) dependence of viscosity (η) for concentrated solutions of disordered polysaccharides (99). All viscosities are expressed as a fraction of the zero-shear viscosity η_0, and shear rates as a fraction of the shear rate ($\dot{\gamma}_{0.1}$) required to reduce viscosity to $\eta_0/10$. Results shown are for guar galactomannan ($\triangle$), λ-carrageenan ($\blacktriangle$), locust bean gum ($\bullet$), alginate ($\blacksquare$), and hyaluronate ($\bigcirc$). From (99), with permission.

The above generalities apply only to concentrated polysaccharide solutions in which the chains are disordered and network rearrangement in response to flow involves only the making and breaking of entanglements. For conformationally ordered molecules (e.g., xanthan), where intermolecular association in concentrated solution appears to occur by a more specific, cooperative mechanism (Section VI), very different solution properties are observed (*180,253*).

VI. Hydrated Networks

A. Influence of Interruptions

In many native polysaccharides, structurally regular homopolymeric or repeating sequences such as those discussed in Sections III and IV are interrupted by the occurrence of other residues (e.g., in carrageenan and agar) or sequences (e.g., in alginate), by partial derivatization (e.g., esterification of galacturonate residues in pectin), or by side chains (e.g., galactose substituents in galactomannans). Such interruptions, even if present as a comparatively small fraction ($\sim 5-10\%$) of the total number of residues, can profoundly alter bulk properties by terminating intermolecular association through structurally and sterically regular junction zones, with consequent exchange of partners to build up a three-dimensional network (*3,134,254*).

Since, as discussed above, cooperative interchain junctions require a finite, specific chain length for stability, the distribution of interruptions, as well as their total number, is of critical importance. For example, in a system in which an uninterrupted sequence of 15 residues is required for cooperativity, regular insertion of anomalous residues equivalent to 10% of the total chain (i.e., every tenth residue) would completely eliminate interchain association, whereas the same residues clustered together in blocks would have comparatively little effect. It is perhaps significant that in pectin, at least from the fruit sources used commercially for its preparation, a critical sequence length of ~ 15 residues has been demonstrated for calcium-induced interchain association of polygalacturonate (*131,138*), and the $(1 \rightarrow 2)$-linked L-rhamnose residues (*155,255–257*) (which are sterically incompatible with the ordered structure) (*258*) are distributed along the chain backbone in such a way that polygalacturonate block length is essentially uniform at ~ 25 residues (*62,138*), close to the optimum spacing for the formation of a stable crosslinked network.

Pectate chains also occur in nature as the partial methyl ester (*155*). The effect of ester sequence on interchain association can be probed by blockwise (enzymatic) and random (chemical) deesterification of essentially fully esterified material to give a range of samples of comparable total ester content but different patterns of esterification. Calcium chelation (as monitored by CD) (66) increases almost linearly (138) with the proportion of unesterified galacturonate present as polygalacturonate blocks (Fig. 33). Randomly inserted unesterified residues, by contrast, show little binding capacity until present as ~50% of the total chain, with a sudden dramatic increase on slight further deesterification. This behavior evidently reflects the statistical probability of consecutive occurrence of sufficient unesterified residues to satisfy the requirements of cooperativity. Intermolecular network formation, characterized by gel yield stress measurements, shows a parallel dependence on esterification pattern (138). Thus, interchain association in calcium pectate gels can be terminated either by the occurrence of L-rhamnosyl insertions in the polymer backbone or by highly esterified chain sequences, as illustrated schematically in Fig. 34.

The primary mechanism of interchain association on calcium-induced gelation of alginate is dimerization (68) of polyguluronate chain sequences

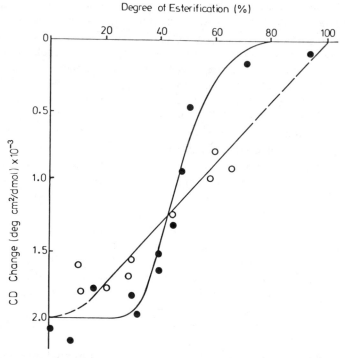

Fig. 33. Calcium-binding capacity, as monitored by CD, for pectin samples with unesterified galacturonate residues distributed randomly (●) and in blocks (○). With permission from (138). *J. Mol. Biol.* **155**, 517 (1982). Copyright by Academic Press, Inc. (London) Ltd.

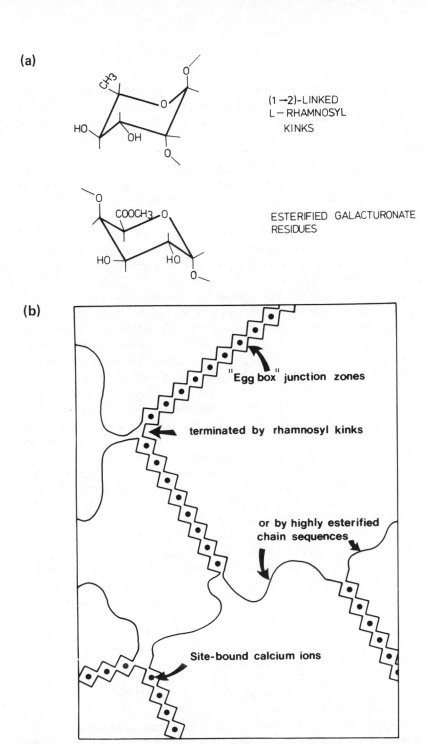

Fig. 34. Calcium pectate gels. (a) Covalent structural features that limit interchain association and solubilize the network. The anomeric configuration of L-rhamnosyl residues present in the polymer backbone is not known; however, neither the α linkage shown nor the alternative β configuration would be compatible (*258*) with incorporation in ordered polygalacturonate junctions. (b) Schematic illustration of network structure. Neutral sugar side chains are not shown, since their distribution and influence on network structure have yet to be established.

(*132*) with interchain chelation of Ca^{2+}. Calcium polyguluronate junctions are closely similar to those of calcium polygalacturonate (Section III,D) but are terminated by the occurrence, in the primary sequence, of D-mannuronate residues present either as homopolymeric blocks (*124,125*) or in mixed sequences with L-guluronate (*126,259*). In contrast to polyguluronate, neither polymannuronate nor alginate mixed sequences show any appreciable capacity for interchain association or strong chelation of Ca^{2+}, as monitored both by CD (*68*) and by ion affinity studies (*132*). In the presence of excess Ca^{2+} uninterrupted polyguluronate chains form a solid precipitate. Alginate chain sequences containing mannuronate residues are therefore essential for the development of an expanded three-dimensional structure, in that they limit the extent of interchain association and promote hydration of the cross-linked network.

Network formation in the carrageenan and agar families of algal polysaccharides is also promoted by departures from regular primary sequence (*254*). In this case interchain association through double helical junction zones is terminated by the occurrence of galactose residues in the normal, unbridged ring form (*260*) in place of the fused anhydride ring structure (Fig. 22a). This has the effect of changing ring geometry from the helix-compatible $^1C_4$ conformation to the normal $^4C_1$ form, thus introducing (*261*) a backbone "kink," which cannot be accommodated within the ordered structure (*29,167*). Both carrageenan- (*262*) and agar-producing algae (*263*) have an enzyme capable of converting residues within the polymer to the anhydride ring form (dekinkase), suggesting that biosynthesis proceeds through a soluble polysaccharide precursor in which anhydride is absent, with subsequent conversion to the helix-forming primary structure. Polysaccharide fractions with the structures and properties appropriate for such a precursor role have, in fact, been isolated (*264*).

Alginate biosynthesis appears to involve analogous enzymatic control of the level of structure-forming chain sequences and hence of *in vivo* physical properties. Thus, alginate extracted from newly synthesized tissue consists predominantly of polymannuronate, with increasing guluronate content on maturation (*64,124,125,265–267*), and enzymatic conversion of D-mannuronate to L-guluronate by C-5 epimerization at the polymer level has been demonstrated for both bacterial (*268*) and algal (*269*) sources. Again, modification of primary structure is reflected in a change in backbone geometry, in this case a conversion from (1→4) diequatorial linkage (in polymannuronate) to (1→4) diaxial in polyguluronate, which radically alters cation-binding behavior, as discussed above (Section III,D).

Modification of backbone structure to inhibit interchain association can also be achieved chemically. Thus, amylose in neutral solution forms gels that, with time, retrograde into an insoluble precipitate. Both gelation and retrogradation can be inhibited by limited chemical modification of the polymer backbone (270), for example, by periodate oxidation. Amylose can be completely solubilized by periodate cleavage of ∼10% of the total number of residues, and substantial retardation of structure formation is observed at ∼5%. The average length of uninterrupted chain sequences decreases rapidly with degree of chemical modification. For example, assuming a random pattern of attack, the fraction of unconverted residues present in sequences of 20 or more would be ∼0.36 for 5% modification, ∼0.12 for 10% modification, and ∼0.01 for 15% conversion. In practice, hemiacetal formation between the aldehyde groups of oxidised residues and hydroxyl groups of the adjacent sugar rings inhibits subsequent attack at these adjacent positions (271), thus further biasing the sequence-length distribution toward short runs of unmodified residues. Recent X-ray fiber diffraction studies (147) suggest that interchain association, at least in the solid state, is through double helices. The observed sensitivity of long-range structure to limited periodate oxidation is therefore consistent with a similar mechanism of interchain association under hydrated conditions, requiring a minimum sequence length for stable junction zone formation comparable to those characterized for other gelling polysaccharides.

Partial substitution of a linear polymer backbone by monosaccharide side chains can also inhibit interchain association, as illustrated by the galactomannan family of plant polysaccharides. The parent β-(1→4)-linked mannan (ivory nut mannan) is insoluble in water, although it may be solubilized by alkali, presumably due to ionization of hydroxy groups (272,273). The known solubilities (273) of relevant oligosaccharides would suggest a critical chain length for stable association of ∼10–15 residues in the mannan and cellulose series. 6-O-Substitution of ∼20% of the backbone residues with α-D-galactose, as found naturally in the galactomannan from *Ceratonia siliqua* (locust bean gum) is sufficient to confer solubility in hot water (272), although a gel network may subsequently develop on standing, or on freezing and thawing (274).

The ability of galactomannans to form a hydrated network structure in both self-association and the mixed interactions outlined in Section VI,B, would suggest a heterogeneity of primary sequence, as in the other gelling polysaccharide systems discussed above. Early studies of the distribution of galactose side chains in a variety of natural galactomannans, by analysis of the products from enzymic hydrolysis (275,276), were interpreted as showing regions of heavy substitution ("hairy" regions) interspersed by stretches of essentially unsubstituted mannan backbone ("smooth" regions), and a subsequent chemical investigation (277) indicated that in the particular case of locust bean gum the length of these unsubstituted stretches might be as great as 85 residues. A block structure of this type could clearly represent one (but not necessarily the only) way of satisfying the need for both stable interchain association and solubilizing structural features in formation of a gel network, as indicated in Fig. 35.

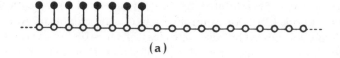

(a)

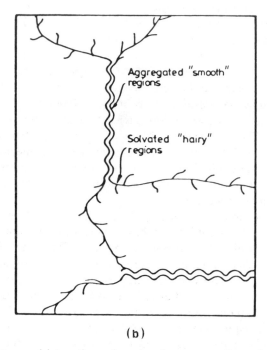

(b)

Fig. 35. Structure and interactions of carob galactomannan (locust bean gum). (a) The reported (*275–277*) "smooth" regions of unsubstituted or sparingly substituted (1→4)-linked β-D-mannopyranose residues (○) and "hairy" regions that are heavily substituted with α-D-galactopyranose residues (●) attached by (1→6) linkages. (b) Schematic representation of proposed network structure in gels formed from carob solutions on freezing and thawing or on standing. From (*274*), with permission.

More recent structural studies using carefully purified enzymes (*278*), however, indicate a more random distribution of side chains, including, in certain galactomannans, regions in which every second mannose residue is substituted (i.e., a trisaccharide repeating structure). Regular alternating sequences of this type have also been proposed from chemical analyses (*277,279*). It would, therefore, seem likely that the differences in patterns of galactose substitution between associating and nonassociating chain segments might be quantitative rather than qualitative, i.e., the model shown in Fig. 35 should be regarded as idealized. It has also been pointed out that, in the preferred twofold conformation of the mannan chain (*116,117*), alternating sequences could have one fully substituted ("hairy") face and one totally unsubstituted ("smooth") face. In terms of this structure, the requirement for sequence heterogeneity in formation of a hydrated network might be satisfied by stable association of "smooth" chain surfaces (*274,278*), solubilized by regions in which the galactose substituents are more randomly distributed.

B. Higher-Order Interactions

Although the adoption of ordered tertiary structures such as carrageenan and agar double helices, xanthan helices, and alginate and pectate "egg box" dimers underlies many of the biologically and technologically important properties of carbohydrate polymers (*3*), it is now becoming evident that higher levels of structural organization are also frequently implicated. This is strikingly illustrated by recent studies (*103,280–282*) of the gelation behavior of carrageenans *in vitro*. As already discussed (Section IV,B) the sol–gel transition of *ι*-carrageenan is accompanied by a random coil–double helix transition, which can be conveniently monitored (*67*) by a discontinuity in the temperature course of optical rotation and NMR relaxation (Fig. 25). A series of different salt forms, however, show essentially the same degree of conformational ordering by optical rotation and NMR, but, despite this, gel formation is extremely sensitive to the nature of the counterion. Thus, optimum gelation is observed with potassium or rubidium, somewhat weaker gels with cesium or ammonium, and no cohesive structure with lithium or sodium, although significant thickening and ultimately precipitation occur at high ionic strength. In the presence of tetramethylammonium as sole counterion there is no evidence of any long-range structure, and solutions remain freely mobile at all temperatures.

The molecular origin of this behavior is deduced from light scattering studies (*93,280–282*) of structurally regular *ι*-carrageenan chain segments (see Section IV,B). Conformational ordering of the tetramethylammonium salt (*282*) is accompanied by an exact doubling of $\bar{M}_w$, entirely consistent with a simple 2 coil–double helix transition. In the presence of potassium ions, however, the molecular weight increase is approximately five- to six-fold (*281*), whereas for sodium ions the value is somewhat in excess of 2 (~ 2.1–2.2), suggesting that helix–helix aggregation occurs to different extents with different cations. For intact (i.e., kinked) *ι*-carrageenan in the nongelling tetramethylammonium salt form, the disorder–order transition is accompanied (*281*) by an approximately 10-fold increase in $\bar{M}_w$. Since (as shown by the results for structurally regular chains) this cannot arise from helix

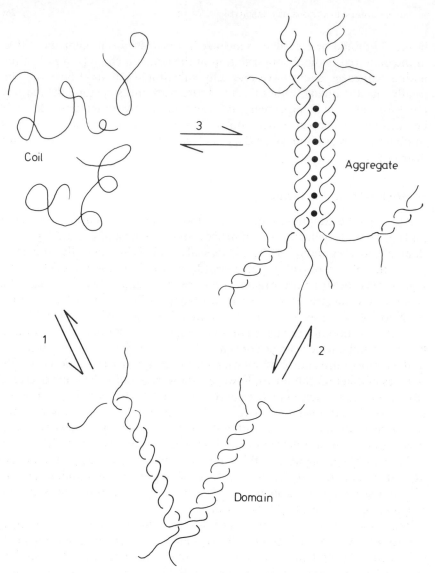

Fig. 36. The domain model of carrageenan gelation. (1) The primary mode of interchain association for ι-carrageenan is the coil–domain transition, which is promoted by cooling and reversed on heating. In an ionic environment that maintains isolation of the individual helices (e.g., Me_4N^+) the domain is the stable ordered state. (2) Below the helix melting point and in the presence of cations (●) that promote gelation (e.g., K^+, Rb^+, Cs^+, NH_4^+, or a high concentration of Na^+), further association occurs by the domain–aggregate transition. Only bound counterions relevant to the model are shown. (3) An alternative mechanism of association for ι-carrageenan in the presence of gel-promoting cations and the sole mechanism for κ-carrageenan appears to be a direct coil–aggregate transition, which occurs at a higher temperature than the coil–domain transition. From (*280*), with permission.

aggregation, we must conclude that there is formation of limited chain clusters or "domains" cross-linked through double helices, with exchange of partners at kink points (*264*). This mechanism is evidently not sufficient for gelation, and further association of domains by helix–helix aggregation (*280,281*) is required to develop a continuous network (Fig. 36). The cation specificity of aggregation suggests that the counterions are site bound between ordered chains, as in alginate and pectate gels (*62,66–68*).

The gelation behavior of the related, but less highly sulfated κ-carrageenan shows a similar dependence on counterion, but conformational ordering (monitored by optical rotation) is now observed only under ionic conditions that promote aggregation in ι-carrageenan (*281*) and is accompanied by very extensive aggregation, as shown by light scattering studies on structurally regular segments. It therefore appears that, under the range of conditions used in this study, the κ-carrageenan double helix is stable only when aggregated. This conclusion is supported by a comparison of the temperature course of conformational change for ι- and κ-carrageenans under different ionic conditions (Fig. 37). As the nonaggregating Me_4N^+ salt, ι-carrageenan shows a single thermally reversible transition with the form expected for a

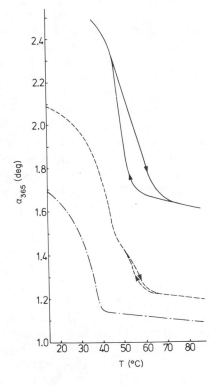

Fig. 37. Thermally induced order–disorder transitions of ι-carrageenan in the Me_4N^+ ($-\cdot-\cdot-$) and K^+ ($---$) salt forms and of K^+ κ-carrageenan (———). Carrageenan concentration (excluding counterion), 1% w/v; cation concentration, 0.12 M. From (*280*), with permission.

simple two-state all-or-none process. κ-Carrageenan in the aggregating K^+ salt form also shows a single transition, but shifted to higher temperature by $\sim 15°C$ and showing substantial hysteresis; this shows that the ordered state is stabilized relative to tetramethylammonium ι-carrageenan and that its formation is now subject to supercooling. The transition for potassium ι-carrageenan is still more complex in that it shows evidence of two conformational transitions, one without hysteresis and centered at the same temperature as the single transition for the nonaggregating Me_4N^+ salt form and the second at a higher temperature, close to that of the κ-carrageenan (K^+) transition, which it further resembles in displaying thermal hysteresis. These two processes are more clearly resolved (103) by scanning calorimetry (Fig. 15) and are attributed to coil–helix and coil–aggregated helix, respectively. Both show second-order transition kinetics (Section II,D), confirming that the rate-limiting step in the coil–aggregate transition involves double helical entities (84,85).

The bulk properties of carrageenan gels are therefore determined by "quaternary" interactions between ordered tertiary structures (double helices), in addition to direct interchain association through double helical junction zones. This two-stage domain mechanism of gelation has so far been demonstrated only for carrageenan. Since it diminishes the obvious topological problems associated with the formation of a continuous network solely through double helical junctions, however, it seems likely that similar structures will emerge for other polysaccharide gels in which the primary mode of interchain association is through multistranded helices (e.g., agar, curdlan, and perhaps amylose). The observed differences in gel strength at constant concentration between helix-forming polysaccharides can then be explained (283) in terms of the different tendencies for domain–domain interaction. Thus, agar, which has a strong tendency to aggregate, gels at low polymer levels, whereas ι-carrageenan, which aggregates only slightly, requires far higher concentrations (~ 10 times) for the development of a continuous network, with κ-carrageenan occupying an intermediate position in both extent of aggregation and concentration requirements for gelation.

Current evidence (Section IV,C) indicates that the primary mechanism of conformational ordering in xanthan is noncovalent association and packing of side chains along the polymer backbone. It is not yet clear whether the resulting "molecular rods" exist in isolation as the fundamental independently moving species in dilute solution and in concentrated solutions under shear or associate into larger, stable entities (by, for example, side-by-side dimerization; Section III,D). There is some evidence from low-angle light scattering (186) and hydrodynamic (185) studies in favor of the latter possibility, but this is far from conclusive and is at variance with results from conventional light scattering (182) and sedimentation velocity (284).

Whatever the detailed nature of the level of organization within the discrete, rodlike entities, however, there seems little doubt that interactions between them are directly responsible for the unusual rheological properties of xanthan solutions (*178–180,253,285*). The most striking of these, which is of considerable technological value, is the capacity of xanthan solutions to hold particles in suspension over a long period, indicating a tenuous but stable intermolecular network that requires a finite force (yield stress) for breakdown, as in the case of true gels. Reestablishment of network properties after shearing, however, is virtually instantaneous, suggesting an activation energy for junction formation only slightly in excess of the mean translational energy of individual rods at ambient temperature. As in the case of permanent gel networks formed by cooperative association of charged polysaccharide chains, yield stress in xanthan solutions increases with increasing ionic strength (*178*), although it remains to be established whether this is due to specific cation chelation to stabilize the structure or to nonspecific suppression of intermolecular electrostatic repulsion. In the presence of urea, network character is destroyed without loss of conformational order (*285*), again consistent with interchain association by noncovalent interactions between the ordered species.

Another consequence of solution yield stress is that, after rupture of the network, the decrease in apparent viscosity with increasing shear rate (shear thinning) is far more pronounced than in the case of random coil polysaccharide solutions (see Fig. 38). Similar extreme shear-thinning behavior is

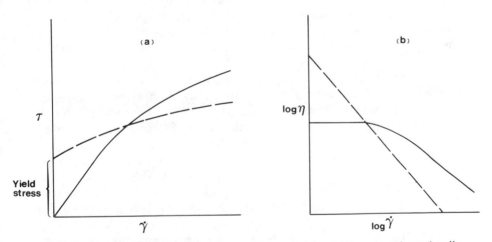

Fig. 38. Effect on flow behavior of quaternary associations between conformationally ordered polysaccharide chains (e.g., xanthan) in solution. (a) Relationship between applied stress (τ) and resulting rate of shear ($\dot{\gamma}$) for random coil polysaccharide solutions (——) and for xanthan (— —). (b) Consequent shear rate dependence of viscosity ($\eta = \tau/\dot{\gamma}$).

observed for the extracellular polysaccharides from certain *Arthrobacter* species (*179,286*), all of which have been shown to be conformationally ordered in solution. The formation of quaternary associations appears, therefore, to be a general property of conformationally rigid polysaccharide tertiary structures. Such behavior is not unexpected since the association of rigid entities involves a smaller loss of conformational entropy compared with flexible chains (*287*), and relatively small increments of interaction energy suffice to stabilize the associations necessary to form the network.

In the presence of excess amounts of monovalent counterions, calcium-induced association of polyguluronate sequences of alginate (*68*) or poly-galacturonate sequences from pectin (*62*) is limited to the formation of chain dimers (see Section III,D). However, in the presence of excess Ca^{2+} as sole or principal counterion, extensive further association occurs (*62,68,102a*), as shown by the precipitation of structurally regular polyguluronate or poly-galacturonate blocks under these conditions. The magnitude of Ca^{2+}-induced CD change (Section III,D) is also greater than in the presence of competing monovalent cations, indicating site binding of calcium ions between dimers in addition to those bound within the dimer structure. In the case of polygalacturonate it has been shown (*102a*) that the relative magnitudes differ by almost exactly a factor of 2. This is consistent (Fig. 14) with interchain association into extensive sheets (with stoichiometric equivalence of Ca^{2+} and uronate) rather than chain dimers, in which Ca^{2+} chelation (and hence CD change) is limited to half the stoichiometric amount. Although such interactions may modify gel properties *in vitro*, they are unlikely to be of importance *in vivo*, at least in the case of algal alginate, since ionic concentrations in seawater (typically, ~ 10 mM Ca^{2+}; 0.5 M Na^+) are close to those used in equilibrium dialysis studies (Sections II,G and III,D) to demonstrate dimerization (*62,68*).

The evident stability of a dimeric structure as a stable intermediate between isolated chains and large aggregates indicates that noncovalent interactions within the dimer are more favorable than those between dimers. One likely contributing factor is the high negative charge density of uncom-plexed polyuronate chains in comparison with that of dimers with an included array of divalent cations. It is also conceivable that optimum chelation of cations within dimers involves some modification of residue geometry, leading to nonequivalence of the interior and exterior faces of the participating chains, or that hydrogen bonding between dimers can occur only at the expense of existing hydrogen bonds within them.

Quaternary associations can also be formed between unlike polysaccha-rides, the most extensively studied being the synergistic interactions of plant galactomannans and related polysaccharides based on a $(1 \rightarrow 4)$-β-D-glycan backbone (e.g., arabinoxylans and glucomannans) with the ordered tertiary structures of carrageenan, agar, xanthan, and the *Arthrobacter* polysaccharides (*57,77,78,174,274*). The capacity of galactomannans to participate in mixed associations shows the same dependence on primary structure as does self-association. Sparingly substituted chains show evidence of more extensive interaction than those with a higher content of galactose substituents, and again there appears to be a specific requirement for unsubstituted backbone sequences or faces. In particular, galactomannans which appear from enzymatic studies to contain a high proportion of regular alternating structure (i.e. one fully substituted and one unsubstituted face in the preferred 2_1 chain conformation) show (*278*) appreciably stronger interaction with xanthan than do other galactomannans with a comparable overall galactose content, but less regular distribution. Polysaccharides with ordered tertiary structures showing a strong tendency toward self-aggregation (e.g. agarose) are more effective in forming mixed associations than are those in which aggregation is limited (e.g., ι-carrageenan, which shows little if any synergism with galactomannans or related molecules). The development of long-range structure through mixed quaternary junctions may range (*274*) from barely detectable enhancement of solution viscosity to the formation of strong, stable gels at concentrations far below those required for the gelation of either component alone.

Such higher-level associations *in vitro* may represent renaturation of functionally relevant interactions *in vivo*, as, for example, in the structure of plant tissues. A biological role for interactions of β-$(1\rightarrow4)$-linked glucans and related chains with the ordered tertiary structure of xanthan (Fig. 39) has been suggested in host–pathogen recognition and adhesion of *Xanthomonas* bacteria within the plant vascular system (*77*). Analogous associations may

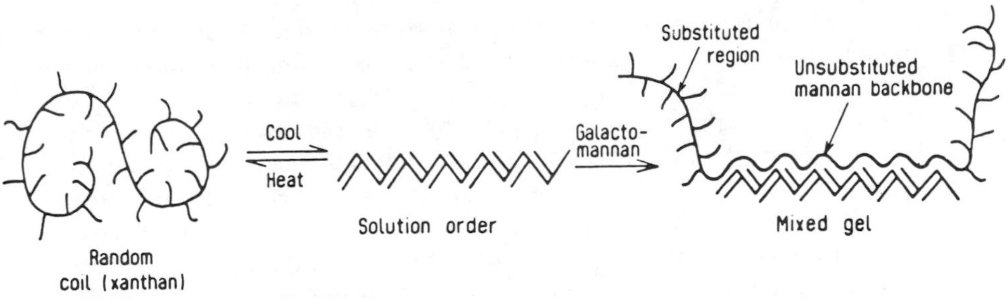

Fig. 39. Schematic representation of the interaction of xanthan with galactomannans and related plant polysaccharides. The stoichiometry of the interaction has not yet been characterized, and it is not known whether the associated form shown here exists as such or as a subunit of some larger aggregate. Analogous models have been proposed for interactions of galactomannans with the double helix conformations of agarose and κ-carrageenan. With permission from (*77*), *J. Mol. Biol.* **110**, 1 (1977). Copyright by Academic Press, Inc. (London) Ltd.

also be of importance in symbiotic interactions of *Arthrobacter* species with rhizoid filaments of plant roots (*78*).

VII. Mixed Interactions

A. Polysaccharide–Protein Interactions

Interaction effects between carbohydrate and protein chains occur widely in both biological systems and industrial applications. This behavior can be of the type that is general for polymer mixtures, which need not depend on specific binding effects, and, as such, has been extensively reviewed (*288–291*). Experimentally it can be observed in solution by thermodynamic measurements or, if the interaction is more marked, as phase separation.

Since entropy of mixing (which favors single-phase systems) is dependent on the number of molecules, it becomes largely insignificant for polymers. The behavior of mixed polymer solutions is therefore determined predominantly by energies of interaction between chains, even when they are small, as in the case of fleeting contacts between disordered chain segments rather than the much stronger cooperative interactions of ordered species discussed in previous sections.

If the interaction between like segments of chains of each type is more favorable than that between segments of the two different chain types in solution, the mixture may separate into two distinct phases with properties resembling immiscible liquids. This is commonly known as polymer incompatibility and can be used to create two immiscible aqueous phases for the separation of particles by partition processes. Statistically, the number of segment–segment contacts increases with both molecular weight and chain flexibility; thus dextran is commonly used as the basis for one of the phases (*291*).

It is found (*292*) that the osmotic pressure of a solution of incompatible polymers exceeds the value expected from the sums of the osmotic pressures of the components. This is interpreted as an exclusion of each polymer from the domain of the other by mutual repulsion, which increases the effective concentration, i.e., the activity coefficient of each. Even though actual phase separation has not been seen in mixtures such as hyaluronic acid and bovine serum albumin that are thought to be good models for natural polysaccharide–protein systems in the extracellular regions of animal tissue, they do show excess osmotic pressures (*293,294*). Another consequence of the changed activity coefficients is that the binding of each polymer to other tissue components may be changed. For example, the apparent equilibrium constants for enzyme–substrate and enzyme–inhibitor binding may be shifted in the presence of an incompatible polysaccharide (295). Moreover, when a flexible polysaccharide is present in sufficient concentration, the steric and repulsive effects of segment–segment contacts may hinder the translational diffusion of globular proteins (but not their rotation) (296–298) while markedly increasing the diffusion of rodlike proteins by restricting and directing their travel in the direction of their long axis (299).

If the interactions between unlike polymer segments are dominant, the two polymers may associate into a liquid- or gel-like phase by a process known as complex coacervation. This is particularly true after the electrostatic binding of proteins and polysaccharides of opposite charges, and it is important in much of today's encapsulation technology.

The molecular features governing electrostatic interactions between polysaccharides and proteins have been investigated in model studies using cationic homopeptides and polyanionic carbohydrate chains. These interactions can induce the conversion of polypeptide sequences from random coil to α-helical chain conformation, which is monitored by CD (300–304). From the stoichiometry and the sharp reversal on heating, it appears that cationic helices are stabilized by regular ionic bonding to the polysaccharides. Relative binding affinities within both polypeptide and polysaccharide series can be assessed from the characteristics of helix formation and melting in the mixed systems. For example, increased exposure of positive charges on the polypeptides facilities the interaction [poly(L-arginine) > poly(L-lysine) > poly(L-ornithine)]. Similarly, increased charge density favors the interaction within sulfated polysaccharides (heparin > chondroitin sulfate > hyaluronic acid), as do stereochemical effects arising from iduronic acid residues (dermatan sulfate > chondroitin 6-sulfate > chondroitin 4-sulfate).

A more specific, but as yet ill-defined interaction with the protein cofactor plasma antithrombin appears to be implicated in the anticoagulant activity of heparin (196). Antithrombin is a glycoprotein of approximate molecular weight 55,000 (305,306), which inactivates virtually all of the serine proteases involved in blood clotting by complexing with them (307–311). In the presence of heparin, complex formation is greatly accelerated, and it has been suggested that this is due to a change in conformation induced by binding of heparin to lysine residues on a single site within the antithrombin molecule (311–313). A loss of anticoagulant activity has been observed on chemical modification of either the carboxyl or N-sulfate groups of heparin (314–316). Further evidence for the specificity of interaction comes from the much lower binding affinity of heparan sulfate (317). Moreover, in studies using a highly purified heparin preparation, only about one-third of the chains were found to bind strongly to antithrombin (318–320). This clearly suggests substantial heterogeneity in residue sequence within and/or between heparin molecules and further indicates a specific sequence requirement for antithrombin binding. Recent structural studies (321,322) show that the antithrombin binding sequence contains a high proportion of sugar residues which are present only as minor overall constituents of heparin. The binding site appears to consist of an octasaccharide sequence, which includes one residue of a unique 3-sulfated glucosamine derivative.

The interactions of carrageenans with milk proteins (323–327) also show evidence of specificity and are of considerable technological importance (328), but they are poorly understood. Casein from bovine milk contains three separate proteins, α_{SI}-, β-, and κ-casein (329). Both α_{SI}- and β-casein are insoluble at the Ca^{2+} level found in milk but are apparently stabilized by complexing with κ-casein, which is soluble over a wide range of calcium ion concentrations. Comparable stabilization of α_{SI}- casein to Ca^{2+}-induced precipitation can be achieved using λ-carrageenan in place of κ-casein, whereas κ- and ι-carrageenan are approximately twice as effective (324). Investigations using carrageenans with different levels of 3,6-anhydrogalactose and different patterns of sulfation suggest that the anhydro ring (and hence helix-forming capacity) is not essential for the interaction but that sulfation at C-6 is antagonistic to stabilization (325). Other charged polysaccharides, including some that are also sulfated, show little if any effect.

In the absence of Ca^{2+} ions both α_{SI}- and β-casein are unaffected (326,327) by carrageenans, but κ-casein forms stable complexes with or without Ca^{2+}. Whether this interaction leads to gel formation appears to depend largely on the gelling capacity of the carrageenan alone, although lower overall carrageenan concentrations are required. Since both carrageenan and κ-casein (isoelectric point pH 4.4) are negatively charged, the formation of a strong mixed complex is surprising. One proposed explanation (327) is that a specific region of the κ-casein molecule (between amino acid residues 97 and 112) carries a net positive charge and interacts electrostatically with the carrageenan polyanion. However, in view of the marked specificity for carrageenan in comparison with other anionic polysaccharides (325) and the relative lack of selectivity among carrageenans of different charge density, further evaluation of the molecular origin of the interaction, including the possibility of stereospecificity, seems to be required.

B. Inclusion Complexes

One of the most familiar properties of amylose is the development of an intense blue color in the presence of iodine. This is due to the adoption of a sixfold hollow helical conformation (V-amylose) (143,144,330) within which an array of iodine–iodide species is accommodated along the helix axis. Interactions with each other and with the ordered polymer modify the electronic transitions of the included molecules to give rise both to a change in UV absorption (i.e., the blue color) and to induced optical activity (59,331–334). The helix is left-handed and is stabilized both by the presence of the "guest" molecules and by hydrogen bonding (Fig. 2) between O-2 and O-3 of neighboring residues and between O-2 and O-6 of residues adjacent on the helix surface but six apart in the primary sequence.

A wide variety of organic molecules can be complexed, provided that they are hydrophobic and have a suitable diameter to fit within the helix. Examples are phenols, aryl halides, tert-butanol, cyclohexane, and other aliphatic molecules. The D-glucopyranose units are so arranged that the axial C—H bonds on C-3 and C-5 as well as one of the H atoms on each C-6 point inward to give some hydrophobic character to the lining of the void—hence, the preference for apolar guest molecules. The bridge oxygens of each connecting linkage are also part of this lining, so that their unshared electron pair orbitals point inward.

The formation of V-amylose complexes is accompanied by a large reduction in D-line optical rotation, which can be correlated (Section II,B) with chain geometry. In the particular case of amylose–iodine, interpretation is complicated by the induced optical activity of the iodine. The iodine–iodide bands, however, can be observed directly in CD, and their contribution to overall ORD calculated (59) by the Kronig–Kramers transform (Section II,B). Hence, by difference, the underlying optical activity of the polymer backbone can be unmasked and is found to be in close quantitative agreement with results for other included species, such as butanol, in which no such spectroscopic difficulties arise.

Like amylose, the Shardinger dextrins are capable of forming complexes with a wide range of nonpolar molecules (335). For example, cyclohexaamylose in its crystalline inclusion complexes assumes a symmetric toroidal form stabilized by hydrogen bonds between O-2 and O-3 of each of the six residues (336), in contrast to the "uncomplexed" ring in the crystalline hexahydrate, in which two water molecules occupy the cavity and one of the glucose residues is rotated about its glycosidic bonds to fill the void (337,338). Solution studies (28) also point to a conformation change when the complex is formed, since aqueous solutions of cyclohexaamylose show shifts in optical rotation when small amounts of alkanoic acids are added (335,339). The magnitude of the shifts increases with the length of the paraffin chain and levels off at high alkyl chain length at a value that is in good agreement with that calculated for the symmetric toroid. Interpretation of the optical rotation before complexation suggests (21) that one residue is twisted about its glycosidic linkages relative to the symmetric form, to adopt a conformation close to an alternative minimum energy conformation for maltose (Section I,C); this twisted residue then fills the cavity.

Biosynthesis of fatty acids by a multienzyme complex from *Mycobacterium phlei* is strongly influenced by carbohydrate chains from the same organism (*340*), and again the formation of polysaccharide–lipid inclusion complexes appears to be implicated. The polysaccharide chains are of two types, based on glucose and mannose, respectively (*341–345*). The glucan contains a linear (1→4)-linked chain of 17 α-D-glucose residues, with a single residue side branch and *O*-methyl, *O*-succinyl, and other *O*-acyl substituents. Although several forms have been identified, they appear to differ only in the level of O-succinylation. The mannan is also predominantly α-(1→4)-linked, and most residues have an *O*-methyl substituent at C-3.

The synthesis of fatty acids by the enzyme systems of *M. phlei* proceeds from the usual substrates, namely, acetyl-CoA and malonyl-CoA. The first stage is growth of the acyl chain to a length of 16 or 18 carbon atoms, after which the chain either is discharged from the enzyme by the action of palmityl transacylase to form palmityl-CoA or, as enzyme-bound palmitate, is elongated to as many as 24 carbon atoms. The course of this synthesis is influenced by the presence of heat-stable substances from extracts of *M. phlei* that are now known (*340*) to correspond to the two polysaccharides. It was found that (a) the synthesis of palmitate is stimulated (*346*) by the addition of either polysaccharide, especially at lower concentrations of acetyl-CoA; (b) this stimulation is accompanied (*346*) by a dramatic lowering of the apparent K_m for acetyl-CoA and a similar but less striking effect on the apparent K_m for malonyl-CoA; (c) among the requirements for fast elongation of palmitate are a high concentration of palmityl-CoA and the addition of polysaccharide (*347*); and (d) the mixture of fatty acid products is shifted in composition by the addition of polysaccharides toward products of shorter chain length (*348*). All these effects were eventually traced to capacity of each of the polysaccharides to form a 1:1 complex with palmityl-CoA and other longer-chain acyl-CoA derivatives (*349*). The two polysaccharides differed little in their influence on the biosynthetic system or in the properties of their complexes, except that (a) the mannan was significantly more effective than the glucan (*346*) in stimulating fatty acid synthesis at low levels of acetyl-CoA and (b) the capacity of mannan to stimulate fatty acid synthesis and to complex with long-chain acyl-CoA derivatives is not affected by further O-methylation, whereas both properties of the glucan are abolished by such treatment (*349*).

Analogies between the two polysaccharides are reinforced by the observation (*346*) that the glucan retains its biological and complexing activities after de-O-acylation followed by digestion with α-amylase to remove a trisaccharide or (less frequently) a tetrasaccharide from the nonreducing terminal. Thus, it is reasonable to propose (*346*) that the important properties are to be traced to the sequence that exists in both polysaccharides of

10 consecutive α-(1→4)-linked mono-O-methylhexose residues. A mechanism by which this sequence could bind palmityl-CoA and its analogues is suggested by a consideration of the conformation and binding behavior of amylose. When the chain of 10 consecutive 6-O-methyl-D-glucose or 3-O-methyl-D-mannose residues is arranged in the V-amylose conformation, the methyl groups project inward to enhance the hydrophobic character of the cavity. This helix matches the palmitate chain closely in length and diameter, and a very plausible inclusion complex, stabilized by hydrophobic interactions, therefore seems possible (*350*).

Recent studies in this laboratory (*270*) and elsewhere (*351,352*) suggest that a similar lipid inclusion complex may exist in native starch granules. On thermal gelatinization of starch a DSC endotherm is observed at $\sim 100^{\circ}$C, which is lost on removal of lipid by methanol extraction but reappears on addition of sodium palmitate to defatted granules. In contrast to the major gelatinization endotherm at $\sim 70^{\circ}$C (*353*), this transition is fully reversible (*270*) and shows substantial thermal hysteresis ($T_m \approx 77^{\circ}$C on cooling). Closely similar thermal behavior, including hysteresis, is observed for aqueous slurries of palmitate and amylose.

As in the case of smaller guest molecules, a substantial reduction in optical rotation is observed (*270*) on formation of complexes between amylose and the salts of fatty acids. The magnitude of this change increases with both the chain length of the fatty acid and its concentration relative to that of amylose, toward a minimum limiting optical rotation value of $[\alpha]_D \approx 100$. This differs considerably from the value calculated (Section II,B) for the sixfold, left-handed V-amylose helix ($[\alpha]_D = 157$) but is in good agreement with calculations for the corresponding right-handed chain geometry ($[\alpha]_D = 104$). This conformation would offer the same possibilities for intramolecular hydrogen bonding and stabilization by hydrophobic interactions between polymer and included species. The proposal of an inversion of helix sense from that commonly accepted from X-ray studies is, however, sufficiently surprising to require extensive investigation.

References

1. E. L. Eliel, N. L. Allinger, S. J. Angyal, and G. A. Morrison, "Conformational Analysis." Wiley, New York, 1965.
2. J. F. Stoddart, "Stereochemisty of Carbohydrates." Wiley, New York, 1971.
3. D. A. Rees, "Polysaccharide Shapes," Outline Studies in Biology. Chapman & Hall, London, 1977.
4. S. J. Angyal, *Aust. J. Chem.* **21**, 2737–2746 (1968).
5. S. J. Angyal, *Angew. Chem., Int. Ed. Engl.* **8**, 157–226 (1969).
6. D. A. Rees and P. J. C. Smith, *J. Chem. Soc., Perkin Trans. 2* pp. 830–835 (1975).

7. L. G. Dunfield and S. G. Whittington, *J. Chem. Soc., Perkin Trans. 2* pp. 654–658 (1977).
8. D. A. Rees, *MTP Int. Rev. Sci.: Org. Chem., Ser. One* **7**, 251–283 (1973).
9. P. J. Flory, "Statistical Mechanics of Chain Molecules." Wiley, New York, 1969.
10. G. N. Ramachandran and V. Sasisekharan, *Adv. Protein Chem.* **23**, 283–438 (1968).
11. J. E. Williams, P. J. Stang, and P. von R. Schleyer, *Annu. Rev. Phys. Chem.* **19**, 531–558 (1968).
12. H. A. Scheraga, *Adv. Phys. Org. Chem.* **6**, 103–184 (1968).
13. H. A. Scheraga, *Chem. Rev.* **71**, 195–217 (1971).
14. W. J. Settineri and R. H. Marchessault, *J. Polym. Sci., Part C* **11**, 253–264 (1965).
15. A. Sarko and R. H. Marchessault, *J. Am. Chem. Soc.* **89**, 6454–6462 (1967).
16. N. S. Anderson, J. W. Campbell, M. M. Harding, D. A. Rees, and J. W. B. Samuel, *J. Mol. Biol.* **45**, 85–99 (1969).
17. S. Melberg and K. Rasmussen, *Carbohydr. Res.* **69**, 27–38 (1979); **71**, 25–34 (1979); **78**, 215–224 (1980).
18. B. Pullman, *Int. J. Quantum Chem., Symp.* **4**, 319–346 (1971).
19. A. Saran and G. Govil, *Indian J. Chem.* **9**, 1095–1097 (1971).
20. M. Giacomini, B. Pullman, and B. Maigret, *Theor. Chim. Acta* **19**, 347–364 (1970).
21. D. A. Rees and P. J. C. Smith, *J. Chem. Soc., Perkin Trans. 2* pp. 836–840 (1975).
22. D. A. Rees and D. Thom, *J. Chem. Soc., Perkin Trans. 2* pp. 191–201 (1977).
23. I. Tanaka, N. Tanaka, T. Ashida, and M. Kakudo, *Acta Crystallogr., Sect. B* **B32**, 155–160 (1976).
24. B. Casu, M. Reggiani, G. G. Gallo, and A. Vigevani, *Tetrahedron* **24**, 803–821 (1968).
25. B. Casu, M. Reggiani, G. G. Gallo, and A. Vigevani, *Carbohydr. Res.* **12**, 157–170 (1970).
26. M. St-Jaques, P. R. Sundararajan, K. J. Taylor, and R. H. Marchessault, *J. Am. Chem. Soc.* **98**, 4386–4391 (1976).
27. J. L. Neal and D. A. I. Goring, *Can. J. Chem.* **48**, 3745–3747 (1970).
28. D. A. Rees, *J. Chem. Soc. B* pp. 877–884 (1970).
29. D. A. Rees and W. E. Scott, *J. Chem. Soc. B* pp. 469–479 (1971).
30. R. U. Lemieux, K. Bock, L. T. J. Delbaere, S. Koto, and V. S. Rao, *Can. J. Chem.* **58**, 631–653 (1980).
31. R. U. Lemieux and S. Koto, *Tetrahedron* **30**, 1933–1944 (1974).
32. D. A. Rees, *MTP Int. Rev. Sci.: Biochemistry Ser. One* **5**, 1–42 (1975).
33. S. Arnott, *in* "Glycosaminoglycan Assemblies in the Extracellular Matrix" (S. Arnott and D. A. Rees, eds.). Humana Press, New York, 1982 (in press).
34. L. Velluz, M. Legrand, and M. Grosjean, "Optical Circular Dichroism: Principles, Measurements and Applications." Academic Press, New York, 1965.
35. C. Djerassi, "Optical Rotatory Dispersion: Applications to Organic Chemistry." McGraw-Hill, New York, 1960.
36. D. W. Sears and S. Beychok, *in* "Physical Principles and Techniques of Protein Chemistry" (S. J. Leach, ed.), Part C, pp. 445–593. Academic Press, New York, 1973.
37. E. S. Pysh, *Annu. Rev. Biophys. Bioeng.* **5**, 63–75 (1976).
38. R. G. Nelson and W. C. Johnson, Jr., *J. Am. Chem. Soc.* **94**, 3343–3345 (1972); **98**, 4290–4295, 4296–4301 (1976).
39. J. S. Balcerski, E. S. Stevens (formerly Pysh), G. C. Chen, and J. T. Yang, *J. Am. Chem. Soc.* **97**, 6274–6275 (1975).
40. J. N. Liang, E. S. Stevens, E. R. Morris, and D. A. Rees, *Biopolymers* **18**, 327–333 (1979).
41. L. A. Buffington, E. S. Stevens, E. R. Morris, and D. A. Rees, *Int. J. Biol. Macromol.* **2**, 199–203 (1980).
42. J. N. Liang, E. S. Stevens, S. A. Frangou, E. R. Morris, and D. A. Rees, *Int. J. Biol. Macromol.* **2**, 204–208 (1980).

43. J. N. Liang and E. S. Stevens, in preparation.
44. D. G. Lewis and W. C. Johnson, Jr., *Biopolymers* **17**, 1439–1449 (1978).
45. A. J. Stipanovic and E. S. Stevens, *Int. J. Biol. Macromol.* **2**, 209–212 (1980).
46. E. R. Morris and S. A. Frangou, *in* "Techniques in Carbohydrate Metabolism" (D. H. Northcote, ed.), B308. Elsevier, Amsterdam, 1981.
47. C. S. Hudson, *J. Am. Chem. Soc.* **31**, 66–86 (1909); **47**, 268–280 (1925); **52**, 1680–1700, 1707–1718 (1930).
48. W. Kauzmann, F. B. Clough, and I. Tobias, *Tetrahedron* **13**, 57–105 (1961).
49. D. H. Whiffen, *Chem. Ind. (London)* pp. 964–968 (1965).
50. J. H. Brewster, *J. Am. Chem. Soc.* **81**, 5475–5483, 5483–5493, 5493–5500 (1959).
51. R. U. Lemieux and A. A. Pavia, *Can. J. Chem.* **46**, 1453–1455 (1968).
52. R. U. Lemieux and J. C. Martin, *Carbohydr. Res.* **13**, 139–161 (1970).
53. D. A. Rees, W. E. Scott, and F. B. Williamson, *Nature (London)* **227**, 390–392 (1970).
54. J. Blackwell, A. Sarko, and R. H. Marchessault, *J. Mol. Biol.* **42**, 379–383 (1969).
55. D. A. Rees, *J. Chem. Soc. B* pp. 217–226 (1969).
56. A. A. McKinnon, D. A. Rees, and F. B. Williamson, *Chem. Commun.* pp. 701–702 (1969).
57. S. Arnott, A. Fulmer, W. E. Scott, I. C. M. Dea, R. Moorhouse, and D. A. Rees, *J. Mol. Biol.* **90**, 269–284 (1974).
58. E. A. Balazs, A. A. McKinnon, E. R. Morris, D. A. Rees, and E. J. Welsh, *J. Chem. Soc., Chem. Commun.* pp. 44–45 (1977).
59. S. A. Frangou, E. R. Morris, D. A. Rees, and E. J. Welsh, *Int. J. Biol. Macromol.* **2**, 178–180 (1980).
60. E. R. Morris, D. A. Rees, G. R. Sanderson, and D. Thom, *J. Chem. Soc., Perkin Trans. 2* pp. 1418–1425 (1975).
61. E. R. Morris, D. A. Rees, and D. Thom, *Carbohydr. Res.* **81**, 305–314 (1980).
62. M. J. Gidley, E. R. Morris, E. J. Murray, D. A. Powell, and D. A. Rees, *J. Chem. Soc., Chem. Commun.* pp. 990–992 (1979).
63. I. Listowsky, S. Englard, and G. Avigad, *Trans. N.Y. Acad. Sci.* [2] **34**, 218–226 (1972).
64. B. Stockton, L. V. Evans, E. R. Morris, and D. A. Rees, *Int. J. Biol. Macromol.* **2**, 176–178 (1980).
65. E. R. Morris, D. A. Rees, and D. Thom, *J. Chem. Soc., Chem. Commun.* pp. 245–246 (1973).
66. G. T. Grant, E. R. Morris, D. A. Rees, P. J. C. Smith, and D. Thom, *FEBS Lett.* **32**, 195–198 (1973).
67. T. A. Bryce, A. A. McKinnon, E. R. Morris, D. A. Rees, and D. Thom, *Faraday Discuss. Chem. Soc.* **57**, 221–229 (1974).
68. E. R. Morris, D. A. Rees, D. Thom, and J. Boyd, *Carbohydr. Res.* **66**, 145–154 (1978).
69. D. Thom, G. T. Grant, E. R. Morris, and D. A. Rees, *Carbohydr. Res.* **100**, 29–42 (1982).
70. E. A. Kabat, K. O. Lloyd, and S. Beychok, *Biochemistry* **8**, 747–756 (1969).
71. P. L. Coduti, E. C. Gordon, and C. A. Bush, *Anal. Biochem.* **78**, 9–20 (1977).
72. C. A. Bush, *in* "Excited States in Organic Chemistry and Biochemistry" (B. Pullman and N. Goldblum, eds.), pp. 209–220. Reidel Publ., Dordrecht, Netherlands, 1977.
73. L. A. Buffington, E. S. Pysh, B. Chakrabarti, and E. A. Balazs, *J. Am. Chem. Soc.* **99**, 1730–1734 (1974).
74. F. Franks and E. R. Morris, *Biochim. Biophys. Acta* **540**, 346–356 (1978).
75. L. D. Melton, E. R. Morris, D. A. Rees, and D. Thom, *J. Chem. Soc., Perkin Trans. 2* pp. 10–17 (1979).
76. A. Darke, E. G. Finer, R. Moorhouse, and D. A. Rees, *J. Mol. Biol.* **99**, 477–486 (1975).
77. E. R. Morris, D. A. Rees, G. Young, M. D. Walkinshaw, and A. Darke, *J. Mol. Biol.* **110**, 1–16 (1977).

78. A. Darke, E. R. Morris, D. A. Rees, and E. J. Welsh, *Carbohydr. Res.* **66**. 133–144 (1978).
79. G. Kotowycz and R. U. Lemieux, *Chem. Rev.* **73**, 669–698 (1973).
80. D. R. Bundle and R. U. Lemieux, *Methods Carbohydr. Chem.* **7**, 79–86 (1976).
81. J. A. Schwarcz and A. S. Perlin, *Can. J. Chem.* **50**, 3667–3676 (1972).
82. K. Bock, I. Lundt, and C. Pedersen, *Tetrahedron Lett.* No. 13, pp. 1037–1040 (1973).
83. B. Casu, M. Reggiani, G. G. Gallo, and A. Vigevani, *Tetrahedron* **22**, 3061–3083 (1966).
84. I. T. Norton, D. M. Goodall, E. R. Morris, and D. A. Rees, *J. Chem. Soc., Chem. Commun.* pp. 515–516 (1978).
85. I. T. Norton, D. M. Goodall, E. R. Morris, and D. A. Rees, *J. Chem. Soc., Chem. Commun.* pp. 988–990 (1979).
86. I. T. Norton, D. M. Goodall, E. R. Morris, and D. A. Rees, *J. Chem. Soc., Chem. Commun.* pp. 545–547 (1980).
87. D. M. Goodall and M. T. Cross, *Rev. Sci. Instrum.* **46**, 391–397 (1975).
88. E. R. Morris, *NATO Adv. Study Inst. Ser., Ser. C* **50**, 379–388 (1979).
89. K. J. Laidler, "Chemical Kinetics," 2nd ed. McGraw-Hill, New York, 1965.
90. D. Margerison and G. C. East, "Introduction to Polymer Chemistry," Pergamon, Oxford, 1967.
91. J. T. Park and M. J. Johnson, *J. Biol. Chem.* **181**, 149–151 (1949).
92. G. N. Richards and W. J. Whelan, *Carbohydr. Res.* **27**, 185–191 (1973).
93. R. A. Jones, E. J. Staples, and A. Penman, *J. Chem. Soc., Perkin Trans. 2* pp. 1608–1612 (1973).
94. J. G. Southwick, M. E. McDonnell, A. M. Jamieson, and J. Blackwell, *Macromolecules* **12**, 305–311 (1979).
95. Y. Einaga, T. Mitani, J. Hashizume, and H. Fujita, *Polym. J.* **11**, 565–574 (1979).
96. J. D. Ferry, "Viscoelastic Properties of Polymers," 3rd ed. Wiley, New York, 1980.
97. D. A. Gibbs, E. W. Merrill, K. A. Smith, and E. A. Balazs, *Biopolymers* **6**, 777–791 (1968).
98. E. R. Morris and S. B. Ross-Murphy, *in* "Techniques in Carbohydrate Metabolism" (D. H. Northcote, ed.), B310. Elsevier, Amsterdam, 1981.
99. E. R. Morris, A. N. Cutler, S. B. Ross-Murphy, D. A. Rees, and J. Price, *Carbohydr. Polym.* **1**, 5–21 (1981).
100. W. P. Cox and E. H. Merz, *J. Polym. Sci.* **28**, 619–622 (1958).
101. E. R. Morris, D. A. Rees, G. Robinson, and G. A. Young, *J. Mol. Biol.* **138**, 363–374 (1980).
102. E. J. Welsh, D. A. Rees, E. R. Morris, and J. K. Madden, *J. Mol. Biol.* **138**, 375–382 (1980).
102a. E. R. Morris, D. A. Powell, M. J. Gidley, and D. A. Rees, *J. Mol. Biol.* **155**, 507–516 (1982).
103. E. R. Morris, D. A. Rees, I. T. Norton, and D. M. Goodall, *Carbohydr. Res.* **80**, 317–323 (1980).
104. G. N. Ramachandran, C. Ramakrishnan, and V. Sasisekharan, *in* "Aspects of Protein Structure" (G. N. Ramachandran, ed.), pp. 121–134. Academic Press, New York, 1963.
105. B. K. Sathyanarayana and V. S. R. Rao, *Biopolymers* **10**, 1605–1615 (1971).
106. R. H. Marchessault and A. Sarko, *Adv. Carbohydr. Chem. Biochem.* **22**, 421–482 (1967).
107. P. R. Sundararajan and R. H. Marchessault, *Can. J. Chem.* **50**, 792–794 (1972).
108. J. Blackwell, *in* "Biomolecular Structure, Conformation, Function and Evolution" (S. Ramachandran, ed.), Vol. 1, pp. 523–535. Pergamon, Oxford, 1979.
109. K. Esau, "Plant Anatomy," 2nd ed. Wiley, New York, 1965.
110. K. H. Gardner and J. Blackwell, *Biopolymers* **13**, 1975–2001 (1974).
111. K. H. Gardner and J. Blackwell, *Biochim. Biophys. Acta* **343**, 232–237 (1974).
112. L. G. Fowle, J. F. W. Juritz, and A. M. Stephen, *S. Afr. Med. J.* **44**, 152 (1970).
113. I. A. Nieduszynski and R. H. Marchessault, *Biopolymers* **11**, 1335–1344 (1972).
114. I. C. M. Dea, D. A. Rees, R. J. Beveridge, and G. N. Richards, *Carbohydr. Res.* **29**, 363–372 (1973).

115. S. M. Gabbay, P. R. Sundararajan, and R. H. Marchessault, *Biopolymers* **11**, 79–94 (1972).
116. E. Frei and R. D. Preston, *Proc. R. Soc. London, Ser. B* **169**, 127–145 (1968).
117. P. Zugenmaier, *Biopolymers* **13**, 1127–1139 (1974).
118. H. Bittiger and R. H. Marchessault, *Carbohydr. Res.* **18**, 469–470 (1971).
119. K. M. Rudall, *J. Polym. Sci., Part C* **28**, 83–102 (1969).
120. E. H. Oldmixon, S. Glauser, and M. L. Higgins, *Biopolymers* **13**, 2037–2060 (1974).
121. H. H. M. Balyuzi, D. A. Reaveley, and R. E. Burge, *Nature (London), New Biol.* **235**, 252–253 (1972).
122. E. Hirst and D. A. Rees, *J. Chem. Soc.* pp. 1182–1187 (1965).
123. D. A. Rees and J. W. B. Samuel, *J. Chem. Soc. C* pp. 2295–2298 (1967).
124. A. Haug, B. Larsen, and O. Smidsrød, *Acta Chem. Scand.* **20**, 183–190 (1966).
125. A. Haug, B. Larsen, and O. Smidsrød, *Acta Chem. Scand.* **21**, 691–704 (1967).
126. J. Boyd and J. R. Turvey, *Carbohydr. Res.* **66**, 187–194 (1978).
127. E. D. T. Atkins, I. A. Nieduszynski, W. Mackie, K. D. Parker, and E. E. Smolko, *Biopolymers* **12**, 1865–1878 (1973).
128. W. Mackie, *Biochem. J.* **125**, 89P (1971).
129. E. D. T. Atkins, I. A. Nieduszynski, W. Mackie, K. D. Parker, and E. E. Smolko, *Biopolymers* **12**, 1879–1887 (1973).
130. R. Kohn and B. Larsen, *Acta Chem. Scand.* **26**, 2455–2468 (1972).
131. R. Kohn, *Pure Appl. Chem.* **42**, 371–397 (1975).
132. O. Smidsrød, *Faraday Discuss. Chem. Soc.* **57**, 263–274 (1974).
133. I. Listowsky, S. Englard, and G. Avigad, *Biochemistry* **8**, 1781–1785 (1969).
134. E. R. Morris, D. A. Rees, D. Thom, and E. J. Welsh, *J. Supramol. Struct.* **6**, 259–274 (1977).
135. K. J. Palmer and M. B. Hartzog, *J. Am. Chem. Soc.* **67**, 2122–2127 (1945).
136. K. J. Palmer, R. C. Merrill, H. S. Owens, and M. Ballantyne, *J. Phys. Chem.* **51**, 710–720 (1947).
137. M. D. Walkinshaw and S. Arnott, *J. Mol. Biol.* **153**, 1055–1073, 1075–1085 (1982).
138. D. A. Powell, E. R. Morris, M. J. Gidley, and D. A. Rees, *J. Mol. Biol.* **155**, 517–531 (1982).
139. C. Sterling, *Biochem. Biophys. Acta* **26**, 186–197 (1957).
140. R. E. Rundle, *J. Am. Chem. Soc.* **69**, 1769–1772 (1947).
141. D. French, A. O. Pulley, and W. J. Whelan, *Staerke* **15**, 349–354 (1963).
142. R. R. Bumb and B. Zaslow, *Carbohydr. Res.* **4**, 98–101 (1967).
143. A. Hybl, R. E. Rundle, and D. E. Williams, *J. Am. Chem. Soc.* **87**, 2779–2791 (1965).
144. A. French and B. Zaslow, *J. Chem. Soc., Chem. Commun.* pp. 41–42 (1972).
145. J. J. Jackobs, R. R. Bumb, and B. Zaslow, *Biopolymers* **6**, 1659–1670 (1968).
146. W. Banks and C. T. Greenwood, *Staerke* **23**, 300–314 (1971).
147. H.-C. H. Wu and A. Sarko, *Carbohydr. Res.* **61**, 7–25, 27–40 (1978).
148. D. French, *in* "Symposium on Foods: Carbohydrates and their Roles" (H. W. Schultz, ed.), pp. 26–54. AVI Publ., Westport, Connecticut, 1969.
149. K. Kainuma and D. French, *Biopolymers* **11**, 2241–2250 (1972).
150. Y. Deslandes, R. H. Marchessault, and A. Sarko, *Macromolecules* **13**, 1466–1471 (1980).
151. T. L. Bluhm and A. Sarko, *Can. J. Chem.* **55**, 293–299 (1977).
152. R. H. Marchessault, Y. Deslandes, K. Ogawa, and P. R. Sundararajan, *Can. J. Chem.* **55**, 300–303 (1977).
153. E. D. T. Atkins and K. D. Parker, *J. Polym. Sci., Part C* **28**, 69–81 (1969).
154. E. R. Ruckel and C. Schuerch, *Biopolymers* **5**, 515–523 (1967).
155. G. O. Aspinall, "Polysaccharides." Pergamon, Oxford, 1970.
156. N. S. Anderson, T. C. S. Dolan, and D. A. Rees, *J. Chem. Soc. C* pp. 596–601 (1968).
157. N. S. Anderson, T. C. S. Dolan, and D. A. Rees, *J. Chem. Soc., Perkin Trans. 1* pp. 2173–2176 (1973).

158. S. Arnott, W. E. Scott, D. A. Rees, and C. G. A. McNab, *J. Mol. Biol.* **90**, 253–267 (1974).
159. E. Percival and R. H. McDowell, "Chemistry and Enzymology of Marine Algal Polysaccharides." Academic Press, New York, 1967.
160. T. C. S. Dolan and D. A. Rees, *J. Chem. Soc.* pp. 3534–3539 (1965).
161. N. S. Anderson, T. C. S. Dolan, C. J. Lawson, A. Penman, and D. A. Rees, *Carbohydr. Res.* **7**, 468–473 (1968).
162. S. T. Bayley, *Biochim. Biophys. Acta* **17**, 194–205 (1955).
163. C. Araki and S. Hirase, *Bull. Chem. Soc. Jpn.* **33**, 291–295 (1960).
164. C. Araki, *J. Chem. Soc. Jpn.* **58**, 1338–1350 (1937).
165. C. Araki, *Proc. Int. Seaweed Symp.* **5**, 3–17 (1966).
166. M. Duckworth and W. Yaphe, *Carbohydr. Res.* **16**, 189–197 (1971).
167. D. A. Rees, I. W. Steele, and F. B. Williamson, *J. Polym. Sci., Part C* **28**, 261–276 (1969).
168. I. J. Goldstein, G. W. Hay, B. A. Lewis, and F. Smith, *Methods Carbohydr. Chem.* **5**, 361–370 (1965).
169. D. A. Rees, *J. Chem. Soc.* pp. 5168–5171 (1961); pp. 1821–1832 (1963).
170. D. S. Reid, T. A. Bryce, A. H. Clark, and D. A. Rees, *Faraday Discuss. Chem. Soc.* **57**, 230–237 (1974).
171. T. A. Bryce, A. H. Clark, D. A. Rees, and D. S. Reid, *Eur. J. Biochem.* **122**, 63–69 (1982).
172. S. Ablett, A. H. Clark, and D. A. Rees, *Macromolecules* **15**, 597–602 (1982).
173. O. Smidsrød, I.-L. Andresen, H. Grasdalen, B. Larsen, and T. Painter, *Carbohydr. Res.* **80**, C11–C16 (1980).
174. I. C. M. Dea, A. A. McKinnon, and D. A. Rees, *J. Mol. Biol.* **68**, 153–172 (1972).
175. P.-E. Jansson, L. Kenne, and B. Lindberg, *Carbohydr. Res.* **45**, 275–282 (1975).
176. L. D. Melton, L. Mindt, D. A. Rees, and G. R. Sanderson, *Carbohydr. Res.* **46**, 245–257 (1976).
177. P. A. Sandford, J. E. Pittsley, C. A. Knutson, P. R. Watson, M. C. Cadmus, and A. Jeanes, *ACS Symp. Ser.* **45**, 192–210 (1977).
178. I. H. Smith, K. C. Symes, C. J. Lawson, and E. R. Morris, *Int. J. Biol. Macromol.* **3**, 129–134 (1981).
179. A. Jeanes, *J. Polym. Sci., Polym. Symp.* **45**, 209–227 (1974).
180. E. R. Morris, *ACS Symp. Ser.* **45**, 81–89 (1977).
181. G. Holzwarth, *Biochemistry* **15**, 4333–4339 (1976).
182. M. Milas and M. Rinaudo, *Carbohydr. Res.* **76**, 189–196 (1979).
183. R. Moorhouse, M. D. Walkinshaw, and S. Arnott, *ACS Symp. Ser.* **45**, 90–102 (1977).
184. G. Holzwarth and E. B. Prestridge, *Science* **197**, 757–759 (1977).
185. G. Holzwarth, *Carbohydr. Res.* **66**, 173–186 (1978).
186. D. A. Rees, *Pure Appl. Chem.* **53**, 1–14 (1981).
187. Y.-M. Choy, F. Fehmel, N. Frank, and S. Stirm, *J. Virol.* **16**, 581–590 (1975).
188. R. Moorhouse, W. T. Winter, S. Arnott, and M. E. Bayer, *J. Mol. Biol.* **109**, 373–391 (1977).
189. M. E. Bayer and H. Thurow, *J. Bacteriol.* **130**, 911–936 (1977).
190. R. H. Marchessault, K. Imada, T. L. Bluhm, and P. R. Sundararajan, *Carbohydr. Res.* **83**, 287–302 (1980).
191. D. H. Isaac, K. H. Gardner, E. D. T. Atkins, U. Elsasser-Beile, and S. Stirm, *Carbohydr. Res.* **66**, 43–52 (1978).
192. E. D. T. Atkins, D. H. Isaac, and H. F. Elloway, *in* "Microbial Polysaccharides and Polysaccharases" (R. C. W. Berkeley, G. W. Gooday, and D. C. Ellwood, eds.), pp. 161–189. Academic Press, New York, 1979.
193. H. F. Elloway, E. D. T. Atkins, and I. W. Sutherland, *Carbohydr. Res.* **76**, 285–289 (1979).
194. H. Muir and T. E. Hardingham, *MTP Int. Rev. Sci.: Biochem. Ser. One* **5**, 153–222 (1975).

195. L. Rodén and M. I. Horowitz, in "The Glycoconjugates" (M. I. Horowitz and W. Pigman, eds.), Vol. 2, pp. 3–71. Academic Press, New York, 1978.
196. U. Lindahl and M. Höök, *Annu. Rev. Biochem.* **47**, 385–417 (1978).
197. V. C. Hascall, *J. Supramol. Struct.* **7**, 101–120 (1977).
198. J. M. Guss, D. W. L. Hukins, P. J. C. Smith, W. T. Winter, S. Arnott, R. Moorhouse, and D. A. Rees, *J. Mol. Biol.* **95**, 359–384 (1975).
199. J. K. Sheehan, K. H. Gardner, and E. D. T. Atkins, *J. Mol. Biol.* **117**, 113–135 (1977).
200. E. D. T. Atkins, D. H. Isaac, I. A. Nieduszynski, C. F. Phelps, and J. K. Sheehan, *Polymer* **15**, 263–270 (1974).
201. E. D. T. Atkins and J. K. Sheehan, *Science* **179**, 562–564 (1973).
202. W. T. Winter and S. Arnott, *J. Mol. Biol.* **117**, 761–784 (1977).
203. W. T. Winter, P. J. C. Smith, and S. Arnott, *J. Mol. Biol.* **99**, 219–235 (1975).
204. R. L. Cleland and J. L. Wang, *Biopolymers* **9**, 799–810 (1970).
205. R. L. Cleland, *Biopolymers* **9**, 811–824 (1970).
206. R. L. Cleland, *Arch. Biochem. Biophys.* **180**, 57–68 (1977).
207. B. Chakrabarti and E. A. Balazs, *J. Mol. Biol.* **78**, 135–141 (1973).
208. J. E. Scott and M. J. Tigwell, *Biochem. Soc. Trans.* **3**, 662–664 (1975).
209. E. A. Balazs, *Fed. Proc., Fed. Am. Soc. Exp. Biol.* **25**, 1817–1822 (1966).
210. J. E. Scott and M. J. Tigwell, *Biochem. J.* **173**, 103–114 (1978).
211. E. R. Morris, D. A. Rees, and E. J. Welsh, *J. Mol. Biol.* **138**, 383–400 (1980).
212. D. Welti, D. A. Rees, and E. J. Welsh, *Eur. J. Biochem.* **94**, 505–514 (1979).
213. S. M. Bociek, A. H. Darke, D. Welti, and D. A. Rees, *Eur. J. Biochem.* **109**, 447–456 (1980).
214. E. D. T. Atkins, D. Meader, and J. E. Scott, *Int. J. Biol. Macromol.* **2**, 318–319. (1980).
215. W. T. Winter, S. Arnott, D. H. Isaac, and E. D. T. Atkins, *J. Mol. Biol.* **125**, 1–19 (1978).
216. J. J. Cael, W. T. Winter, and S. Arnott, *J. Mol. Biol.* **125**, 21–42 (1978).
217. S. Arnott, J. M. Guss, D. W. L. Hukins, and M. B. Mathews, *Biochem. Biophys. Res. Commun.* **54**, 1377–1383 (1973).
218. E. D. T. Atkins and D. H. Isacc, *J. Mol. Biol.* **80**, 773–779 (1973).
219. A. Mitra, S. Arnott, E. D. T. Atkins, and D. H. Isaac, in preparation.
220. G. Gatti, B. Casu. G. Torri, and J. R. Vercellotti, *Carbohydr. Res.* **68**, C3–C7 (1979).
221. A. S. Perlin, B. Casu, G. R. Sanderson, and L. F. Johnson, *Can. J. Chem.* **48**, 2260–2268 (1970).
222. G. K. Hamer and A. S. Perlin, *Carbohydr. Res.* **49**, 37–48 (1976).
223. L.-Å. Fransson, *Carbohydr. Res.* **36**, 339–348 (1974); **62**, 235–244 (1978).
224. S. Arnott, J. M. Guss, D. W. L. Hukins, I. C. M. Dea, and D. A. Rees, *J. Mol. Biol.* **88**, 175–184 (1974).
225. M. B. Mathews and I. Lozaityte, *Arch. Biochem. Biophys.* **74**, 158–174 (1958).
226. S. M. Partridge, H. F. Davis, and G. S. Adair, *Biochem. J.* **79**, 15–26 (1961).
227. D. Heinegård, *J. Biol. Chem.* **252**, 1980–1989 (1977).
228. E. D. T. Atkins and I. A. Nieduszynski, *Fed. Proc., Fed. Am. Soc. Exp. Biol.* **36**, 78–83 (1977).
229. G. Gatti, B. Casu, G. K. Hamer, and A. S. Perlin, *Macromolecules* **12**, 1001–1007 (1979).
230. A. S. Perlin, N. M. K. Ng Ying Kin, S. S. Bhattacharjee, and L. F. Johnson, *Can. J. Chem.* **50**, 2437–2441 (1972).
231. B. Casu, G. Gatti, N. Cyr, and A. S. Perlin, *Carbohydr. Res.* **41**, C6–C8 (1975).
232. G. Gatti, B. Casu, and A. S. Perlin, *Biochem. Biophys. Res. Commun.* **85**, 14–20 (1978).
233. B. Casu, Results presented at the Third International Symposium on Glycoconjugates, Brighton, England, 1975.
234. J. Boyd. F. B. Williamson, and P. Gettins, *J. Mol. Biol.* **137**, 175–190 (1980).

235. A. S. Perlin, *in* "Proceedings of the International Symposium of Macromolecules" (E. B. Mano, ed.), pp. 337–348. Elsevier, Amsterdam, 1975.
236. S. Arnott and W. T. Winter, *Fed. Proc., Fed. Am. Soc. Exp. Biol.* **36**, 73–78 (1977).
237. H. F. Elloway and E. D. T. Atkins, *Biochem. J.* **161**, 495–498 (1977).
238. P. Hovingh and A. Linker, *Carbohydr. Res.* **37**, 181–192 (1974).
239. D. A. Brant and W. L. Dimpfl, *Macromolecules* **3**, 655–664 (1970).
240. R. L. Cleland, *Biopolymers* **10**, 1925–1948 (1971).
241. E. R. Morris, D. A. Rees, E. J. Welsh, L. G. Dunfield, and S. G. Whittington, *J. Chem. Soc., Perkin Trans. 2*, pp. 793–800 (1978).
242. V. S. R. Rao, N. Yathindra, and P. R. Sundararajan, *Biopolymers* **8**, 325–333 (1969).
243. S. G. Whittington, *Biopolymers* **10**, 1481–1489 (1971).
244. P. J. Flory, "Principles of Polymer Chemistry." Cornell Univ. Press, Ithaca, New York, 1953
245. C. Tanford, "Physical Chemistry of Macromolecules." Wiley, New York, 1961.
246. J. R. van Wazer, J. W. Lyons, K. Y. Kim, and R. E. Colwell, "Viscosity and Flow Measurement: A Laboratory Handbook of Rheology." Wiley, New York, 1963.
247. M. Fixman, *J. Chem. Phys.* **41**, 3772–3778 (1964).
248. O. Smidsrød and A. Haug, *Biopolymers* **10**, 1213–1227 (1971).
249. K. Šolc, *J. Chem. Phys.* **55**, 335–344 (1971).
250. W. W. Graessley, *Adv. Polym. Sci.* **16**, 1–179 (1974).
251. J. L. Doublier and B. Launay, *Ind. Miner.* **4**, 191–198 (1977).
252. J. L. Doublier and B. Launay, *Proc. Int. Congr. Rheol., 7th, 1976* pp. 532–533 (1976).
253. P. J. Whitcomb and C. W. Macosko, *J. Rheol.* **22**, 493–505 (1978).
254. D. A. Rees and E. J. Welsh, *Angew. Chem., Int. Ed. Engl.* **16**, 214–224 (1977).
255. G. O. Aspinall, I. W. Cottrell, S. V. Egan, I. M. Morrison, and J. N. C. Whyte, *J. Chem. Soc. C* pp. 1071–1080 (1967).
256. G. O. Aspinall, J. W. T. Craig, and J. L. Whyte, *Carbohydr. Res.* **7**, 442–452 (1968).
257. V. Zitko and C. T. Bishop, *Can. J. Chem.* **43**, 3206–3214 (1965).
258. D. A. Rees and A. W. Wight, *J. Chem. Soc. B* pp. 1366–1372 (1971).
259. A. Haug, B. Larsen, O. Smidsrød, and T. Painter, *Acta Chem. Scand.* **23**, 2955–2962 (1969).
260. A. Penman and D. A. Rees, *J. Chem. Soc., Perkin Trans. 1* pp. 2182–2187, 2191–2196 (1973).
261. D. A. Rees, *Biochem. J.* **126**, 257–273 (1972).
262. C. J. Lawson and D. A. Rees, *Nature (London)* **227**, 392–393 (1970).
263. D. A. Rees, *Biochem. J.* **81**, 347–352 (1961).
264. D. A. Rees, *Adv. Carbohydr. Chem. Biochem.* **24**, 267–332 (1969).
265. A. Haug, B. Larsen, and O. Smidsrød, *Carbohydr. Res.* **32**, 217–225 (1974).
266. A. Haug, S. Myklestad, B. Larsen, and O. Smidsrød, *Acta Chem. Scand.* **21**, 768–778 (1967).
267. B. Stockton, L. V. Evans, E. R. Morris, D. A. Powell, and D. A. Rees, *Bot. Mar.* **23**, 563–567 (1980).
268. B. Larsen and A. Haug, *Carbohydr. Res.* **17**, 287–296 (1971); A. Haug and B. Larsen, *ibid.* pp. 297–308.
269. J. Madgwick, A. Haug, and B. Larsen, *Acta Chem. Scand.* **27**. 3592–3594 (1973).
270. P. V. Bulpin, E. J. Welsh and E. R. Morris, *Staerke*, in press.
271. O. Smidsrød, B. Larsen, and T. Painter, *Acta Chem. Scand.* **24**, 3201–3212 (1970).
272. I. C. M. Dea and A. Morrison, *Adv. Carbohydr. Chem. Biochem.* **31**, 241–312 (1975).
273. R. L. Whistler, *Adv. Chem. Ser.* **117**, 242–255 (1973).
274. I. C. M. Dea, E. R. Morris, D. A. Rees, E. J. Welsh, H. A. Barnes, and J. Price, *Carbohydr. Res.* **57**, 249–272 (1977).
275. J. E. Courtois and P. le Dizet, *Carbohydr. Res.* **3**, 141–151 (1966).

276. J. E. Courtois and P. le Dizet, *Bull. Soc. Chim. Biol.* **52**, 15–22 (1970).
277. C. W. Baker and R. L. Whistler, *Carbohydr. Res.* **45**, 237–243 (1975).
278. B. V. McCleary, *Carbohydr. Res.* **71**, 205–230 (1979).
279. T. J. Painter, J. J. González, and P. C. Hemmer, *Carbohydr. Res.* **69**, 217–226 (1979).
280. G. Robinson, E. R. Morris, and D. A. Rees, *J. Chem. Soc., Chem. Commun.* pp. 152–153 (1980).
281. E. R. Morris, D. A. Rees, and G. Robinson, *J. Mol. Biol.* **138**, 349–362 (1980).
282. I. T. Norton, D. M. Goodall, E. R. Morris, and D. A. Rees, *J. Chem. Soc., Faraday Trans. 1* (submitted for publication).
283. D. A. Rees, *Chem. Ind. (London)* pp. 630–636 (1972).
284. M. Rinaudo and M. Milas, *Biopolymers* **17**, 2663–2678 (1978).
285. S. A. Frangou, E. R. Morris, D. A. Rees, R. K. Richardson, and S. B. Ross-Murphy, *J. Polym. Sci., Polym. Lett. Ed.* (in press).
286. A. R. Jeanes, C. A. Knutson, J. E. Pittsley, and P. R. Watson, *J. Appl. Polym. Sci.* **9**, 627–638 (1965).
287. P. J. Flory, *Proc. R. Soc. London, Ser. A* **234**, 73–89 (1956).
288. P. J. Flory and J. Rehner, Jr., *J. Chem. Phys.* **11**, 512–520 (1943).
289. H. Morawetz, "Macromolecules in Solution." Wiley, New York, 1965.
290. R. Koningsveld, *Adv. Colloid Interface Sci.* **2**, 151–215 (1968).
291. P.-Å. Albertsson, *Adv. Protein Chem.* **24**, 309–341 (1970).
292. E. Edmond and A. G. Ogston, *Biochem. J.* **109**, 569–576 (1968).
293. T. C. Laurent and A. G. Ogston, *Biochem. J.* **89**, 249–253 (1963).
294. B. N. Preston, M. Davies, and A. G. Ogston, *Biochem. J.* **96**, 449–471 (1965).
295. T. C. Laurent, *Eur. J. Biochem.* **21**, 498–506 (1971).
296. T. C. Laurent and B. Öbrink, *Eur. J. Biochem.* **28**, 94–101 (1972).
297. B. N. Preston, B. Öbrink, and T. C. Laurent, *Eur. J. Biochem.* **33**, 401–406 (1973).
298. B. Öbrink and T. C. Laurent, *Eur. J. Biochem.* **41**, 83–90 (1974).
299. G. J. Cumming, C. J. Handley, and B. N. Preston, *Biochem. J.* **181**, 257–266 (1979).
300. R. A. Gelman, W. B. Rippon, and J. Blackwell, *Biopolymers* **12**, 541–558 (1973).
301. R. A. Gelman, D. N. Glaser, and J. Blackwell, *Biopolymers* **12**, 1223–1232 (1973).
302. R. A. Gelman and J. Blackwell, *Biopolymers* **12**, 1959–1974 (1973).
303. R. A. Gelman and J. Blackwell, *Biochim. Biophys. Acta* **297**, 452–455 (1973).
304. R. A. Gelman and J. Blackwell, *Arch. Biochem. Biophys.* **159**, 427–433 (1973).
305. K. Kurachi, G. Schmer, M. A. Hermodson, D. C. Teller, and E. W. Davie, *Biochemistry* **15**, 368–373 (1976).
306. B. Nordenman, C. Nyström, and I. Björk, *Eur. J. Biochem.* **78**, 195–203 (1977).
307. P. S. Damus, M. S. Hicks, and R. D. Rosenberg, *Nature (London)* **246**, 355–357 (1973).
308. J. S. Rosenberg, P. W. McKenna, and R. D. Rosenberg, *J. Biol. Chem.* **250**, 8883–8888 (1975).
309. K. Kurachi, K. Fujikawa, G. Schmer, and E. W. Davie, *Biochemistry* **15**, 373–377 (1976).
310. E. T. Yin, S. Wessler, and P. J. Stoll, *J. Biol. Chem.* **246**, 3712–3719 (1971).
311. R. D. Rosenberg and P. S. Damus, *J. Biol. Chem.* **248**, 6490–6505 (1973).
312. G. B. Villanueva and I. Danishefsky, *Biochem. Biophys. Res. Commun.* **74**, 803–809 (1977).
313. E. H. H. Li, J. W. Fenton, II, and R. D. Feinman, *Arch. Biochem. Biophys.* **175**, 153–159 (1976).
314. J. A. Cifonelli, *Carbohydr. Res.* **37**, 145–154 (1974).
315. J. Riesenfeld, M. Höök, I. Björk, U. Lindahl, and B. Ajaxon, *Fed. Proc., Fed. Am. Soc. Exp. Biol.* **36**, 39–43 (1977).
316. I. Danishefsky, *Fed. Proc., Fed. Am. Soc. Exp. Biol.* **36**, 33–35 (1977).
317. A. N. Teien, U. Abildgaard, and M. Höök, *Thromb. Res.* **8**, 859–867 (1976).

318. L.-O. Andersson, T. W. Barrowcliffe, E. Holmer, E. A. Johnson, and G. E. C. Sims, *Thromb. Res.* **9**, 575–583 (1976).
319. L. H. Lam, J. E. Silbert, and R. D. Rosenberg, *Biochem. Biophys. Res. Commun.* **69**, 570–577 (1976).
320. M. Höök, I. Björk, J. Hopwood, and U. Lindahl, *FEBS Lett.* **66**, 90–93 (1976).
321. U. Lindahl, G. Bäckström, L. Thunberg, and I. G. Leder, *Proc. Natl. Acad. Sci. U.S.A.* **77**, 6551–6555 (1980).
322. B. Meyer, L. Thunberg, U. Lindahl, O. Larm, and I. G. Leder, *Carbohydr. Res.* **88**, C1–C4 (1981).
323. J. Grindrod and T. A. Nickerson, *J. Dairy Sci.* **51**, 834–841 (1968).
324. P. M. T. Hansen, *J. Dairy Sci.* **51**, 192–195 (1968).
325. C. F. Lin and P. M. T. Hansen, *Macromolecules* **3**, 269–274 (1970).
326. T. A. J. Payens, *J. Dairy Sci.* **55**, 141–150 (1972).
327. Th. H. M. Snoeren, T. A. J. Payens, J. Jeunink, and P. Both, *Milchwissenschaft* **30**, 393–396 (1975).
328. M. Glicksman, "Gum Technology in the Food Industry." Academic Press, New York, 1969.
329. H. M. Farrell, Jr. and M. P. Thompson, *in* "Fundamentals of Dairy Chemistry"(B. H. Webb, A. H. Johnson, and J. A. Alford, eds.), 2nd ed., pp. 442–473. AVI Publ., Westport, Connecticut, 1974.
330. W. T. Winter and A. Sarko, *Biopolymers* **13**, 1447–1460 (1974).
331. R. S. Stein and R. E. Rundle, *J. Chem. Phys.* **16**, 195–207 (1948).
332. B. Pfannemüller, H. Mayerhöfer, and R. C. Schulz, *Biopolymers* **10**, 243–261 (1971).
333. B. Pfannemüller, *Carbohydr. Res.* **61**, 41–52 (1978).
334. T. Handa and H. Yajima, *Biopolymers* **18**, 873–886 (1979).
335. J. A. Thoma and L. Stewart, *in* "Starch: Chemistry and Technology" (R. L. Whistler and E. F. Paschall, eds.), Vol. 1, pp. 209–249. Academic Press, New York, 1965.
336. A. Hyble, R. E. Rundle, and D. E. Williams, *J. Am. Chem. Soc.* **87**, 2779–2788 (1965).
337. P. C. Manor and W. Saenger, *Nature (London)* **237**, 392–393 (1972).
338. P. C. Manor and W. Saenger, *J. Am. Chem. Soc.* **96**, 3630–3639 (1974).
339. H. Schlenk and D. M. Sand, *J. Am. Chem. Soc.* **83**, 2312–2320 (1961).
340. M. Ilton, A. W. Jevans, E. D. McCarthy, D. Vance, H. B. White, and K. Bloch, *Proc. Natl. Acad. Sci. U.S.A.* **68**, 87–91 (1971).
341. C. E. Ballou, *Acc. Chem. Res.* **1**, 366–373 (1968).
342. G. R. Gray and C. E. Ballou, *J. Biol. Chem.* **246**, 6835–6842 (1971).
343. W. L. Smith and C. E. Ballou, *J. Biol. Chem.* **248**, 7118–7125 (1973).
344. M. H. Saier, Jr., and C. E. Ballou, *J. Biol. Chem.* **243**, 992–1005, 4332–4341 (1968).
345. J. M. Keller and C. E. Ballou, *J. Biol. Chem.* **243**, 2905–2910 (1968).
346. D. E. Vance, O. Mitsuhashi, and K. Bloch, *J. Biol. Chem.* **248**, 2303–2309 (1973).
347. D. E. Vance, T. W. Esders, and K. Bloch, *J. Biol. Chem.* **248**, 2310–2316 (1973).
348. H. Knoche, T. W. Esders, K. Koths, and K. Bloch, *J. Biol. Chem.* **248**, 2317–2322 (1973).
349. Y. Machida and K. Bloch, *Proc. Natl. Acad. Sci. U.S.A.* **70**, 1146–1148 (1975).
350. C. E. Ballou, *Pure Appl. Chem.* **53**, 107–112 (1981).
351. J. W. Donovan and C. J. Mapes, *Staerke* **32**, 190–193 (1980).
352. M. Kugimiya, J. W. Donovan, and R. Y. Wong, *Staerke* **32**, 265–270 (1980).
353. G. Hollinger, L. Kuniak, and R. H. Marchessault, *Biopolymers* **13**, 879–890 (1974).

6

Immunology of Polysaccharides

C. T. BISHOP AND H. J. JENNINGS

I.	Introduction	292
II.	Molecular Architecture of Bacteria, Yeasts, and Fungi	293
III.	The Immune Response to Polysaccharides	296
	A. General Description of the Immune Response	296
	B. Cellular Response to Polysaccharide Antigens	299
	C. Measurement of Antibody	304
IV.	Structural Aspects of the Immune Response to Polysaccharides	312
	A. Molecular Weight	312
	B. Determinants and Repeating Units	313
	C. Noncarbohydrate Determinants in Polysaccharides	316
	D. Conformation	317
	E. Cross-Reactions	319
V.	Polysaccharide Antigens in Human Disease	320
	A. Polysaccharides and Virulence	321
	B. Protective Vaccines	321
	C. Diagnosis	324
	D. Natural Immunity	325
	References	325

THE POLYSACCHARIDES, VOL. 1

I. Introduction

The history of the role of polysaccharides in immunology can be said to date from 1917 with the report by Dochez and Avery (1) of a "specific soluble substance" secreted by pneumococci during growth. This substance gave precipitates with antibodies to the bacterium of the pneumococcus type from which it was derived and thus constituted the type-specific serological antigen of the bacterium. Subsequent studies by Heidelberger, Avery, and Goebel showed that the "specific soluble substances" of the pneumococci were polysaccharides, i.e., complex polymers of sugars linked together through glycosidic bonds. This was the first time that any material other than protein had been shown to be antigenic. Heidelberger (2) has given an interesting and entertaining account of those important times in the history of immunology. The demonstrated protective effect of antibodies raised to pneumococcal polysaccharides against pneumococcal infections provided proof of the importance of these substances in immunology (3,4). These results showed that polysaccharides are true immunogens in that they induce an immune response and the generation of specific antibodies (serum globulins). Then followed the concept of determinant groups when it was shown that only a relatively small portion of a polysaccharide is the major site of antibody specificity. The presence of the same determinant group in several polysaccharides was also shown to be responsible for their serological cross-reactivity; that is, the capacity of a polysaccharide from one species of bacterium to precipitate the polysaccharide-specific antibodies of another. Cross-reactions have been used extensively in the immunochemical analysis of polysaccharides, and by this means Heidelberger (5) was able to predict the presence of structural features before they were verified chemically. It is a vindication of the exquisite specificity of immune reactions that all of those predictions proved to be correct.

In more recent years the use of polysaccharides as antigens and immunogens has contributed greatly to the classification and identification of bacteria, to a better understanding of the immune response, to the definition of the active site in antigen–antibody interactions, and to the detection and prevention of human disease caused by invasive microorganisms. This chapter cites relevant and significant contributions in each of those areas. It is not intended to be exhaustive but rather to present a state of the art.

Of course, glycose moieties occur in other biopolymers such as glycoproteins, glycolipids, and nucleic acids and have been shown to function as the serological specific site in some of them, e.g. the blood group glycoproteins (6) and membrane glycolipids (7). However, these molecules cannot be classified as polysaccharides and are therefore beyond the scope of this review.

II. Molecular Architecture of Bacteria, Yeasts, and Fungi

The overwhelming majority of immunologically significant polysaccharides are of microbial origin. It is therefore appropriate to begin this chapter with a description of the anatomy of those microorganisms and in particular to summarize that which is known about the location of polysaccharide antigens.

A simplified picture of the bacterial cell is shown in Fig. 1. The plasma membrane that surrounds the cytoplasm is common to all cells. It consists of phospholipid bilayers, lipids, and proteins joined by hydrophobic forces into a fluid, dynamic structure. The plasma membrane regulates the movement of ions and molecules into and out of the cell and also houses enzymes and proteins that are specifically required by the cell.

Outside the plasma membrane is the cell wall, which provides form and rigidity. In bacteria the cell walls are of two general types: one that has an outer membrane over a peptidoglycan layer and one that lacks the outer

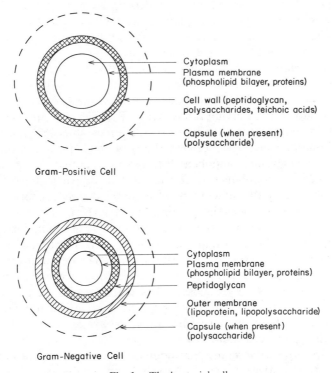

Fig. 1. The bacterial cell.

membrane but has additional components within the peptidoglycan layer. These two wall types can in general be related to the gram designation of bacteria, as shown in Fig. 1. This broad classification of bacteria was introduced by Gram (8) in 1884 on the basis of a color stain. Briefly, the bacteria are treated with a solution of gentian violet followed by an aqueous solution of iodine in potassium iodide. Excess reagents are then washed off with water, leaving all bacterial cells colored a deep purple. A subsequent wash with ethanol removes the color from some bacteria, called gram-negative, and leaves others still colored; these are designated gram-positive. Considerable work has been done in attempts to explain the chemical basis of Gram's stain but without definitive conclusions. Salton (9) has reviewed this topic and suggests that retention of the color in gram-positive organisms is due to a change in the permeability of the cell wall induced by the 95% ethanol used for decolorization. More recent investigations of bacterial cell walls by electron microscopy have shown that the correspondence of wall type with Gram's stain breaks down in some instances in which the peptidoglycan layer is either very thin or very thick (10). As a consequence Costerton (11) has proposed that electron microscopy is the more definitive method for determining cell wall type and that new descriptive terms are required. For the purposes of this chapter the significant distinction to be made is between bacteria with an outer membrane and those without, and these are generally gram-negative and gram-positive, respectively.

A common feature of both types of cell wall is the peptidoglycan layer (Fig. 1). This consists of a backbone of alternating units of 2-acetamido-2-deoxy-D-glucose (N-acetylglucosamine) and 2-acetamido-3-O-carboxyethyl-2-deoxy-D-glucose (N-acetylmuramic acid). Peptides are attached to this backbone by amide formation between L-alanine units and the carboxyl groups of muramic acid residues, and the peptide chains are in turn cross-linked by a polyglycine component through the ε-amino groups of lysine. This complex heteropolymer is further modified, particularly in gram-positive bacteria, by the attachment of oligo- and polysaccharides and of teichoic acids (see below). The peptidoglycan layer is therefore a highly cross-linked mosaic, and this structure undoubtedly accounts for the properties of strength and rigidity that are necessary for its function of containing and protecting the bacterial cell. The peptidoglycan layer is destroyed during lysis of bacterial cells; the term "lysis" is derived from the enzyme lysozyme, which is a β-(1→4)-N-acetylglucosaminidase and cleaves the glucosamine–muramic acid backbone. The reader is referred to two excellent reviews (9,12) for more details about the developments that have led to this description of the peptidoglycan cell wall.

In gram-positive bacteria the peptidoglycan layer also contains two types of carbohydrate antigens: one a relatively simple polysaccharide, and the other a teichoic acid. The polysaccharides, which contain only two or three different monosaccharides, are usually extractable and are the major group or type-specific antigens of the bacterium. They must therefore be located at the surface of the cell wall, where they can be recognized by the immunocompetent cells. The teichoic acids are complex structures of polyglycerol or polyribitol phosphate with glycosyl or alanyl residues attached variably at some of the polyol hydroxyl groups (13,14). They may also exist as copolymers of glycosyl units with glycerol or ribitol phosphate (15,16). The teichoic acids have been shown to be antigenic determinants and the phage receptors of certain species (17). Their location at the outer surface of the cell wall has been established by electron microscopy (18). In that study the cells were treated with concanavalin A, which interacted with terminal glucosyl units on the teichoic acids. The electron micrograph showed a series of protuberances around the outer surface, and these were interpreted as being several chains of teichoic acid bound together by the multivalent lectin. The teichoic acid then must be present as protruding chains perpendicular to the cell surface.

In gram-negative bacteria the outer membrane contains strong antigens, the main component being lipopolysaccharide, which constitutes 10–15% of the dry cell wall. In cases in which the lipopolysaccharide contains a hydrophilic, high molecular weight polysaccharide chain (O antigen) (19), it constitutes a principal antigen of the bacteria and is functionally equivalent to capsular polysaccharides (see below). Studies with specific ferritin-conjugated antibodies have made it possible to determine the location of the lipopolysaccharides by electron microscopy (20). The results showed that newly formed lipopolysaccharide appeared in about 220 patches per cell and was evenly spread over the entire cell surface after 2–3 min of growth. It was also found that 86% of the patches of newly formed lipopolysaccharide were located over sites where the cytoplasmic and outer membranes adhered to each other. On this basis it was suggested that new lipopolysaccharide emerges from those sites and then spreads evenly over the cell surface by lateral movement.

Many bacteria, either gram-positive or gram-negative, produce extracellular polysaccharides (exopolysaccharides). These polysaccharides, together with the O antigens of the lipopolysaccharides, constitute the principal immunogens and antigens of the bacteria because of their location at the extreme outer surface of the cell. In this position they participate in most of the interactions between the bacteria and the immune mechanisms of a host. The exopolysaccharides may exist in the form of a discrete capsule surrounding the bacterial cell or in the form of a loose slime, unattached to the cell surface. Capsules can be recognized by the India ink staining technique (21) or by electron microscopy after reaction with specific antibody (22) or ferritin-conjugated antibody (23). Capsular polysaccharides usually have a high content of acidic constituents such as uronic acids, phosphate groups, or pyruvate ketals and are found most often in pathogenic bacteria. A protective function can be assigned to the capsules because they render bacteria resistant to phagocytosis and to the action of complement (24,25).

The cell walls of yeasts and fungi are less well understood than those of bacteria. The polymeric components of yeast cell walls include mannans, glucans, chitin, glycoproteins, proteins, and lipids, the relative proportions of which may vary widely among different genera. Attempts to demonstrate discrete layers by electron microscopy have not given definitive results. Phaff (26) has reviewed this topic and suggests that "the apparent continuous nature of the wall components is probably caused by the presence of protein throughout the wall and the partial overlapping of wall layers." However, it has been shown that mannans or mannan–protein complexes occur in the outer regions of the cell wall (27), and observations that mannans are the major antigens of many species of yeasts would support their location at the cell surface (28–31). The cell walls of hyphal fungi contain protein, glucans, and chitin, with the latter located in the innermost regions merging into proteinaceous material that also contains admixed glucans (32). Like the bacteria, many species of yeasts produce exopolysaccharides that may occur in the form of a discrete capsule or may be released into the culture medium (26).

III. The Immune Response to Polysaccharides

A. General Description of the Immune Response

The immune system consists of a highly complex network of cells and molecules whose special characteristic is pattern recognition. The cells (lymphocytes and macrophages) and molecules (antibody and complement) circulate through the bloodstream and the lymphatic system, and their function is to patrol the body and destroy or inactivate any foreign invaders. All of the cells of the immune system are derived from a common, blood-forming stem cell. In some unknown way, the stem cell differentiates to a progenitor cell, which in turn differentiates into one of the different blood cells. Differentiation of progenitor cells to lymphocytes occurs in the thymus and in the bone marrow, leading to T lymphocytes and B lymphocytes, respectively. These lymphocytes are then exported to the spleen and the lymph nodes, where they are functional. Macrophages, derived from the stem cell by differentiation (probably in the bone marrow), are concentrated mainly in the spleen and the peritoneum.

These three kinds of cells, T and B lymphocytes and macrophages, cooperate and interact in a highly complex way to produce an immune response. It would be impossible to present all of the evidence for these interactions because of space limitations and because many points have not yet been clarified. Instead, we present the currently accepted view of the immune response and refer the reader to some books and reviews for further detail (33–37).

The introduction of a foreign material (antigen or immunogen) into an animal induces the formation of *specific* immunoglobulin protein molecules (antibodies) and of *specific* reactive cells. These antibodies and reactive cells circulate in the blood and react with their specific antigen, the result being that the foreign invader is inactivated, killed, or degraded. Those responses that can be transferred from one animal to another by means of serum are classified as humoral responses (i.e., due to antibody). Those that can be transferred by means of sensitized cells but not by serum are termed cell-mediated responses. The T lymphocytes have been shown to be responsible for the cell-mediated response, whereas the B lymphocytes produce antibodies and are thus responsible for the humoral response. Both kinds of lymphocytes require the presence of macrophages to give a functional immune response. Macrophages are large, mononuclear, phagocytic cells that can be distinguished by their capacity to adhere to glass or plastic surfaces. The macrophage appears to be the first major cell type to react with an antigen in most immune responses, and its major role seems to be one of "presenting" the antigen to the lymphocytes. The macrophage has several types of surface receptors that enable it to carry out this function. There are nonspecific membrane binding sites to which antigens may adhere; there are also surface receptors for immunoglobulin and receptors for the third component of complement (C3). The activity of complement depends on the operation of nine protein components acting in a sequence, which is referred to as the complement cascade. It is now recognized that this cascade can be activated by two pathways. The classical pathway is initiated by the antigen–antibody complex, to which complement then binds and assists in phagocytosis by the macrophage. The alternate pathway does not require specific antibody and can be activated at the C3 step by polysaccharides such as zymosan and lipopolysaccharides. Sialic acid is an inhibitor of the alternate pathway of complement activation (38). An antigen that encounters a macrophage may therefore be bound to its surface specifically (through attachment to the immunoglobulin) or nonspecifically (at a membrane binding site), and the addition of complement is a preliminary step to the ultimate engulfment and destruction of an invasive organism by phagocytosis (a term that simply means the destruction of one cell by another through enzymatic degradation). Although phagocytosis is a major activity of macrophages, it seems clear that at least some of the antigen is retained on the surface of some of these cells, which then function as accessory cells to stimulate the lymphocytes. All lymphocytes display their specific immunoglobulins on their surfaces, and these act as receptor sites for an antigen as suitably "presented" on the surface of the macrophage. The interaction of "presented antigen" with these immunoglobulin receptors stimulates the lymphocytes to differentiate along a number of pathways depending on their function.

The T lymphocytes serve a number of functions when activated by an antigen, and the mechanisms of differentiation into those different subsets are unknown. Some T cells function as killer cells and are responsible for such cell-mediated immune responses as delayed hypersensitivity, allograft immunity, and graft-versus-host responsiveness. Other T cells function as controls over the response of B cells and have a helper or suppressor role. There is as yet no consensus on the mechanisms by which these subsets of T cells exert their functions.

The B lymphocytes are responsible for the production of antibody (immunoglobulins). These cells have large quantities of immunoglobulin on their surface, and for each individual cell this immunoglobulin is identical to the antibody that the cell or its progeny will secrete. Upon activation by antigen reacting with the surface receptor immunoglobulin, the B lymphocyte differentiates into an antibody-secreting cell producing molecules with the same specificity as those on the surface of the precursor cell. As indicated above, for some antigens this activation of a B lymphocyte requires the cooperation of a T lymphocyte as well as a macrophage.

In addition to the effector responses of T and B lymphocytes as described above, two other events may occur when these cells meet an antigen. The first is paralysis or tolerance, and the second is the generation of memory cells. Immune tolerance, loss of the capacity to react to an antigenic stimulus, may be caused by an initial exposure to either very low or very high concentrations of antigen, although B cells seem to be paralyzed only at high antigen concentrations. Thus, the dose range in immunization can be critical. If an animal does make an immune response, subsequent exposure to the same antigen will usually give a larger and faster response. This is called immunological memory and is due to the persistence of memory cells generated by the initial response.

It should be emphasized that cellular immunology is one of the most active areas of science with an enormous body of literature. The foregoing description of the immune response is of necessity a simplification of a highly complex, controlled system with each type of cell and response modulating the others.

B. Cellular Response to Polysaccharide Antigens

Most current knowledge about cellular responses to polysaccharide antigens has been obtained from studies with mice and simple polysaccharides, the one most frequently used being the capsular polysaccharide of type 3 *Streptococcus pneumoniae*. Although it is still too early to be certain about the application of these results to human beings, it is clear that the use of polysaccharides as model antigens has contributed greatly to our understanding of cellular responses in the immune system.

1. T-Cell Independence

As explained in the previous section, the activation of a B cell to secrete immunoglobulin generally requires the cooperation of T cells. However, a few antigens have been shown to be T-cell independent, and these include polysaccharides (39), polyvinylpyrrolidine (40), and poly(D-amino acids) (41). These antigens have in common the properties of high molecular weight and repeating sequences of antigenic units. Thus, an athymic nude mouse or a thymectomized adult mouse will make a substantial humoral response to those antigens, and the response may equal or even exceed that given by normal mice. The antibody produced by thymectomized mice is of the IgM class (39–41) (see Section III,B,3 for a description of classes of antibody), which suggests that those B cells which excrete IgM are much less dependent on T-cell cooperation than are those that excrete IgG. Alternatively, there is evidence that in a nonimmunized animal nearly all antigen-binding cells have IgM receptors (42). When stimulated by an antigen, in cooperation with T lymphocytes, these IgM-bearing precursors may then differentiate along two pathways to give both IgM- and IgG-producing B cells. The necessity for T-cell cooperation in this antigen-driven switch would account for the exclusive presence of IgM in the absence of T lymphocytes. An important property of the IgM response to a T-cell-independent antigen, such as a purified polysaccharide, is that memory cells do not appear to be generated. Thus, an immunologically naive animal that is immunized with a T-cell-independent antigen will not raise a secondary response on subsequent reexposure to the same antigen. This has obvious, important consequences in considering the immunization of infants with purified polysaccharides. In an immunologically mature individual there are already present both IgM and IgG precursor cells, generated by natural exposure. In that instance the polysaccharide is intimately associated with other bacterial antigens in the complete microorganism. These other antigens activate T cells, which also cooperate in the stimulation of B cells; the latter secrete both IgM and IgG, some of which are specific for the polysaccharide. In this way a substantial pool of polysaccharide-specific, IgG-bearing precursor cells is established, providing an immunological memory and the capacity to give a secondary response subsequently on reexposure to either the purified polysaccharide or the whole bacterium.

Finally, although polysaccharides are T-cell-independent antigens, Baker and colleagues (35) have shown that when T cells are present they do participate in regulating the response to polysaccharide antigens. The use of anti-lymphocyte serum to deplete T cells in a normal mouse caused a marked enhancement in the response to type 3 pneumococcal polysaccharide. This result was interpreted as being due to the removal of a group of T cells that normally suppress the antibody response. Baker *et al.* also found that the response of athymic nude mice was only slightly higher than that of their normal littermates, whereas one would have predicted that, lacking the suppressor T cells, the nude mice would give a greatly enhanced response. On this basis Baker (35) proposed the presence of two kinds of T cells to regulate the immune response, amplifiers and suppressors, and recently there has been a direct demonstration of the existence of specific suppressor T cells (43).

2. Heterogeneity

The immunoglobulin molecules generated in an immune response are heterogeneous with respect to a number of parameters. Analysis of serum antibodies by ultracentrifugation shows that they have molecular weights ranging from 150,000 to over 1,000,000. Electrophoresis shows heterogeneity by charge, and they are also heterogeneous in their reaction with the same antigen and as antigens themselves. This diversity can be accounted for by the structure of the immunoglobulin molecule, which is a protein made up of four polypeptide chains: two identical light chains and two identical heavy chains, the four joined together by disulfide bridges (Fig. 2). There can also be intrachain disulfide bridges creating loops or domains in both the

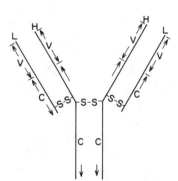

Fig. 2. Antibody molecule: two heavy (H) and two light (L) chains with variable (V) and constant (C) regions.

light and heavy chains. The light chains consist of about 214 amino acids, whereas the heavy chains have about twice that number. Antibody molecules are alike except for a sequence of about 50 amino acids at the ends of both the heavy and light chains, known as the variable regions. These variable regions together constitute the antigen-binding sites, and it will be noted that there are two identical sites per molecule. If there were a choice between just two amino acids for each of the 50 positions, the number of potentially different molecules would be 10^{30}. Thus, in any antiserum there will be a complex population of antibodies that react to various extents with the same antigen.

There are also constant regions in the polypeptide chains of immunoglobulins, and here it has been found that there are five major types of heavy chains in human beings. These are responsible for the *class* of immunoglobulin designated IgG, IgA, IgD, and IgE. The class distinction is an antigenic one based on the fact that immunoglobulins of one species are antigenic in another. Thus, the class of immunoglobulin is established serologically, and the antigenic differences reside in the heavy chains (*36, 44,45*).

There are two natural circumstances under which antibodies of restricted heterogeneity or even homogeneity may be produced. The first of these was the discovery that intensive immunization of rabbits with bacteria that presented a polysaccharide antigen on their surface sometimes resulted in the production of antibodies with restricted heterogeneity as determined by antigen-binding properties and by electrophoresis (*46–48*). Furthermore, those rabbits that produced uniform antibody were also high responders, giving antisera that contained up to 60.0 mg/ml of antibody. This led to a search for possible genetic factors that influence the immune response, and it was shown by breeding experiments (*48*) that a line of progeny could be established that gave high responses of restricted heterogeneity. However, it was also shown (*49*) that 90% of outbred rabbits will give antibodies of restricted heterogeneity to type 3 and 8 *S. pneumoniae*. In these experiments there was no correlation between degree of heterogeneity and antibody concentration. There is no solid explanation for this phenomenon of antibodies with restricted heterogeneity. Krause (*34*) postulated that such responses may occur "because of the selection of a most favored plasma cell clone which proliferates at the expense of all others." However, the reason for the selective advantage of such a clone remains obscure. Whatever the reason, it is clear that polysaccharide antigens, *when presented on the surface of a bacterial cell*, can induce the formation of homogeneous antibodies. This has been observed with groups A (*50*), A variant (*51*), B (*52*) and C (*52*) *Streptococcus*, with type 3 and 8 *S. pneumoniae* (*53*), with type A

Neisseria meningococcus (*53*), and with *Salmonella typhi* (*54*). Restricted heterogeneity of antibody has not been observed with purified, soluble polysaccharides as antigens.

The second source of homogeneous antibody to polysaccharides is myeloma proteins. Myelomas can be induced in mice by intraperitoneal application of mineral oil, and these plasma cell tumors excrete a protein that is a homogeneous immunoglobulin, generally of the IgA class. Many of these myeloma proteins have been shown to react specifically with microbial polysaccharides including lipopolysaccharides, teichoic acid, and dextrans (*44,55,56*). Glaudemans (*57*) has discussed the known anticarbohydrate myeloma proteins, their specificities, and binding properties. The origins of these specificities are unknown but may lie in the stimulation of a lymphocyte by an endogenous antigen before the transformation to malignancy. Evidence in favor of this concept is the absence of polysaccharide-specific myeloma proteins in germ-free mice (*58*).

Myelomas can be used in another way to produce homogeneous antibody of predetermined specificity, and that is by cell fusion with antigen-sensitized spleen cells (*59*). Thus, spleen cells from a mouse that has been immunized with a specific antigen are fused with myeloma cells, and the product of cell fusion, a hybridoma, produces a myeloma immunoglobulin with specificity for the immunizing antigen. Such hybridomas can be transplanted into mice and thus provide a continuing source of specific, monoclonal antibody. This discovery is of profound significance because it has provided a means for obtaining homogeneous antibody to virtually any antigen (*60*).

3. Types of Antibodies

Reference has already been made to the classification of immunoglobulins based on antigenic differences in their heavy chains. It now seems certain that these differences in the heavy chains are responsible for the different biological activities of the immunoglobulin classes and subclasses. Often the biological activity is expressed only after the immunoglobulin molecule has reacted with an antigen and probably arises through an allosteric change in conformation consequent to the reaction at the combining site. Thus, the different classes of immunoglobulins may be regarded as a system that provides antibodies with a range of biological activities following combination with antigen.

The most prevalent immunoglobulin in human serum is IgG, which constitutes 75–80% of the total. It exists in the monomeric form (i.e., one unit of the four-peptide chain molecule, Fig. 2) and diffuses readily throughout the body. Its primary function is to neutralize bacterial toxins and to bind microorganisms to macrophages to enhance their phagocytosis. It can cross the placenta and thus provides the major defense against infection during the early weeks of life. Four subclasses of IgG have been identified that show variations in these biological properties.

Immunoglobulin A is the next most prevalent immunoglobulin, and it constitutes about 13% of the total immunoglobulin content in serum. It appears selectively in mucous secretions such as saliva, tears, nasal fiuids, colostrum, and secretions from the lung and gastrointestinal tract. Clearly, it represents the body's first line of defense against attack by microorganisms in these exposed surfaces. It is present in serum as the monomeric, four-peptide chain unit but is prone to form dimers. In secretions it is usually found as a dimer together with two other proteins: a secretory component and a J (joining) chain. Immunoglobulin A is more resistant to the normal digestive proteolytic enzymes than other classes of immunoglobulins, and this may be of particular advantage for neutralizing antigens in food. Two subclasses of IgA have been identified, but they have not been assigned specific biological roles.

The IgM class is present to the extent of 6% of total immunoglobulins in human serum. This immunoglobulin is a pentamer of the four-peptide chain unit. It contains 10 heavy chains, 10 light chains, and 1 joining chain. There are therefore 10 combining sites that are identical, and this polyvalency accounts for the property of this class as being very effective agglutinating agents. Molecules of IgM appear early in the response to infection or to primary immunization and are confined largely to the bloodstream; they are therefore of importance in cases of bacteremia. Immunoglobulins comparable to mammalian IgM have been found in all classes of vertebrates, and it seems likely that IgM represents the primordial antibody from which the other classes evolved.

The IgD class occurs to the extent of about 1% of total immunoglobulin content, and its biological function is obscure. It occurs on the surface membranes of lymphocytes, often in association with IgM, and may serve as the receptor site for activation of the lymphocyte by antigen.

The IgE class of immunoglobulins is present in the smallest amount, and the concentration in serum of healthy adults is 250–300 ng/ml. This concentration may be 20 times higher in patients with asthma and hay fever, in those with parasitic infections including schistosomiasis and hookworm, and in those with certain types of dermatosis including eczema. In the body, IgE is fixed to the mast cells in the skin, and contact with an antigen leads to the liberation by those cells of pharmacologically active substances such as histamine. These substances induce vasodilation and smooth muscle contraction, leading to the symptoms of hay fever or asthma. It may be that the vasoactive amines that are released by the mast cells also facilitate the ejection of parasites. Thus, the prime role of IgE seems to be that of antigen receptor in allergic responses.

Most of the foregoing description of immunoglobulin classes is drawn from three sources (36,44,45), which can be consulted for further details.

4. Adjuvants

Adjuvants are materials that cause an enhanced immune response, and a host of substances have been identified as having adjuvant activity. The effects measured include increased tumor immunity, conversion of a non-antigenic substance to an effective antigen, increased levels of circulatory antibody, increase in cell-mediated immunity, and more effective protective immunity. The majority of materials that have been tested and used as adjuvants are microbial products derived from a large variety of mycobacteria or related species. Undoubtedly, the most frequently used material is Freund's complete adjuvant, which consists of water-in-oil emulsions with added mycobacteria or nocardia (61). Numerous attempts were made to identify those parts of the mycobacterium that were responsible for adjuvant activity. It was found that whole organisms could be replaced by a liposoluble wax, by purified cell walls, or by water-soluble fractions of lysozyme-treated cell walls. These earlier investigations have been reviewed (62,63) and will not be discussed in detail here. It was found subsequently that adjuvant activity resided in the peptidoglycan portion of the cell wall (64), and further chemical breakdown showed that the minimal active unit was a mono-saccharide tripeptide (65). Finally, the monosaccharide dipeptide (N-acetyl-muramyl-L-alanyl-D-isoglutamine) was synthesized and shown to be the minimal active unit (65,66). Various analogues of the muramyl dipeptide have been tested for adjuvanticity, antigenicity, and pharmacological prop-erties when administered with a number of antigens and by different routes (63,67). It is noteworthy that none of these adjuvants appears to potentiate immune responses to T-cell-independent antigens such as polysaccharides, and this has been interpreted as indicating that the cellular mechanisms of adjuvanticity operate primarily through macrophages and T cells (68).

Finally, it should be noted that lipopolysaccharides are, by themselves, both potent antigens (69) and potent adjuvants for many protein antigens (68).

C. Measurement of Antibody

The existence and extent of a humoral immune response require some method to detect and measure the specific antibody that is produced. Gill (70) has published a summary of all methods used to detect antibody, including the lower limits of detection. This section describes the methods that are most widely used with polysaccharide antigens. It includes only the basic principles; experimental details can be found in the references.

1. Precipitins

The precipitin reaction is the fundamental antigen–antibody interaction on which immunochemistry is based. When a soluble antigen is added to serum that contains specific antibodies to that antigen, a precipitate forms. Heidelberger and Kendall (71), by using a nitrogen-free polysaccharide antigen, were able to quantify the amount of antibody in the precipitate simply by measuring the amount of nitrogen. Thus, for the first time the precipitin reaction was given a quantitative basis. It was found that incremental additions of antigen to its homologous antiserum caused increasing amounts of antigen–antibody complex to be precipitated up to a maximum level. Addition of a slight excess of antigen caused no change in the amount of complex precipitated, but further additions brought about dissolution of some of the precipitate. The theory proposed (71) to explain these results was that a number of equilibrium reactions occurred simultaneously between the multivalent antigen and an antibody that was at least bivalent. These reactions would lead to the cross-linking of antibody molecules by the multivalent (multiple determinants) antigen until all of the antibody-combining sites were filled (point of maximum precipitation). Additional antigen would then compete for the binding sites in the complex, resulting in its partial dissolution. The precipitin reaction gives an absolute quantitative measure of the serum antibody to a specific antigen. The use of colorimetric methods to determine antibody nitrogen makes it possible to detect as little as 10–20 μg of antibody nitrogen per milliliter. The quantitative precipitin reaction can be used to measure the extent of cross-reactions (see Section IV,E) and to determine the effectiveness of inhibitors. Inhibition studies are important in establishing the structures of determinant groups (see Section IV,B). The procedure consists of addition of the test inhibitor (usually a soluble oligosaccharide in the case of polysaccharide antigens) to the antiserum before addition of the antigen in the amount that normally gives maximum precipitation. The test inhibitor competes with the antigen for occupation of the binding sites on the antibody and thereby inhibits precipitation of the antigen–antibody complex. The degree to which the precipitin reaction is inhibited is a measure of the structural similarities between the test inhibitor and the actual determinant on the antigen. Experimental details for these procedures have been assembled by Kabat and Mayer (72) in their classical book. A variation of the quantitative precipitin test was introduced by Farr (73), who used the binding of a radioactive antigen as a measure of antibody concentration. This method has been used to quantify the immune response to polysaccharide vaccines (74,75) and with proper standardization can determine antibody concentrations of 0.05–0.1 μg/ml. The precipitin reaction can also be used qualitatively in the serological classification of bacteria, for the identification of bacteria, and in testing for antibody. Possibly the most common qualitative test is the ring test, in which a solution of antigen is carefully layered over a sample of antiserum in a narrow glass tube. The appearance of turbidity at the interface is a positive test. An early example of this application of the precipitin reaction was the differentiation of specific types of hemolytic streptococci, group B (76).

2. Immunodiffusion

Immunodiffusion represents a further refinement of the precipitin reaction. Oudin (77) introduced the technique of single diffusion of an antigen into an agar gel that contains antiserum. As the antigen diffuses into the gel, an opaque precipitin band appears at the advancing front, where the concentrations of antigen and antibody are optimal for precipitation. As antigen concentration increases behind the precipitin band, the complex is dissolved and the band migrates through the gel. With mixtures of antigens and antibodies several precipitin bands can be obtained based on the different diffusion constants of the antigens. The number of bands indicates the *minimum* number of antigen–antibody complexes in the system. A variation of this technique is to allow both antigen and antibody to diffuse toward each other from opposite ends of a gel (78); this may reveal more precipitin complexes because of different diffusion constants of specific antibodies as well as those of the antigens.

The most extensively used immunodiffusion method is that of double diffusion in two dimensions introduced by Ouchterlony (79,80). This method is advantageous for the direct identification of antigens and for revealing cross-reactions. The diffusion is done in a flat layer (3–6 mm thick) of agar gel contained in a petri dish. Holes are cut in the gel to serve as wells for solutions of antigen and antibody. The usual pattern of wells is a central well surrounded by six circumferential wells so that all wells are an equal distance (ca. 10 mm) from each other. Antiserum is placed in the central well and various antigens in the surrounding wells or vice versa. As diffusion of antigen and antibody takes place an opaque, white precipitin band forms between the two wells where antigen and antibody are in the proportions for maximum precipitation. When there are mixtures of several antibodies and several antigens, a number of bands may be formed, which is indicative of the minimum number of antigen–antibody complexes present. If two antigens are placed in adjacent wells and diffused against a single antiserum, one of three results may be obtained. (a) A band is formed against only one antigen, showing that the two antigens are different. (b) Bands are formed against both antigens, and the two bands fuse where they intersect to form a continuous precipitin line, an indication that the two antigens are identical or have identical determinants. (c) Both antigens give bands, but there is only partial fusion at the intersection with a spur extending beyond the point of fusion; this indicates that the two antigens are cross-reactive but not fully identical. An example of an immunodiffusion experiment is shown in Fig. 3.

With complex mixtures improved resolution can be obtained by electrophoresis of the antigens in the gel before double diffusion against antiserum at a right angle to the direction of electrophoretic migration (81). This technique of immunoelectrophoresis can be used to estimate the minimum number of antigens in a complex mixture. The foregoing diffusion methods detect antigen–antibody complexes with antiserum that contains 20 μg/ml or more of antibody. The detection and measurement of lower concentrations of antibody require one of the more sensitive methods described below.

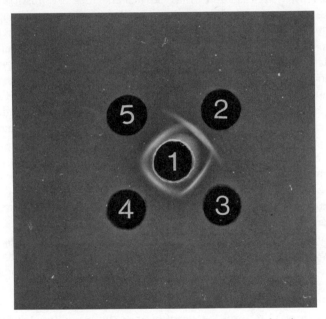

Fig. 3. Immunodiffusion in agar of type III group B streptococcal antiserum (well 1) with the native type III polysaccharide (well 2), reduced native type III polysaccharide (well 3), and the desialylated type III polysaccharide (wells 4 and 5). A line of identity is formed between wells 4 and 5 because each contains the identical desialylated antigen. Spurring occurs between wells 2 and 5 because, in addition to having the determinants of the desialylated antigen, the native antigen also possesses a unique determinant controlled by the presence of its terminal sialic acid residues. The latter determinant is destroyed by reduction of the carboxylate groups of the sialic acid residues as indicated by spurring between wells 2 and 3, whereas the line of identity between wells 3 and 4 indicates that despite the presence of reduced sialic acid residues the reduced native antigen still possesses the same determinants as the desialylated antigen (137).

3. Hemagglutination

Agglutination, or the clumping of cells (bacteria or erythrocytes), may be regarded as a precipitin reaction that occurs at the surface of the cell. Thus, the specific antibody in homologous antiserum will react with the antigen on the surface of a cell, causing cross-linking of these cells and resulting in visible agglutination. Because the antigen is spread out over the cell surface, much less antibody is required to bring about enough cross-linking to give a visible response than is needed to precipitate a soluble antigen. Agglutination methods are therefore much more sensitive than precipitin reactions, and immune sera can usually be diluted several thousand times and still cause a detectable response.

Agglutination methods became important in the immunology of polysaccharides when it was found that polysaccharides are adsorbed on the surface of erythrocytes (red blood cells) (82). The majority of polysaccharides are adsorbed without pretreatment simply by incubation with erythrocytes at 37°C in isotonic saline. With some lipopolysaccharides, a pretreatment with alkali has been found to improve their capacity to coat the erythrocytes (83). Periodate oxidation of polysaccharides produces reactive aldehyde groups that couple to side-chain amino groups of proteins, and this method is used to couple polysaccharides to erythrocytes (84). The hemagglutination test is done by making serial, twofold dilutions of antiserum, usually in the wells of a microtiter plate, and then adding a fixed volume of suspended erythrocytes previously coated with the test antigen. The reagents are mixed and incubated for 1–4 h, and each well is then read on a 1 + to 4 + scale. The method is an empirical dilution procedure that is useful for detecting antibody and for measuring the relative strengths of different antisera to the same antigen. Artenstein et al. (85) compared a number of assays for antibody to determine the responses of human beings to infection by N. meningitidis and to vaccination by the specific polysaccharides. They found that hemagglutination was highly sensitive and serogroup specific for antibodies raised by both infection and vaccination. Other examples of the application of this method are the measurement of antibody response to pneumococcal polysaccharides in rats (86) and the serotyping of the lipopolysaccharides of N meningitidis (87). It will be obvious that, like all antibody–antigen reactions, the hemagglutination test can be used for inhibition studies. This simply involves the addition of the test inhibitor to the antiserum before the addition of the coated erythrocytes. Comparison of inhibitory activity provides a useful measure of the purity of different preparations of the same antigen. When carefully standardized with antiserum of known concentration, the hemagglutination method detects antibody in concentrations down to 0.03 μg/ml.

4. Radioimmunoassay

The use of radioactive isotopes to measure antigen–antibody interactions was introduced by Farr (73) in a study of the equilibrium between bovine serum albumin and antibody globulin. The technique is based on the fact that unbound antigen is soluble in 50% ammonium sulfate, whereas the antigen–antibody complex is precipitated. Thus, by using a radioactively labeled antigen it is possible to measure all of the bound antigen and to characterize the antiserum by its antigen-binding capacity. The advantages of using this procedure are that *all* of the antigen–antibody complexes are measured and the use of radioactive labels provides for greater sensitivity and precision than the methods described earlier.

Brandt *et al.* (75) modified Farr's original method to detect antibody response to polysaccharide vaccines of *N. meningitidis*. These workers prepared ^{14}C-labeled polysaccharides by growing the organisms in the presence of sodium [1-^{14}C]acetate. A standard volume of a solution of the purified, labeled polysaccharide was then added to the test antiserum and, after incubation at 4°C for 16–18 h, the solution was mixed with an equal volume of ammonium sulfate. The resulting precipitate was centrifuged, washed, redissolved, and counted together with a standard of the same amount of labeled polysaccharide in normal serum. The results were expressed as percentage of antigen bound. This method was refined to measure low levels of antibody response of adults, children, and infants to several preparations of group A and group C polysaccharides from *N. meningitidis* (74). The refinement consisted of double labeling of the antigen to increase the sensitivity of the method. Thus, the ^{14}C-labeled polysaccharide was activated with cyanogen bromide and then reacted with tyramine, the product then being iodinated with ^{125}I. In order to perform binding tests on very small quantities of sera, ^{22}Na was added to the antigen solution as a volume marker. By using antisera with a known concentration of precipitating antibody, it was found that there was a linear relationship between the percentage of antigen bound and the logarithm of the antibody concentration, and the latter could be measured down to 0.1 μg/ml. The radioactive antigen-binding assay was used to investigate the role of maternal antibody in neonatal group B streptococcal infection (88). In this study the polysaccharide antigen was labeled with tritium by growing the organism in the presence of ^{3}H-labeled sodium acetate. The results showed that maternal antibody deficiency could be an important determinant in the pathogenesis of group B streptococcal infections in infants. Schiffman and Austrian (89) reported a radioimmunoassay system for 18 types of *Streptococcus pneumoniae*. The system was based on ^{14}C labeling of the polysaccharides by biosynthesis, and antigen binding was related to antibody content by the use of standardized antisera. By means of this method it was possible to measure 1–10 ng of antibody nitrogen in samples of 50 μl, and the procedure could be adapted to inhibition studies. The capsular polysaccharide of *Hamophilus influenzae* type b has been radioactively labeled by the cyanogen bromide–^{125}I-tyramine route described above (*74*) and used in a radioimmunoassay to measure antibodies in normal and immunized human beings down to concentrations of 0.04 μg of antibody per milliliter (*90*).

In summary, radioimmunoassay has become firmly established as a technique for detecting and measuring antibodies to polysaccharide antigens. Polysaccharides can be labeled biosynthetically with ^{14}C or ^{3}H or chemically by substitution with tyramine and iodination with ^{125}I.

5. Enzyme-Linked Immunosorbent Assay

The enzyme-linked immunosorbent assay (ELISA) was introduced by Engvall and Perlmann (*91–93*) and has become widely accepted as the method of choice for quantifying antigen–antibody reactions. The principle of the method is the detection and estimation of antibody by reacting it with a specific antiimmunoglobulin to which an enzyme has been attached. The amount of enzyme, which can be determined colorimetrically by its action on a substrate, is then a measure of bound antibody.

Application of the ELISA method involves the following steps (see Fig. 4):

1. *Adsorption of antigen on a solid support.* It was found that proteins and some other antigens are passively adsorbed on plastic surfaces (*93*), thus permitting the use of polystyrene tubes or microtiter plates. It is essential that conditions for optimal adsorption be established because these vary with the nature of the antigen, and in particular polysaccharides show considerable variability. Lipopolysaccharides adhere directly to polystyrene tubes, with optimum coating obtained with solutions of 1–5 μg/ml at pH 9–10 (*94*). The capsular polysaccharide of group C *N. meningitidis* is also adsorbed directly, but the group A polysaccharide is not (*95*). When polysaccharides are not adsorbed passively, two alternatives are available. The polysaccharide can be coupled covalently to a protein to give an adsorbable conjugate (*96*), or the container wells can be precoated with an antiserum that is specific for the polysaccharide (*97*), thus providing a specific adsorptive surface. The coupling of a polysaccharide to a protein can be done with cyanuric chloride (*98*), and this method has been used to conjugate capsular polysaccharides of *Streptococcus pneumoniae* and group B *Streptococcus* to poly(L-lysine) to make adsorbable antigens for ELISA tests (*96*). The alternative procedure employed single type-specific rabbit antipneumococcal serum to precoat the wells of a microtiter plate. Addition of the polysaccharide for which the antiserum was specfic then resulted in an antigen-coated well (*97*).

A. Antigen adsorbed
on surface.

B. Antibody bound
to adsorbed antigen.

C. Enzyme-labeled
immunoglobulin
bound to antibody.

D. Enzyme substrate (□) added.
Colored product (■) measured
colorimetrically.

Fig. 4.   Steps in enzyme-linked immunosorbent assay.

2. *Binding of test antiserum.* The antiserum to be tested is added in serial dilutions to the antigen-coated wells and incubated in the presence of 0.5% Tween 20 (a wetting agent) to prevent nonspecific binding. The incubation period should be determined separately; maximum binding may require anywhere from 30 min to 16 h. The cells are then washed to remove any unbound antibody.

3. *Preparation of enzyme-labeled antiimmunoglobulin.* An antiimmunoglobulin is selected that has been raised against the species of immunoglobulin in the test serum. Thus, if the serum to be tested is from human beings, a rabbit antihuman immunoglobulin may be selected. The antiimmunoglobulin is conjugated with the enzyme by cross-linking using 0.2% glutaraldehyde at room temperature for 2 h. The enzyme used most frequently is alkaline phosphatase, and the conjugates are stable with respect to both enzyme and immune reactivity for more than 1 year (*93*).

4. *Binding of enzyme-labeled antiimmunoglobulin.* The enzyme–antiimmunoglobulin conjugate, at a predetermined concentration, is added to the washed cells from step 2 that contain the adsorbed antigen with bound antibody from the test serum. After incubation the excess conjugate is washed away, leaving only the conjugate that is bound to the antibody, which is in turn bound to the adsorbed antigen.

5. *Enzyme–substrate reaction.* When substrate for the conjugated enzyme is added to the cells from step 4, the enzyme–substrate reaction is a measure of the enzyme remaining and hence also a measure of the antibody to which the conjugate is bound. With alkaline phosphatase as the enzyme, the substrate is *p*-nitrophenyl phosphate, and the yellow color produced by the release of *p*-nitrophenol is measured colorimetrically at 405 nm.

With the ELISA method antibody can be detected with the same sensitivity as with the radioimmunoassay (less than 1 ng/ml); the method is simple to perform, uses stable reagents, and does not require counting equipment. It is more versatile than other methods and offers several additional advantages, e.g., the detection of antigenic determinants on cell surfaces and the capacity to determine separately the various classes of immunoglobulin (IgG, IgM, etc.) by using the appropriate antiimmunoglobulin serum (see *94*). With careful standardization of each step this method can provide relative quantitative comparisons of antisera, but it does not give the same direct measurement of antibody–antigen complexes as the radioimmunoassay.

IV. Structural Aspects of the Immune Response to Polysaccharides

A. Molecular Weight

It has been known for a long time that molecular weight is a critical parameter in the antigenicity of a molecule. Thus, proteins with a molecular weight of less than 10,000 do not stimulate the formation of antibody (*99*). Insulin, with a molecular weight of 20,000, does not give rise to antibodies but can cause hypersensitivity, so it has some antigenicity. Ovalbumin, with a molecular weight of about 40,000, is a good antigen, and the serum globulins at MW 150,000 are highly antigenic (*2*). A similar relationship between molecular weight and antigenicity has been found for polysaccharide anti-antigens. Kabat and Bezer (*100*) showed that dextrans with an average molecular weight of 90,000 or higher were immunogenic in man as determined by increases in precipitating antibody; dextrans with an average molecular weight of 50,000 or lower were not immunogenic. Similarly, Howard and colleagues (*101*) showed a decreasing immune response in mice to the capsular polysaccharide of type 3 *Streptococcus pneumoniae* with decreasing molecular weight. The serogroup-specific polysaccharides of *N. meningitidis* are good immunogens in man at molecular weights of 50,000–130,000, but the response is much weaker with preparations that have a molecular weight of 30,000 (74,102). Lipopolysaccharides are highly immunogenic (103) and tend to form aggregates that have molecular weights in the millions. It has been shown that disaggregation of lipopolysaccharide by heating reduces its immunogenicity in rabbits (104). The O-antigenic polysaccharide chains of lipopolysaccharides have molecular weights of 10,000–20,000 when separated from lipid A and in this form are not immunogenic. The accumulated evidence points to the generality that 45,000 ± 5000 is the molecular weight above which polysaccharides are immunogenic and below which their immunogenicity falls off rapidly.

B. Determinants and Repeating Units

The immunological specificity of polysaccharide antigens resides in their structures; i.e., antibodies are formed that will recognize only those structural features present in the polysaccharide that induced the formation of that antibody. The chemical structures of microbial polysaccharides are described elsewhere in this work (Chapter 5, Volume II,) and are not discussed in detail here. However, it is important to recognize how the enormous variety of antigenic specificities can be accounted for in terms of polysaccharide structure.

Polysaccharides may be linear or branched; the branches may be single-unit side chains, side chains of intermediate length (four to six units), or simply a joining point for two identical linear chains, e.g., glycogen and dextrans. Within these structural forms a further variation exists in the number of different sugars that are present. The polysaccharide may be a homopolymer (composed of a single type of monosaccharide) such as dextran (D-glucose), levan (D-fructose), or mannan (D-mannose), or it may be a heteropolymer (composed of two or more different monosaccharide residues). The monosaccharides most often found in polysaccharide antigens are D-glucose, D-galactose, D-mannose, D-glucuronic acid, D-galacturonic acid, D-mannuronic acid, L-rhamnose, L-fucose, D-glucosamine, and D-galactosamine. Another variation in structure arises from a uniform or nonuniform assembly of the monosaccharide residues in the polysaccharide molecule. There is now abundant evidence that the majority of bacterial polysaccharides are composed of relatively small repeating units (one to six monosaccharide residues) (105,106) as distinct from plant polysaccharides, in which linkages may be consistent but repeating units are not evident. The use of ^{13}C-NMR spectroscopy has demonstrated vividly the basic repeating unit construction of bacterial polysaccharides (107). However, repetitive patterns may not be as readily apparent for substituent groups (e.g., O-acetyl, phosphate) that are not present in every repeating unit. It must be remembered that all analytical methods are limited to the detection of possible structural variation in approximately 1 out of 10 repeating units (*108*). Finally, there is the variety introduced by the position and anomeric configuration of glycosidic linkages between the monosaccharide residues and the presence of hydroxyl substituents such as acetate, phosphate, and ketal groups. The combination of all these variations in polysaccharide structure provides for an enormous range of immunological specificities.

   Immunological studies of polysaccharides are aimed at establishing which part of the polysaccharide is responsible for its immunological specificity, and that part is known as the determinant group. A determinant group may comprise several monosaccharide residues, one of which contributes most to the specificity; that monosaccharide residue is termed the immunodominant sugar. Since it is responsible for immunospecificity, a determinant group is that part of the antigen that enters the combining site of the antibody. It was known from the pioneering work of Landsteiner (*109*) that the combining site in antibody was directed toward only a small part of a macromolecular antigen. The size and specificity of the combining site could therefore be studied by inhibition of an antigen–antibody reaction (see section III) with low molecular weight compounds that were chemically defined and identical with or very similar to the determinant group. This procedure has been used extensively to define determinant groups and immunodominant sugars in polysaccharide antigens. The classical studies on this method are those of Kabat (*110*) on the inhibition of the dextran–antidextran precipitin reaction by a series of oligosaccharides of the isomaltose series [α-D-Glc-(1→6)-α-D-Glc———]. With human antidextran sera it was found that the inhibitory power of the isomaltose oligosaccharide series increased with increasing size up to the hexasaccharide. From the relative inhibitory powers of each oligosaccharide it was possible to calculate the contributions of each D-glucose residue to the binding energy. The results showed that the nonreducing terminal residue contributed about 40% and the next two residues together about 60%. Additional residues each contribute less, and the sixth contributes only some 2–3% of the total binding energy (*111*). Although the nonreducing terminal D-glucose unit may be considered immunodominant in this system, the specificity must extend farther than that because the series of maltose [α-D-Glc$p$-(1→4)-α-D-Glc$p$———] oligosaccharides were much poorer inhibitors. These results indicate that in a linear polysaccharide the immunological specificity resides primarily in the terminal sugar residue and extends along the polysaccharide chain. The antiserum generated to such a polysaccharide contains many antibody molecules with different sizes of combining sites, each directed toward the same determinant region. Gelzer and Kabat (*112*) were able to demonstrate this heterogeneity of antibody by fractionating antidextran serum into mainly two subpopulations of antibody molecules: one in which the combining sites were mostly complementary to the trisaccharide (iso-maltotriose and the other in which they were mostly complementary to the hexasaccharide (isomaltohexaose). It is likely that any linear antigenic polysaccharide will generate antibodies with a range of specificities extending from two to about six sugar residues.

The situation is somewhat different for branched polysaccharide antigens in which the immunodominant sugars are invariably those that form the branches. This was demonstrated conclusively by Ballou ($31,113$) for yeast mannans in which side chains ranging from di- to pentasaccharides were attached to an α-(1→6)-linked mannan backbone. The side chains consisted of α-(1→2)-linked mannopyranose residues terminated at the nonreducing end by an α-(1→3)-linked residue. It was found that the tetrasaccharide side chain gave 100% inhibition of the homologous precipitin reaction. Further-more, the isolated backbone of α-(1→6)-linked mannose residues neither precipitated antibody nor inhibited the precipitation. Further evidence of the immunological importance of the side chains in this family of polysaccharides was provided by the discovery of a yeast mannan that contained 2-acetamido-2-deoxy-D-glucose. The amino sugar was located as a subbranch on one of the mannose oligosaccharide side chains and was shown to be a very strong immunodeterminant (114).

In contrast to the yeast mannans that have oligosaccharide branches, the majority of microbial polysaccharides have single-unit branches. These single-unit side chains invariably are immunodominant when they occur in antigenic polysaccharides. Possibly the best illustration of this principle is to be found in the lipopolysaccharides, which are the somatic antigens of gram-negative bacteria and hence form the chemical basis of serological classification (115). Extensive investigations of lipopolysaccharides have been the subject of several reviews ($19,55,116-118$). These investigations have shown that lipopolysaccharides are composed of three regions: (a) a lipid portion, referred to as lipid A, which is the point of attachment in the outer membrane; (b) a basic core, which is termed the R antigen and consists of a short polysaccharide chain, and (c) the O-antigen-specific polysaccharide chains. The last-named chains carry the serological specificities in the form of oligosaccharide repeating units, a majority of which contain one or two single monosaccharide side chains. These single-unit side chains have been shown to be immunodominant and responsible for the O factors that to-gether account for the specificity of the organism. Many of the single-unit side chains consist of an unusual class of sugars, the 3,6-dideoxyhexoses, which are not found elsewhere and contribute to the very large number of serological classes within the Enterobacteriaceae.

There is less variation within the basic core region of lipopolysaccharides, the R-antigenic region. Within the same genera the core structures are very similar, if not identical, and contribute little to serological specificity. The core polysaccharides generally contain the common sugars 2-acetamido-2-deoxy-D-glucose, D-glucose, and D-galactose joined together in a pentasaccharide unit. Variations in the structures of the core polysaccharide arise from the location and identity of single-unit side chains in the pentasaccharide, thus providing for variations across genera. Some natural mutants of the Enterobacteriaceae lack the O chains in their lipopolysaccharides and are known as *rough* mutants. In those instances the R chains, the core polysaccharide portions of the molecule, become the predominant antigens and serve to distinguish one R mutant from another. For example, variations in core structure lead to serotype antigens of *N. meningitidis* (*119,120*). Bacteria with complete lipopolysaccharides, including O chains, are said to be in the *smooth* form.

Immunochemical studies on exopolysaccharides such as those from *Streptococcus pneumoniae* (*106,121*) and *Klebsiella* species (*122*) have yielded information about carbohydrate determinants similar to that exemplified by the lipopolysaccharides. The general principles of the immune response to polysaccharide antigens may be stated as follows. (a) Antibodies are generated against small portions (determinant groups) of the polysaccharide molecule ranging from two to four monosaccharide units; (b) a monosaccharide residue at the nonreducing end of a linear molecule, or as a single-unit side chain, or the terminal residue on an oligosaccharide side chain is the immunodominant sugar of the determinant group; (c) in polysaccharides with oligosaccharide side chains the serological specificity resides in the side chains and not in the backbone; (d) in polysaccharides with single-unit side chains the specificity extends to include the two or three neighboring residues in the backbone or main chain; (e) antipolysaccharide immunoglobulins raised in animals are heterogeneous and consist of subpopulations of molecules with specificities toward different parts or different sides of the polysaccharide.

The inference from these conclusions is that it is the most exposed parts of a polysaccharide antigen that are seen and recognized by the antibody molecule in its role as either a receptor on a lymphocyte or as a serum immunoglobulin.

C. Noncarbohydrate Determinants in Polysaccharides

Certain noncarbohydrate groups may function as antigenic determinants *when present as substituents in polysaccharides.* Jann and Westphal (*55*) summarized earlier work showing that *O*-acetyl groups are essential parts of the determinant regions in some *Salmonella* lipopolysaccharides. More recently, *O*-acetyl groups have been found in some of the capsular polysaccharide antigens of *N. meningitidis* (*107*) and shown to be immunologically significant in the one from serogroup *C* (*95*).

When present in antigenic polysaccharides, pyruvate ketals are determinant groups because their removal leads to weaker precipitin reactions in homologous antisera (*123–125*). These pyruvate ketals have three structural features that affect immunological specificity: (a) the configuration of the monosaccharide residue that bears the ketal, (b) the positions of the two hydroxyl groups that are bridged by the ketal, and (c) the configuration of the methyl and carboxyl groups at the asymmetric ketal carbon atom. The importance of the latter structural feature has been proved for the pyruvate ketal in the polysaccharide from type 27 *Streptococcus pneumoniae* (*126*), and the importance of the first two can be inferred from studies of cross-reactions (see Section IV,E).

It is more difficult to demonstrate immunological significance for the phosphate esters that occur widely in polysaccharides. The teichoic acids (*13–16*), polymers of glycerol phosphate and ribitol phosphate, usually contain monosaccharides attached glycosidically to the polyol, and these monosaccharides are the immunodominant groups (*122*). Glycerol phosphate was found to play only a minor role in the specificity of the polysaccharide from type 28 *Streptococcus pneumoniae* (*127*) but was an immunodominant substituent in a polysaccharide from *Escherichia coli* (*128*). Phosphorylcholine was a better inhibitor than choline of the homologous precipitin reaction of the polysaccharide from type 27 *Streptococcus pneumoniae*, indicating some involvement of the phosphate group (*125*).

The discovery (*129*) that carrageenans, sulfated polygalactans present in marine algae, are immunogenic has led to immunochemical studies to establish the effective determinants (*129–132*). These studies have shown the presence of antibodies directed toward some structural feature associated with the 6-sulfate group of the (1→4)-linked D-galactose (*131,132*). These seem to be the only reports showing involvement of a sulfate substituent in an immune reaction.

D. Conformation

In addition to the structural determinants that have been discussed in the previous two sections, it is likely that polysaccharides present conformational determinants that depend on the secondary structure or shape of the molecule. Sela *et al.* (*133*) demonstrated such conformational determinants for proteins by showing that antibodies to a sequential peptide determinant did not recognize the same sequence when present in a rigid helical structure. Such a clear-cut demonstration has not yet been obtained with polysaccharides, but there is some evidence that conformational determinants may be important in this class of antigens. Rees (*134*) showed that many plant polysaccharides have an ordered structure in solution arising from intra- and interchain interactions and from association with unlike chains. Of particular interest was the subsequent observation that an exopolysaccharide from *Xanthomonas campestris* not only had an ordered structure in solution but interacted with certain plant cell wall polysaccharides, suggesting a possible role in the infection of plants by this plant pathogen (*135*). Similar studies were done on the capsular polysaccharides of some *Klebsiella* species, and temperature-dependent order–disorder transitions were observed (*136*). In that study the serotype K8 polysaccharide was found to have a two-stage transition. This polysaccharide has the repeating unit

D-GlcUA

1

↓

4

$----\to 3)$-β-D-Gal-$(1\to 3)$-α-D-Gal-$(1\to 3)$-β-D-Glc-$(1---$

and the first transition was interpreted as a cooperative change in orientation of the uronic acid, which resulted in a reordering of the backbone into another helical conformation. If this interpretation can be substantiated, it would represent an example of true allosterism in polysaccharides. More direct evidence for participation of a substituent group in conformational control has been provided by Jennings, Lugowski, and Kasper (*137*) in studies of the type III group B streptococcal polysaccharide. This important antigen has the following repeating unit:

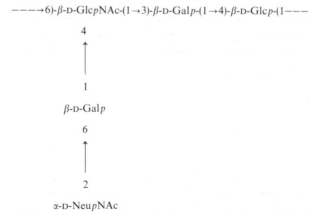

$----\to 6)$-β-D-Glc$p$NAc-$(1\to 3)$-β-D-Gal$p$-$(1\to 4)$-β-D-Glc$p$-$(1---$

4

↑

1

β-D-Gal$p$

6

↑

2

α-D-Neu$p$NAc

On structural grounds the *N*-acetylneuraminosylgalactose side chain would be expected to be immunodominant; however, serotransferrin, which contains this disaccharide unit, did not inhibit the homologous precipitin reaction. Furthermore, removal of C-8 and C-9 from the glycol chain of the N-acetylneuraminic acid residues by controlled oxidation with periodate did not alter the serological properties. However, since reduction of the carboxyl group or hydrolytic removal of the N-acetylneuraminic acid altered the serological properties of the antigen, it was clear that the N-acetyl-neuraminosyl residue played some role in establishing specificity. Chemical shift displacements in ^{13}C-NMR spectra before and after removal of the N-acetylneuraminosyl residues showed that these residues were responsible for conformational control over the determinant and that this change in conformation was restricted to the branches and did not occur in the backbone of the polysaccharide. Thus, the N-acetylneuraminosyl residues control the orientation of the branch β-D-galactopyranosyl residues in relation to the backbone to create a conformation that is immunodominant and responsible for the specificity of the native polysaccharide. It was proposed that this conformational control could be due to interaction of the N-acetyl-neuraminic acid with the backbone of the polysaccharide, possibly by hydrogen bonding of the carboxylate group to the 3-OH of the N-acetyl-D-glucosaminosyl residue. It is possible and perhaps even likely that other substituents such as O-acetyl groups, pyruvate ketals, and phosphate and sulfate esters confer immunological specificity by similar mechanisms.

E. Cross-Reactions

An early premise of immunology was that an antigen generates its own antibody and that the two are mutually specific. Two subsequent developments have shown that the specificity of antigen–antibody reactions is not totally exclusive. The first was the realization that antibodies raised by immunization consisted of subpopulations of antibody molecules with different degrees of complementarity for the antigen. More recently, it has been shown that a relatively simple determinant, the type 3 *Streptococcus pneumoniae* polysaccharide, can be bound by several different complementary sites (*138*). The second development was the concept of determinant groups and the recognition that only a relatively small portion of a polysaccharide antigen is the major site of antibody specificity. These two developments led to an explanation of cross-reactivity, i.e., in which one antigen reacts with some of the antibodies raised to another antigen. It should be emphasized that cross-reactivity of two antigens in the same antiserum *does not* mean that the antigens are identical. It *does* mean that the two antigens probably have some structural features in common.

Cross-reactions abound in bacterial polysaccharides and can be used to establish the structures of immunodeterminant groups. Thus, if antiserum that has been raised against a polysaccharide of known structure is available, it can be used to screen other polysaccharides for cross-reactivity, the extent of which is an indication of structural similarity in determinants. Heidelberger (5) has used this approach extensively, and the results have been included in several reviews (5,121,139–141). Only a few examples are given here to demonstrate the principles involved. Cross-reactivity of a number of polysaccharides of known structure in antipneumococcal type 7 sera showed that the principal determinant in the type VII polysaccharide must include a terminal β-D-galactopyranose residue (142). The cross-reactivity of some *Rhizobia* exopolysaccharides in antipneumococcal type 27 sera was traced to the common presence of pyruvate groups (123). Studies of cross-reactions of more than 60 capsular, type-specific polysaccharides of *Klebsiella* in 26 specific types of antipneumococcal sera permitted the assignment of several structural features such as nonreducing terminal residues of D-glucuronic acid, L-rhamnose, and D-galacturonic acid and the linkage of some monosaccharides within the polysaccharide chains (143,144). The results of cross-reactions must always be interpreted with caution. In particular, a positive cross-reaction is good evidence of a common determinant, but lack of a cross-reaction does not necessarily indicate that common determinants are not present. For example, the de-O-acetylated polysaccharide antigen from *Tremella mesenterica* contains single-unit, nonreducing terminal residues of D-glucuronic acid but does not give a cross-reaction with anti-type 2 pneumococcal serum (145), for which D-glucuronic acid is a strong determinant. This lack of reactivity was traced to the presence, in the *Tremella* polysaccharide, of single-unit D-xylopyranosyl residues located in the vicinity of the uronic acid residues so that the latter were "hidden" or "screened" and thus prevented by steric hindrance from reacting with the antibody.

V. Polysaccharide Antigens in Human Disease

The significance of polysaccharide antigens in human disease was first recognized in the early studies on the pneumococci (*Streptococcus pneumoniae*) already cited (1–4). Those studies were initiated because at that time pneumococcal pneumonia was the most common and most serious of all acute infections, with a mortality that exceeded the combined rate from measles, scarlet fever, diphtheria, whooping cough, typhoid fever, and dysentery; Osler (146) referred to pneumonia as "Captain of the men of death." With increasing knowledge about the immune system the role of polysaccharides in human disease became clearer. In this section we discuss how polysaccharides affect bacterial virulence, their use in protective vaccines and for diagnosis, and finally their possible role in passive or naturally acquired immunity.

A. Polysaccharides and Virulence

The development of a bacterial infection depends on complex interactions between the host and the organism. Variation in host resistance is one factor in this complex relationship, and the virulence of the infecting organism is another. Bacterial virulence may be regarded as the capacity to resist the host's defense mechanisms so that the microorganism can multiply and eventually damage or kill the host. There is now abundant evidence that capsular polysaccharides are important virulence factors in many bacterial infections including those caused by *Streptococcus pneumoniae, Neisseria meningitidis, Hemophilus influenzae, Escherichia coli, Salmonella typhosa, Klebsiella pneumoniae*, and *Staphylococcus aureus (1–4,55,147–149)*. The primary role of capsular polysaccharides in bacterial virulence is to protect the organism from destruction by the phagocytic cells of the body. Phagocytic cells, macrophages and neutrophils, originating in the bone marrow, circulate in the bloodstream and migrate, by chemotactic response, to the site of infection. There they ingest and kill the invading organism by hydrolytic and oxidative enzymatic processes. Ingestion of the microbial cell usually requires that it be opsonized, i.e., coated with specific IgG antibody and sometimes the C3 component of complement to cover antiphagocytic groups on the surface and to form a ligand between the bacterial cell and the IgG and complement receptors on the phagocyte. When capsule-specific antibody is not present, the bacterium will multiply, resulting in an acute infection that may be fatal. This accounts for the success of serum therapy *(150)* and for the repeated observation that the amount of circulating capsular polysaccharide in patients is indicative of the severity of the infection *(1,151–155)*. In noncapsulated gram-negative bacteria the surface lipopolysaccharides appear to have antiphagocytic properties. Studies with an *E. coli* strain showed that two mutants, in which the O-antigenic chains of the lipopolysaccharide were incomplete, were less virulent and more easily phagocytized than the parent strain, in which the O-antigenic strains were complete *(156)*. Noncapsulated organisms are generally phagocytized readily, because of the ability of other exposed surface molecules (e.g. teichoic acid) to directly effect complement induced phagocytosis. The role of microbial surface components in phagocytosis has been reviewed *(157)*, as have the mechanisms by which phagocytic cells function *(149)*.

B. Protective Vaccines

The first capsular polysaccharide vaccine arose from the early work on *Streptococcus pneumoniae (1–4)* when a multivalent vaccine containing the polysaccharides of six types showed 100% efficacy in preventing pneumonia caused by those organisms when tested in U.S. Army recruits (4). As a result of these trials a license was awarded to the E. R. Squibb Company in 1945 to manufacture and distribute this vaccine. With the introduction of antibiotics the company withdrew its license in 1947. The complacency engendered by the success of antibiotic therapy resulted in the discontinuation of accurate diagnosis by typing sera and a general lack of awareness of the importance of pneumococcal infection. However, the clinical and epidemiological studies of Austrian (150,158–160) showed that pneumococcal pneumonia occurred with the same frequency and with the same fatality (18–25%) as in the preantibiotic era. This coupled with the emergence of multiple antibiotic resistance (161) led to a reconsideration of prevention rather than cure and resulted in the licensing of a 14-valent pneumococcal polysaccharide vaccine in 1977 (162). At about the same time it was found that N. meningitidis had developed resistance to sulfadiazine (163), and outbreaks or epidemics of meningitis occurred in several countries. This led to the rapid development and licensing of meningococcal polysaccharide vaccines (164–166). These pneumococcal and meningococcal vaccines are the only ones based on capsular polysaccharides that are generally available, although, as shown in Table I, there are several other encapsulated bacterial species that cause serious diseases. Infectious diseases are the fifth most common cause of death in the United States and place a heavy burden on health care delivery systems throughout the world. This together with the emergence of widespread, multiple antibiotic resistance has given rise to renewed interest in the prevention of these diseases by immunization. Approximately 130 million individuals have been immunized with capsular polysaccharides, resulting in a high degree of disease protection and no fatalities or significant adverse effects from the vaccination. The immunity generated by these vaccines appears to be long lasting (2,147), and a cost effectiveness analysis of

TABLE I

Diseases Caused by Encapsulated Bacteria

Bacterial species	Diseases
Streptococcus pneumoniae	Pneumococcal pneumonia, meningitis, otitis media
Neisseria meningitidis	Meningitis
Hemophilus influenzae type b	Meningitis, epiglottitis, septic arthritis, and pneumonitis with empyema
Salmonella typhosa	Typhoid fever
Group B *Streptococcus*	Meningitis in newborns
Escherichia coli	Meningitis, neonatal septicemia, urinary tract infections
Staphylococcus aureus	Abscesses, septicemia
Klebsiella pneumoniae	Pneumonia, meningitis

pneumococcal vaccination has shown that provision of the vaccine to the elderly and to convalescents merits attention as a strategy to moderate medical care expenditures (*167*). The promising outlook for polysaccharide vaccines has given rise to considerable research in this field, most of which has already been summarized (*147,168,169*). In addition to the bacterial diseases listed in Table I, work has been proceeding toward possible immunization against infections by anaerobes such as *Bacteroides fragilis* (*170*) and *Bacteroides asaccharolyticus* (*171*) and by *Pseudomonas aeruginosa* in burns (*172*) and in pulmonary tissue (*173*).

The difficulties of preparing capsular polysaccharides in an immunogenic form and in good yield have been largely overcome; however, two other problems associated with the development of these vaccines should be mentioned. The first is that antibody raised in response to these antigens is not always protective. There are several possible reasons for this phenomenon. (a) The newer methods (such as RABA and ELISA) for detecting antibodies are extremely sensitive, and, although the presence of specific antibody may be demonstrated, the level of response may not be high enough to confer protection; (b) antibodies generated by different antigens vary in their avidity (binding efficiency), and some may not be sufficiently avid to opsonize the bacterial cell in preparation for phagocytosis; (c) some organisms may be eliminated by a cell-mediated response rather than a humoral response. It is therefore essential that antisera be tested not only for the presence of specific antibody but also for its bactericidal effect on the target organism. Bactericidal tests are easily done by serial dilution of the test serum in microtiter plates. To the serum samples are then added a source of complement and a small number (10–20) of bacterial cells (pregrown and in the log phase). After a brief period of incubation (30–60 min) the contents of each well are inoculated onto culture plates, which are then incubated, and the number of colonies is counted (*174*).

The second and more serious problem in developing polysaccharide vaccines is that infants and young children (under 2 years), unlike adults, do not develop protective levels of serum antibodies (*175,176*). For most of the bacterial infections listed in Table I the highest attack rate, mortality, and morbidity occur in infants and young children. In young children polysaccharide vaccines give rise to only IgM antibodies and there is no anamnestic (memory) response, whereas protein vaccines induce both IgM and IgG antibodies and an immunological memory (*177*). This difference in immune response is thought to be due to the T-cell independence of polysaccharide antigens, the participation of T cells being essential for the induction of IgG antibodies and memory cells (*178*). There have been several attempts to overcome this problem by conjugating polysaccharides to an antigenic protein in order to form a T-cell-dependent antigen (*179–181*). Possibly the most promising approach is the coupling through a single, terminal, free aldehyde group, generated in the polysaccharide, to an acceptable protein antigen, tetanus toxoid (95). This method offers the advantages of a defined linkage that does not affect the antigenic structure and is innocuous for purposes of human vaccination. As discussed earlier (Section III,B,4), adjuvants do not appear to be useful in potentiating immune responses to T-cell-independent antigens such as polysaccharides.

C. Diagnosis

Polysaccharides on the surface of a microbial cell are the serological determinants of that organism and therefore represent a highly specific means of identification. With the development of multiple and variable antibiotic resistance, the accurate identification of the causative organism in an infection has become important in deciding on the appropriate therapy. Thus, the use of specific typing sera, common in the preantibiotic era, is once again receiving attention (159). This means of identification is invaluable in following the course of epidemics, in charting endemic diseases, and in detecting the presence of bacteria that are otherwise difficult to identify. A few examples will serve to illustrate these uses. There are approximately 83 distinct serotypes of *Streptococcus pneumoniae*, but most pneumococcal disease is caused by about 14 capsular types. A study of the changes in occurrence of different serotypes in disease isolates over a period of 40 years showed that there was considerable variation both with time and with geographic location (182). A more recent report (183) showed that only 70% of isolates from cases of pneumococcal disease were of the specific types included in the polyvalent vaccine. The results of these two studies emphasize the need for continuing surveillance as a guide to vaccine formulation. Similarly, *N. meningitidis* has at least seven serogroups, of which groups A, B, and C cause most meningococcal disease. The severe epidemic in Brazil from 1972 to 1974 was caused originally by group C but then changed to group A, and there have been outbreaks of group B in Belgium, Norway, and England (147). The other groups have so far shown a low disease attack rate, but continuous monitoring is desirable in view of our limited knowledge of the epidemiology of this organism. An example of the identification of bacteria is the recent development of diagnostic tests for *Neisseria gonorrhoeae* by the use of highly specific antiserum raised in hens to the lipopolysaccharide (184,185). Reference has already been made to the use of capsular polysaccharides for measuring antibody content in immunized or infected patients (74,75,85) and to the significance of circulating polysaccharide antigen as evidence of severe infection (1,151–155).

D. Natural Immunity

It has long been known that adult animal sera contain antibodies to a variety of polysaccharide antigens including those of pathogenic bacteria. The result is that an animal may show immunity to a pathogenic organism even though it was not exposed to it previously. This passive, or natural, immunity was found to be based on cross-reactions among the polysaccharide antigens (*186*). Thus, many nonpathogenic bacteria, normal inhabitants of the intestine or pharynx, have capsular polysaccharides that are sufficiently similar to those of some pathogens that protective antibodies are induced. Robbins (*147*) has given a summary of the antigens from nonpathogens that cross-react with type 1 and 3 *Streptococcus pneumoniae*, *N. meningitidis* groups A, B, and C, *Hemophilus influenzae* type b, and *Salmonella typhosa*. As proof of this cross-reaction protection Schneerson and Robbins (*187*) showed that the feeding to human volunteers of a nonpathogenic *E. coli* strain that was cross-reactive with *H. influenzae* type b resulted in intestinal colonization by the *E. coli* and the induction of antibodies to the *H. influenzae* type b capsular polysaccharide. The possibility therefore exists of deliberate intestinal colonization with cross-reacting *E. coli* strains as a method of immunization against some pathogens. Of course, cross-reacting polysaccharides may also find use in vaccines.

Finally, it is likely that natural immunity to some pathogenic bacteria arises from a mild or asymptomatic infection by another pathogen due to the presence of cross-reactive antigens. Cross-reactions among major pyogenic bacteria have been shown to occur frequently, and this topic has been the subject of intense investigation (*168*).

References

1. A. R. Dochez and O. T. Avery, *J. Exp. Med.* **26**, 477–493 (1917).
2. M. Heidelberger, "Lectures in Immunochemistry." Academic Press, New York, 1956.
3. C. M. MacLeod, R. G. Hodges, M. Heidelberger, and W. G. Bernhard, *J. Exp. Med.* **82**, 445–465 (1945).
4. M. Heidelberger, C. M. MacLeod, J. J. Kaiser, and B. Robinson, *J. Exp. Med.* **83**, 303–320 (1946).
5. M. Heidelberger, *Res. Immunochem. Immunobiol.* **3**, 1–40 (1978).
6. E. A. Kabat, "Blood Group Substances." Academic Press, New York, 1956.
7. B. Siddiqui and S. Hakomori, *Biochim. Biophys. Acta* **330**, 147–155 (1973).
8. C. Gram, *Fortschr. Med.* **2**, 185–189 (1884).
9. M. R. J. Salton, "The Bacterial Cell Wall." Elsevier, Amsterdam, 1964.
10. K-J. Cheng and J. W. Costerton, *J. Bacteriol.* **129**, 1506–1512 (1977).
11. J. W. Costerton, *Annu. Rev. Microbiol.* **23**, 459–479 (1979).

12. H. J. Rogers, H. R. Perkins, and J. B. Ward, "Microbial Cell Walls and Membranes." Chapman & Hall, London, 1980.
13. N. Shaw and J. Baddiley, *Biochem. J.* **93**, 317–321 (1964).
14. J. Baddiley, J. G. Buchanan, R. O. Martin, and U. L. Rajbhandary, *Biochem. J.* **85**, 49–56 (1962).
15. A. R. Archibald, J. Baddiley, and D. Button, *Biochem. J.* **95**, 8c–11c (1965).
16. J. Baddiley, *Proc. R. Soc. London, Ser. B* **170**, 331–348 (1968).
17. M. M. Burger, *Proc. Natl. Acad. Sci. U.S.A.* **56**, 910–917 (1966).
18. D. C. Birdsell, R. J. Doyle, and M. Morgenstern, *J. Bacteriol.* **121**, 726–734 (1975).
19. O. Lüderitz, O. Westphal, A. M. Staub, and H. Nikaido, *in* "Microbial Toxins" (G. Weinbaum, S. Kadis, and S. J. Ajl, eds.), Vol. 4, pp. 145–233. Academic Press, New York, 1971.
20. P. F. Mühlradt, J. Menzel, J. F. Golecki, and V. Speth, *Eur. J. Biochem.* **35**, 471–481 (1973).
21. E. M. Bott, C. W. Bonynge, and R. L. Joyce, *J. Infect. Dis.* **58**, 5–9 (1936).
22. M. E. Bayer and H. Thurow, *J. Bacteriol.* **130**, 911–936 (1977).
23. D. L. Kasper and C. J. Baker, *J. Infect. Dis.* **139**, 147–151 (1979).
24. S. Schwarzmann and J. R. Boring, *Infect. Immun.* **3**, 762–767 (1971).
25. A. A. Glynn, "Microbial Pathogenicity in Man and Animals." Cambridge Univ. Press, London and New York, 1972.
26. H. J. Phaff, *in* "The Yeasts" (A. T. Rose and J. S. Harrison, eds.), pp. 133–210. Academic Press, New York, 1971.
27. B. Mundkur, *Exp. Cell Res.* **20**, 28–42 (1960).
28. H. F. Hasenclever and W. O. Mitchell, *J. Bacteriol.* **82**, 570–573 (1961).
29. H. F. Hasenclever and W. O. Mitchell, *J. Immunol.* **93**, 763–771 (1964).
30. S. Suzuki, H. Sunayama, and T. Saito, *Jpn. J. Microbiol.* **12**, 19–24 (1968).
31. C. E. Ballou, *J. Biol. Chem.* **245**, 1197–1202 (1970).
32. J. H. Burnett and A. P. J. Trinci, "Fungal Walls and Hyphal Growth." Cambridge Univ. Press, London and New York, 1979.
33. M. C. Raff, *Nature (London)* **242**, 19–23 (1973).
34. R. M. Krause, *DHEW Publ. (NIH) (U.S.)* **NIH-74-533**, 16–25 (1973).
35. P. J. Baker and B. Prescott, *Dev. Immunol.* **2**, 67–104 (1979).
36. E. S. Golub, "The Cellular Basis of the Immune Response." Sinauer Assoc., Sunderland, Massachusetts, 1977.
37. E. Diener and R. E. Langman, *Prog. Allergy* **18**, 6–42 (1975).
38. H. J. Müller-Eberhard, *Adv. Immunol.* **8**, 1–80 (1968).
39. J. G. Howard, G. H. Christie, B. M. Courtenay, E. Leuchars, and A. J. S. Davies, *Cell. Immunol.* **2**, 614–626 (1971).
40. B. Andersson and H. Blomgren, *Cell. Immunol.* **2**, 411–424 (1971).
41. G. F. Mitchell, F. C. Grumet, and H. D. McDevitt, *J. Exp. Med.* **135**, 126–135 (1972).
42. N. L. Warner, P. Byrt, and G. L. Ada, *Nature (London)* **226**, 942–943 (1970).
43. H. Braley-Mullen, *J. Immunol.* **125**, 1849–1864 (1980).
44. A. Nisonoff, J. E. Hopper, and S. B. Spring, "The Antibody Molecule." Academic Press, New York, 1975.
45. I. Roitt, "Essential Immunology." Blackwell, Oxford, 1971.
46. H. G. Kunkel, M. Mannik, and R. C. Williams, *Science* **140**, 1218–1219 (1962).
47. D. G. Braun, K. Eichmann, and R. M. Krause, *J. Exp. Med.* **129**, 809–830 (1969).
48. K. Eichmann, D. G. Braun, and R. M. Krause, *J. Exp. Med.* **134**, 48–65 (1971).
49. F. W. Chen, A. D. Strosberg, and E. Haber, *J. Immunol.* **110**, 98–106 (1973).

50. K. Eichmann and R. W. Krause, *Fed. Proc., Fed. Am. Soc. Exp. Biol.* **28**, 695 (1969).
51. C. K. Osterland, E. J. Miller, W. W. Karakawa, and R. M. Krause, *J. Exp. Med.* **123**, 599–614 (1966).
52. E. J. Miller, C. K. Osterland, J. M. Davie, and R. M. Krause, *J. Immunol.* **98**, 710–715 (1967).
53. J. H. Pincus, J. C. Jaton, K. J. Bloch, and E. Haber, *J. Immunol.* **104**, 1143–1148, 1149–1154 (1970).
54. J. Oudin and M. Michel, *C.R. Hebd. Seances Acad. Sci.* **257**, 805–808 (1963).
55. K. Jann and O. Westphal, *in* "The Antigens" (M. Sela, ed.), Vol. 3, pp. 1–125. Academic Press, New York, 1975.
56. M. Potter, S. Rudikoff, E. A. Padlan, and M. Vrana, *in* "Antibodies in Human Diagnosis and Therapy" (E. Haber and R. M. Krause, eds.), pp. 9–28. Raven, New York, 1977.
57. C. P. J. Glaudemans, *Adv. Carbohydr. Chem. Biochem.* **31**, 313–346 (1975).
58. K. R. McIntire and G. Princler, *Immunology* **17**, 481–487 (1969).
59. G. Köhler and C. Milstein, *Nature (London)* **256**, 495–497 (1975).
60. C. Milstein and G. Köhler, *in* "Antibodies in Human Diagnosis and Therapy" (E. Haber and R. M. Krause, eds.), pp. 271–284. Raven, New York, 1977.
61. J. Freund, *Annu. Rev. Microbiol.* **1**, 291–308 (1947).
62. E. Lederer, A. Adam, R. Ciorbaru, J. F. Petit, and J. Wietzerbin, *Mol. Cell. Biochem.* **7**, 87–104 (1975).
63. L. Chédid and F. Audibert, *in* "Microbiology–1977" (D. Schlessinger, ed.), pp. 388–394. Am. Soc. Microbiol. Washington, D.C., 1977.
64. A. Adam, R. Ciorbaru, J. F. Petit, and E. Lederer, *Proc. Natl. Acad. Sci. U.S.A.* **69**, 851–854 (1972).
65. F. Ellouz, A. Adam, R. Ciorbaru, and E. Lederer, *Biochem. Biophys. Res. Commun.* **59**, 1317–1324 (1974).
66. C. Merser, P. Sinaÿ, and A. Adam, *Biochem. Biophys. Res. Commun.* **66**, 1316–1322 (1975).
67. L. Chédid, C. Carelli, and F. Audibert, *J. Reticuloendothel., Soc.* **26**, 631–641 (1979).
68. R. G. White, *Annu. Rev. Microbial.* **30**, 579–600 (1976).
69. J. A. Rudbach, *in* "The Role of Immunological Factors in Infectious, Allergic, and Autoimmune Processes" (R. F. Beers, Jr. and E. G. Bassett, eds.), pp. 29–41. Raven, New York, 1976.
70. T. J. Gill, III, *Immunochemistry* **7**, 997–1000 (1970).
71. M. Heidelberger and F. E. Kendall, *J. Exp. Med.* **50**, 809–823 (1929).
72. E. A. Kabat and M. M. Mayer, "Experimental Immunochemistry." Thomas, Springfield, Illinois 1961.
73. R. S. Farr, *J. Infect. Dis.* **103**, 239–262 (1958).
74. E. C. Gotschlich, M. Rey, R. Triau, and K. J. Sparks, *J. Clin. Invest.* **51**, 89–96 (1972).
75. B. L. Brandt, F. A. Wyle, and M. S. Artenstein, *J. Immunol.* **108**, 913–920 (1972).
76. R. C. Lancefield, *J. Exp. Med.* **59**, 441–458 (1934).
77. J. Oudin, *C.R. Hebd. Seances Acad. Sci.* **222**, 115–116 (1946).
78. J. R. Preer, Jr., *J. Immunol.* **77**, 52–60 (1956).
79. O. Ouchterlony, *Acta Pathol. Microbial. Scand.* **25**, 186–191 (1948).
80. O. Ouchterlony, *Prog. Allergy* **6**, 30–154 (1962).
81. P. Grabar and C. A. Williams, *Biochim. Biophys. Acta* **17**, 67–84 (1955).
82. E. V. Keogh, E. A. North, and M. F. Warburton, *Nature (London)* **160**, 63 (1947).
83. E. Neter, O. Westphal, O. Lüderitz, and E. A. Gorzynski, *Ann. N.Y. Acad. Sci.* **66**, 141–161 (1956).

84. C. J. Sanderson and D. V. Wilson, *Immunochemistry* **8**, 163–168 (1971).
85. M. S. Artenstein, B. L. Brandt, E. C. Tramont, W. C. Brauche, Jr., H. D. Fleet, and R. L. Cohen, *J. Infect. Dis.* **124**, 277–288 (1971).
86. G. R. Hodges, S. E. Worley, C. E. Degener, and G. M. Clark, *Infect. Immun.* **28**, 832–836 (1980).
87. R. E. Mandrell and W. D. Zollinger, *Infect. Immun.* **16**, 471–475 (1977).
88. C. J. Baker and D. L. Kasper, *N. Engl. J. Med.* **294**, 753–756 (1976).
89. G. Schiffman and R. Austrian, *Fed. Proc., Fed. Am. Soc. Exp. Biol.* **30**, 658 (1971).
90. J. B. Robbins, J. C. Parke, Jr., R. Schneerson, and J. K. Whisnant, *Pediatr. Res.* **7**, 103–110 (1973).
91. E. Engvall and P. Perlmann, *Immunochemistry* **8**, 871–874 (1971).
92. E. Engvall, K. Jonsson, and P. Perlmann, *Biochim. Biophys. Acta* **251**, 427–434 (1971).
93. E. Engvall and P. Perlmann, *J. Immunol.* **109**, 129–135 (1972).
94. H. E. Carlsson, A. A. Lindberg, and S. Hammarström, *Infect. Immun.* **6**, 703–708 (1972).
95. H. J. Jennings and C. Lugowski, *J. Immunol.* **127**, 1011–1018 (1981).
96. B. M. Gray, *J. Immunol. Methods* **28**, 187–192 (1979).
97. D. J. Barrett, A. J. Ammann, S. Stenmark, and D. W. Wara, *Infect. Immun.* **27**, 411–417 (1980).
98. R. J. Fielder, C. T. Bishop, S. F. Grappel, and F. Blank, *J. Immunol.* **105**, 265–267 (1970).
99. D. Givol, S. Fuchs, and M. Sela, *Biochim. Biophys. Acta* **63**, 222–224 (1962).
100. E. A. Kabat and A. E. Bezer, *Arch. Biochem. Biophys.* **78**, 306–318 (1958).
101. J. G. Howard, H. Zola, G. H. Christine, and B. M. Courtenay, *Immunology* **21**, 535–546 (1971).
102. B. L. Brandt, M. S. Artenstein, and C. D. Smith, *Infect. Immun.* **8**, 590–596 (1973).
103. J. A. Rudbach, *J. Immunol.* **106**, 993–1001 (1971).
104. E. Netu, H. Y. Whang, and H. Mayer, *in* "Bacterial Lipopolysaccharides" (E. H. Kass and S. M. Wolff, eds.), pp. 48–51. Univ. of Chicago Press, Chicago, Illinois, 1973.
105. I. W. Sutherland, *Adv. Microb. Physiol.* **8**, 143–213 (1972).
106. O. Larm and B. Lindberg, *Adv. Carbohydr. Chem. Biochem.* **33**, 295–322 (1976).
107. H. J. Jennings, A. K. Bhattacharjee, D. R. Bundle, C. P. Kenny, A. Martin, and I. C. P. Smith, *J. Infect. Dis.* **136**, 578–583 (1977).
108. H. J. Jennings and I. C. P. Smith, *in* "Methods in Enzymology" (V. Ginsburg, ed.), Vol. 50, Part C, pp. 39–50. Academic Press, New York, 1978.
109. K. Landsteiner, "The Specificity of Serological Reactions." Harvard Univ. Press, Cambridge, Massachusetts, 1945.
110. E. A. Kabat, *J. Immunol.* **84**, 82–85 (1960).
111. E. A. Kabat, *J. Immunol.* **77**, 377–385 (1956).
112. J. Gelzer and E. A. Kabat, *Immunochemistry* **1**, 303–316 (1964).
113. C. E. Ballou, P. N. Lipke, and W. C. Raschke, *J. Bacteriol.* **117**, 461–467 (1974).
114. W. C. Raschke and C. E. Ballou, *Biochemistry* **11**, 3807–3816 (1972).
115. F. Kauffmann, "The Bacteriology of Enterobacteriaceae." Munksgaard, Copenhagen, 1966.
116. D. A. R. Simmons, *Bacteriol. Rev.* **35**, 117–148 (1971).
117. O. Lüderitz, A. M. Staub, and O. Westphal, *Bacteriol. Rev.* **30**, 192–255 (1966).
118. I. Orskov, F. Orskov, B. Jann, and K. Jann, *Bacteriol. Rev.* **41**, 667–710 (1977).
119. H. J. Jennings, A. K. Bhattacharjee, L. Kenne, C. P. Kenny, and G. Calver, *Can. J. Biochem.* **58**, 128–136 (1980).
120. W. D. Zollinger and R. E. Mandrell, *Infect. Immun.* **18**, 424–433 (1977).
121. M. Heidelberger, *Res. Immunochem. Immunobiol.* **3**, 1–40 (1973).

122. I. W. Sutherland, *in* "Immunochemistry: An Advanced Textbook" (L. E. Glynn and M. W. Steward, eds.), pp. 399–443. Wiley, New York, 1977.

123. W. F. Dudman and M. Heidelberger, *Science* **164**, 954–955 (1969).

124. M. Heidelberger, W. F. Dudman, and W. Nimmich, *J. Immunol.* **104**, 1321–1328 (1970).

125. L. G. Bennett and C. T. Bishop, *Can. J. Chem.* **55**, 8–16 (1977).

126. L. G. Bennett and C. T. Bishop, *Immunochemistry* **14**, 693–696 (1977).

127. S. Estrada-Parra and M. Heidelberger, *Biochemistry* **2**, 1288–1294 (1963).

128. B. Jann, K. Jann, G. Schmidt, I. Orskov, and F. Orskov, *Eur. J. Biochem.* **23**, 515–522 (1970).

129. K. H. Johnston and E. L. McCandless, *J. Immunol.* **101**, 556–562 (1968).

130. V. DiNinno and E. L. McCandless, *Carbohydr. Res.* **66**, 85–93 (1978).

131. V. DiNinno and E. L. McCandless, *Carbohydr. Res.* **67**, 235–241 (1978).

132. V. DiNinno and E. L. McCandless, *Immunochemistry* **15**, 273–274 (1978).

133. M. Sela, B. Schechter, I. Schechter, and F. Borek, *Cold Spring Harbor Symp. Quant. Biol.* **32**, 537–545 (1967).

134. D. A. Rees, *Biochem. J.* **126**, 257–273 (1972).

135. E. R. Morris, D. A. Rees, G. Yound, M. D. Walkinshaw, and A. Darke, *J. Mol. Biol.* **110**, 1–16 (1977).

136. C. Wolf, U. Elsässer-Beile, S. Stirm, G. G. S. Dutton, and W. Burchard, *Biopolymers* **17**, 731–748 (1978).

137. H. J. Jennings, C. Lugowski, and D. L. Kasper, *Biochemistry* **20**, 4511–4518 (1981).

138. D. A. Holowka, A. D. Strosberg, J. W. Kimball, E. Haber, and R. E. Cathou, *Proc. Natl. Acad. Sci. U.S.A.* **69**, 3399–3403 (1972).

139. M. Heidelberger, *Fortschr. Chem. Org. Naturst.* **18**, 503–536 (1960).

140. M. Heidelberger, *Annu. Rev. Biochem.* **36**, 1–12 (1967).

141. M. Heidelberger and M. E. Slodki, *Carbohydr. Res.* **24**, 401–407 (1972).

142. J. M. Tyler and M. Heidelberger, *Biochemistry* **7**, 1384–1392 (1968).

143. M. Heidelberger and W. Nimmich, *J. Immunol.* **109**, 1337–1344 (1972).

144. M. Heidelberger and W. Nimmich, *Immunochemistry* **13**, 67–80 (1976).

145. C. G. Fraser, H. J. Jennings, and P. Moyna, *Can. J. Biochem.* **51**, 225–230 (1973).

146. W. Osler, "The Principles and Practice of Medicine," 7th ed. Appleton, New York, 1909.

147. J. B. Robbins, *Immunochemistry* **15**, 839–854 (1978).

148. R. Bortolussi, P. Ferrieri, B. Björksten, and P. G. Quie, *Infect. Immun.* **25**, 293–298 (1979).

149. P. Densen and G. L. Mandell, *Rev. Infect. Dis.* **2**, 817–838 (1980).

150. R. Austrian, *Rev. Infect. Dis., Suppl.* **3**, S1–S17 (1981).

151. S. C. Bukantz, P. F. Gara, and J. G. M. Bullowa, *Arch. Intern. Med.* **69**, 191–212 (1942).

152. J. D. Coonrod and M. W. Rytel, *J. Lab. Clin. Med.* **81**, 778–786 (1973).

153. T. A. Hoffman and E. A. Edwards, *J. Infect. Dis.* **126**, 636–644 (1972).

154. G. E. Kenny, B. B. Wentworth, R. P. Beasley, and H. M. Foy, *Infect. Immun.* **6**, 431–437 (1972).

155. M. Pollack, *Infect. Immun.* **13**, 1543–1548 (1976).

156. D. N. Medearis, Jr., B. M. Camitta, and E. C. Heath, *J. Exp. Med.* **128**, 399–414 (1968).

157. H. Smith, *Bacteriol. Rev.* **41**, 475–500 (1977).

158. R. Austrian and J. Gold, *Ann. Intern. Med.* **60**, 759–776 (1964).

159. R. Austrian, *J. Infect. Dis.* **131**, 474–484 (1975).

160. R. Austrian, *J. Infect. Dis.* **136**, 538–542 (1977).

161. P. C. Appelbaum, A. Bhamjee, J. N. Scragg, A. F. Hallet, A. Bowen, and R. Cooper, *Lancet* **2**, 995–997 (1977).

162. P. Smit, D. Oberholzer, S. Hayden-Smith, H. J. Koornhof, and M. R. Hilleman, *JAMA, J. Am. Med. Assoc.* **233**, 2613–2620 (1977).

163. R. Gold and M. L. Lepow, *Adv. Pediatr.* **23**, 71–93 (1976).
164. I. Goldschneider, E. C. Gotschlich, and M. Artenstein, *J. Exp. Med.* **129**, 1307–1323 (1969).
165. M. S. Artenstein, R. Gold, J. G. Zimmerly, F. A. Wyle, H. Schneider, and C. Harkins, *N. Engl. J. Med.* **282**, 417–420 (1970).
166. K. H. Wong, O. Barrera, A. Sutton, J. May, D. H. Hockstein, J. D. Robbins, J. B. Robbins, P. D. Parkman, and E. B. Seligman, Jr., *J. Biol. Stand.* **5**, 197–215 (1977).
167. J. S. Willems, C. R. Sanders, M. A. Riddiough, and J. C. Bell, *N. Engl. J. Med.* **303**, 553–559 (1980).
168. J. B. Robbins, R. E. Horton, and R. M. Krause, eds., "New Approaches for Inducing Natural Immunity to Pyogenic Organisms," DHEW Publ. No. (NIH) 74–553. USDHEW, Public Health Serv., Rockville, Maryland, 1973.
169. R. M. Krause, *J. Infect. Dis.* **135**, 318–329 (1977).
170. D. L. Kasper, A. B. Onderdonk, J. Crabb, and J. G. Bartlett, *J. Infect. Dis.* **140**, 724–731 (1979).
171. B. L. Mansheim and D. L. Kasper, *J. Infect. Dis.* **140**, 945–951 (1979).
172. J. W. Alexander and M. W. Fisher, *J. Infect. Dis.* **130**, S152–S158 (1974).
173. J. E. Pennington, *J. Infect. Dis.* **140**, 73–80 (1979).
174. C. E. Frasch and S. S. Chapman, *Infect. Immun.* **5**, 98–102 (1972).
175. R. Gold, M. L. Lepow, I. Goldschneider, T. L. Draper, and E. C. Gotschlich, *J. Clin. Invest.* **66**, 1536–1547 (1975).
176. R. Gold, M. L. Lepow, I. Goldschneider, and E. C. Gotschlich, *J. Infect. Dis.* **136**, S31–S35 (1977).
177. P. J. Baker and P. W. Stashak, *J. Immunol.* **103**, 1342–1348 (1969).
178. H. Braley-Mullen, *Immunology* **40**, 521–527 (1980).
179. W. E. Paul, D. H. Katz, and B. Benacerraf, *J. Immunol.* **107**, 685–688 (1971).
180. E. C. Beuvery, F. Miedema, R. W. Van Delft, and J. Nagel, Seminars in Infectious Disease, Bacterial Vaccines. Vol. 4. ed. by J. B. Robbins, J. C. Hill, and J. C. Sadoff. Thieme-Stratton Inc., New York. (1982).
181. R. Schneerson, O. Barrera, A. Sutton, and J. B. Robbins, *J. Exp. Med.* **152**, 361–376 (1980).
182. M. Finland and M. W. Barnes, *J. Clin. Microbiol.* **5**, 154–166 (1977).
183. C. V. Broome, R. R. Facklam, J. R. Allen, and D. W. Fraser, *J. Infect. Dis.* **141**, 119–123 (1980).
184. B. B. Diena, F. E. Ashton, A. Ryan, R. Wallace, and M. B. Perry, *Can. J. Microbial.* **24**, 117–123 (1978).
185. R. Wallace, R. E. Ashton, A. Ryan, B. B. Diena, C. Malysheff, and M. B. Perry, *Can. J. Microbiol.* **24**, 124–128 (1978).
186. J. B. Robbins, R. Schneerson, M. P. Glode, W. Vann, M. S. Schiffer, T.-Y. Liu, J. C. Parke, and C. Huntley, *J. Allergy Clin. Immunol.* **56**, 141–151 (1975).
187. R. S. Schneerson and J. B. Robbins, *N. Engl. J. Med.* **292**, 1093–1096 (1975).

Index

A

2-Acetamido-4-amino-2,4,6-trideoxy-D-galactopyranose residues, deamination in polysaccharides, 121, 123

2-Acetamido-2-deoxy-D-mannurono-D-glucan, from *Micrococcus lysodeikticus,* deamination, 122

5-Acetamido-3,5-dideoxy-D-*glycero*-D-*galacto*-nonulosonic acid, Fischer convention, 7

Acetolysis, polysaccharide, 63

N-Acetylneuraminic acid, Fischer convention, 7

Adjuvants, 304

Agarose
  carbon-13 nuclear magnetic resonance spectroscopy and chemical structure, 158
  mercaptolysis and methanolysis, 67, 68
  oxolane ring formation, 95–97
  repeating units, 238
  structure, 240, 242, 245
  vacuum ultraviolet circular dichroism, 208, 209

Alditol acetates, methylated, primary fragment ions, 52–56

Alditols, permethylated oligosaccharide, mass spectral fragmentation, 76–79

Aldobiouronic acid, from rye flour arabinoxylan, 92

Aldononitrile acetates, methylated, primary fragment ions, 55

Alginic acid
  carbon-13 nuclear magnetic resonance spectroscopy and chemical structure, 158
  conformation and nuclear magnetic resonance spectroscopy, 160
  infrared-Raman spectroscopy, 178
  network formations, 266
  ribbon sequences and cation binding, 228–232
  structure, 21
  vacuum ultraviolet circular dichroism, 208
  viscosity and solution concentration, 260, 262

D-Allose, Fischer convention, 5, 6

Alkaline degradations, 100–112

D-Altrose, Fischer convention, 5, 6

2-Amino-2-deoxy-3-O-[(R)-1-carboxyethyl]-D-glucose, *see* Muramic acid

2-Amino-2-deoxy-D-galactose, *see* D-Galactosamine

2-Amino-2-deoxy-D-glucopyranosides, deamination and cleavage, 120

2-Amino-2-deoxy-D-glucose, *see* D-Glucosamine

2-Amino-2-deoxyglycosides, acid hydrolysis, 63

Aminodeoxyglycosidic linkages, deamination cleavage, 121

5-Amino-3,5-dideoxy-D-*glycero*-D-*galacto*-nonulosonic acid, *see* Neuraminic acid,

Aminoglycosidic linkages, selective cleavage by deamination, 118–124
Amylopectin, structure, 22, 24
Amylose
  inclusion complexes, 278, 279, 281
  linkage conformations, 198
  network formations, 266, 267
  nuclear magnetic resonance spectroscopy, conformations and relaxation characteristics, 161, 169
Antibodies
  measurement, 304–312
    enzyme-linked immunosorbent assay, 310–313
    hemagglutination, 308
    immunodiffusion, 306, 307
    precipitin reaction, 305, 306
    radioimmunoassay, 309, 310
Antigens
  polysaccharide, cellular response, 299–304
  in human disease, 320–325
Antithrombin, mixed interactions with heparin and heparan sulfate, 277
D-Apiose, structure, 9, 10
D-Arabinose
  Fischer convention, 5, 6
  structure, 9, 10
L-Arabinose, structure, 9, 10
Arabinoxylan
  oxidation of rye flour, 92, 93
  structure, 22, 23

B

Bacteria
  chitin structure, ribbon sequences, 227
  molecular architecture, 293–296
Barry degradation, of polysaccharide, 89
Base-catalyzed fragmentations, 100–112

C

Calcium pectate chains, interruptions, 263–266
Carbohydrate chains, see also Polysaccharides
  chiroptical techniques, 206–214
  conformational entropy, 203, 204

conformational principles, 195–204
differential scanning calorimetry, 223, 224
disordered, in concentrated solution, 259–263
  in dilute solution, 255–259
higher-order interactions, 269–276
hydrated networks, effect of interruptions, 263–269
interresidue linkages, 197–200
nuclear magnetic resonance spectroscopy, 214–216
shapes and interactions, 195–290
X-ray fiber diffraction, 204–206
Carbon-13 nuclear magnetic resonance spectroscopy
  polysaccharide, 134, 140
  proton-coupled and proton-decoupled, 143–145
Carboxymethylamylose, viscosity and solution concentration, 260
O-(Carboxymethyl)cellulose, carbon-13 nuclear magnetic resonance spectroscopy, 167, 168
ι-Carrageenan
  nuclear magnetic resonance spectroscopy, conformations, 165
  structure, 240, 241, 244
κ-Carregeenan
  mercaptolysis or methanolysis, 67
  structure, 240
λ-Carrageenan
  structure, 240
  viscosity and solution concentration, 260, 262
Carrogeenans
  carbon-13 nuclear magnetic resonance spectroscopy and chemical structure, 158
  chain interruptions, 266
  gelation behavior, 269–272
  interactions with milk proteins, 278
  oxidative hydrolysis, 68
  oxolane ring formation, 95–97
  repeating units, 238
Casein, interactions with carrageenans, 278
Cation binding
  ribbon sequences, 228–235
  stoichiometry, in carbohydrate chains, 222, 223

Cellobiose, conformational analysis, 200
Cellulase, in polysaccharide hydrolysis, 69, 70
Cellulose
  alkaline degradation, 103
  infrared-Raman spectroscopy, 183
  infrared-Raman spectra of base paper, 174
  linkage conformations, 198
  nuclear magnetic resonance spectroscopy, 166
  primary structure, 225
  solvents, nuclear magnetic resonance spectroscopy, 172
  structure, ribbon sequences in plant cell walls, 226
Cellulose derivatives, nuclear magnetic resonance spectroscopy, 167–169
Ceratocystis brannea glucomannan, structural modification, 97
Chitin
  primary structure, 225
  structure, ribbon sequences, 227, 228
Chondroitin, acid hydrolysis of carboxyl-reduced, 63
Chondroitin sulfate
  carbon-13 nuclear magnetic resonance spectroscopy and chemical structure, 157
  conformational behavior, 249, 250, 252–254
  desulfation, 98
  infrared-Raman spectroscopy, 178, 181
  periodate oxidation, 85
  proton-coupled and proton-decoupled carbon-13 nuclear magnetic resonance spectra, 143–145
Chromatography, ion-exchange, polysaccharide fractionation, 29
Circular dichroism, *see also* Vacuum ultraviolet circular dichroism
  conformational analysis of carbohydrate chains, 206–214
  polyuronates, and cation binding, 229–235
Configuration
  anomeric, in multiresidue polysaccharides, 158–160
  polysaccharide, infrared-Raman spectroscopy, 176–180

Conformation
  carbohydrate chains, 196–204
    chiroptical techniques, 206–214
    nuclear magnetic resonance spectroscopy, 214–216
    x-ray fiber difraction, 204–206
  homopolysaccharides, 224–226
  immune response determinants, 317–319
  of individual polysaccharide residues, 160
  polysaccharide, differential scanning calorimetry, 223, 224
  relaxation characteristics and nuclear magnetic resonance spectroscopy, 161–166
Copolysaccharides, regular
  algal, 240–245
  bacterial, 245–248
  glycosaminoglycans, 248–255
  repeating units, 238, 239
Crustaceans, chitin structure, ribbon sequences, 227
Curdlan, nuclear magnetic resonance spectroscopy, conformations, 163, 164
Curtius rearrangement, glycosiduronamide, 115
Cyclohexaamylose, inclusion complexes, 279

D

Deacylation, polysaccharide, 98, 99
Deamination, selective cleavage of amino-glycosidic linkages, 118–124
Degradation, *see* Barry degradation; Depolymerization; Hofmann-Weermann degradation; Smith degradation
3-Deoxy-D-glucose, nomenclature, 5
3-Deoxy-D-*arabino*-hexose, nomenclature, 5
3-Deoxy-D-*ribo*-hexose, Fischer convention, 7
3-Deoxy-D-*manno*-octulosonic acid
  Fischer convention, 7
  structure, 9, 11
2-Deoxy-D-*erythro*-pentose
  Fischer convention, 7
  nomenclature, 5
2-Deoxy-D-ribose, nomenclature, 5

Depolymerization
  alkaline degradation from reducing
    groups, 100–103
  base-catalyzed cleavage of phosphodies-
    ter linkages, 112
  base-catalyzed β-elimination from hex-
    uronic acid residues, 103–107
  degradation from sulfone derivatives,
    107
  degradations preceded by oxidation,
    109–112
  by hydrolysis and related reactions, 56–
    81
  methylated polysaccharides, 52–56
  permethylated polysaccharides, 71–73
  selective cleavage of uronic acid link-
    ages, 112–118
Dermatan
  carbon-13 nuclear magnetic resonance
    spectroscopy and chemical struc-
    ture, 157
  conformation and nuclear magnetic reso-
    nance spectroscopy, 160, 171
Dermatan sulfate
  conformational behavior, 252, 253
  infrared-Raman spectroscopy, 178
Desulfation, polysaccharide, 98
Dextran
  linkage conformations, 198
  nuclear magnetic resonance spectros-
    copy, 165, 166
    and chemical structure, 153, 154
  viscosity and solution concentration, 260
Dextrins, inclusion complexes, 279
Differential scanning calorimetry, carbohydrate chain conformations, 223, 224
Diplococcus pneumoniae type 14, capsular
  polysaccharide, degradation, 93
Disaccharides, conformational analysis,
  200–203
Diseases, see also Immunity; Vaccines;
  Virulence factors
  diagnosis, 324
  encapsulated bacteria infections, 322, 323
  polysaccharide antigens, 320–325

E

Electron spin resonance spectroscopy,
  polysaccharide, 140–142

Endoglycanases, in polysaccharide hydrolysis, 69, 70
Enzyme-linked immunosorbent assay, antibody, 310–313
Enzymes, in polysaccharide hydrolysis,
  69–71
D-Erythrose, Fischer convention, 5, 6
Escherichia coli 069, lipopolysaccharide,
  deamination cleavage, 122
Escherichia coli serotype K29
  capsular polysaccharide, repeating units,
    239
  structure, 246
Exoglycanases, polysaccharide hydrolysis,
  69–71
Exoglycosidases, in polysaccharide hydrolysis, 69
Exopolysaccharides, immune response,
  316, 318

F

Fatty acids, biosynthesis, inclusion com-
  plexes, 280
Fischer convention, 5, 6
Freund's complete adjuvant, 304
D-Fructose
  Fischer convention, 7
  nomenclature, 5
  structure, 9, 10
L-Fucose, structure, 9, 10
Fungi, molecular architecture, 296
Furcellaran
  repeating units, 238
  structure, 240

G

Galactoarabinoxylans, structure, 23, 24
Galactomannans
  higher-order interactions, 274–276
  hydrated network structure, 267–269
  hydrolysis of permethylated, 48
  nuclear magnetic resonance spectros-
    copy and chemical structure, 156
  structure, 22, 23
  sulfone degradation, 109
  viscosity and solution concentration, 262
D-Galacto-D-mannan, shorthand nomenclature, 16

D-Galactosamine
  Fischer convention, 7
  nomenclature, 5
  structure, 9, 10
D-Galactose, Fischer convention, 5, 6
D-Galactose, structure, 9, 10
L-Galactose, structure, 9, 10
D-Galacturonic acid, structure, 9, 11
Gangliosides, brain, acetolysis, 65
Gels, polysaccharide, hydrodynamic and
    rheological techniques, 219–222
α-D-Glucan, structure, 21
β-D-Glucan
  branched, Smith degradation, 86, 87
  carbon-13 nuclear magnetic resonance
    spectroscopy, 140, 141
  nuclear magnetic resonance spectros-
    copy, conformations, 163
  Smith degradation of oat, 86–88
Glucans
  carbon-13 nuclear magnetic resonance
    spectroscopy and structure, 156,
    157
  nuclear magnetic resonance spectros-
    copy, 165, 166
α-D-Glucofuranose, structure, 5, 8, 9
β-D-Glucofuranose, structure, 5, 8, 9
Glucomannan
  Ceratocystis brannea, structural mod-
    ification, 97
  structure, 21
α-D-Glucopyranose, Haworth and chair
    formulas, 5, 8, 9
β-D-Glucopyranose, Haworth and chair
    formulas, 5, 8, 9
D-Glucosamine, structure, 9, 10
D-Glucose, Fischer convention, 5, 6
D-Glucose, structure, 9, 10
β-D-Glucose, pyranose ring conformations,
    197
D-Glucuronic acid, structure, 9, 11
Glycan, definition, 14
Glyceraldehyde, Fischer convention, 5, 6
Glycoamylase, nuclear magnetic resonance
    spectroscopy, 165
Glycogen, structure, 22, 24
Glycoproteins, proton nuclear magnetic
    resonance spectroscopy, 159–160
Glycopyranosides, permethylated, mass
    spectral fragmentation pathways, 77

Glycosaminoglycans
  conformational behavior, 248–254
  stereochemistry of action by nuclear
    magnetic resonance spectroscopy,
    169, 170
  and chemical structure, 157
Glycosidases, nuclear magnetic resonance
    spectroscopy, 170, 171
Glycosides, linkage formation, 13
Glycosidic linkages
  chemical shifts of carbon, 148, 149
  chemical shifts of protons, 149
  configuration, conformation, 149–151
  hydrolysis rates, 62–64
Glycosiduronic acid linkages, selective
    cleavage, 112–118
Glycosiduronic acids, hydrolysis, 90–94
Glycuronans, methylated, depolymeriza-
    tion, 104
D-Gulose, Fischer convention, 5, 6
L-Guluronic acid, structure, 9, 11

                        H

Hakomori methylation, polysaccharide, 49,
    50, 99
Hemagglutination, 308
Hemiacetals, cyclic, 5
Heparan sulfate
  conformational behavior, 254, 255
  conformation and nuclear magnetic reso-
    nance spectroscopy, 160
  mixed interactions with antithrombin,
    277
  proton nuclear magnetic resonance spec-
    tra, 136, 138
Heparin
  carbon-13 nuclear magnetic resonance
    spectroscopy, 157
  conformational behavior, 254, 255
  conformation and nuclear magnetic reso-
    nance spectroscopy, 160, 170, 171
  desulfation, 98
  infrared-Raman spectroscopy, 179
  mixed interactions with antithrombin,
    277
  proton nuclear magnetic resonance spec-
    troscopy, 137–140
  conformations, 162
  selective cleavage by deamination, 118

L-*glycero*-D-*manno*-Heptose
Fischer convention, 7
structure, 9, 11
Heterogeneity
concepts, 20–26
polysaccharide, sequences and structure, 152–158
Heteroglycans, definition, 15
Heteropolysaccharides, 21–23
branched, 22–24
Hex-5-enopyranosides, formation and selective hydrolysis of, from uronic acids, 117
D-*arabino*-Hexulose, *see* D-Fructose
Hofmann-Weermann degradation, glycosiduronamide linkages, 113–115
Homogeneity
concepts, 20–26
polysaccharide, criteria, 31, 32
Homoglycans, definition, 15
Homopolysaccharides, 21–23
branched, 22–24
conformational types, 224–226
flexible coils and linkages, 237
hollow helices, with energy reserve functions, 235–237
with structural functions, 237
ribbon sequences, and cation binding, 228–235
in plant cell walls, 226, 227
Hyaluronic acid
carbon-13 nuclear magnetic resonance spectroscopy and chemical structure, 157
conformational behavior, 248–252
infrared-Raman spectroscopy, 178
nuclear magnetic resonance spectroscopy, conformations, 163, 170
shorthand nomenclature, 16
structure, 21
viscosity and solution concentration, 260, 262
Hybridoma, 302
Hydration, hydrogen-bonding and infrared-Raman spectroscopy, 182–184
Hydrogen bonding, infrared-Raman spectroscopy, 182–184
Hydrolysis, *see also* Depolymerization; Oxidative hydrolysis

enzyme-catalyzed, polysaccharide, 69–71
partial, of glycosidic linkages, 62–64

I

D-Idose, Fischer convention, 5, 6
L-Iduronic acid, structure, 9, 11
Immune response
polysaccharide, 296–312
conformational determinants, 317–319
cross-reactions, 319, 320
determinants and repeating units, 313–316
molecular size effect, 312, 313
noncarbohydrate determinants, 316, 317
structural aspects, 312–320
Immunity, natural, 325
Immunodiffusion, antigens and antibodies, 306, 307
Immunoelectrophoresis, 307
Immunoglobulins
classification, 302–304
heterogeneity, 300–302
Immunology
polysaccharide, 291–330
history, 292
Inclusion complexes, polysaccharide, 278–281
Infrared-Raman spectroscopy
analytical aspects, 180–182
methodology, instrumentation and techniques, 172–176
polysaccharide, 134, 172–184
Insects, chitin structure, ribbon sequences, 227
6'-Iodophenyl α-maltoside, conformational analysis, 201

K

Keratan sulfate, conformational behavior, 254
Klebsiella, capsular polysaccharides, nuclear magnetic resonance spectroscopy and anomeric configuration, 158–160

Klebsiella type 18
  capsular polysaccharide, enzymic hydrolysis, 71
    partial hydrolysis of methylated, 71, 72
Klebsiella type 28, capsular polysaccharide, degradation, 107
Klebsiella type 33, capsular polysaccharide, methylation, 99
Klebsiella type 47, capsular polysaccharide, degradation, 105, 106, 107
Klebsiella type 59, capsular polysaccharide, Svensson oxidation alkaline degradation, 111, 112
Klebsiella type 63, capsular polysaccharide, methylation, 99

L

Laminaran, nuclear magnetic resonance spectroscopy, 163, 164, 170
Lead tetraacetate
  carbohydrate oxidation, 86
  oxidative decarboxylation of uronic acid linkages, 115–117
Leiocarpan A
  base-catalyzed fragmentation, 106, 107
  partial acid hydrolysis, 91
  selective cleavage of glucopyranosiduronic acid linkages, 116
Lentinan, nuclear magnetic resonance spectroscopy, conformations, 163, 164
Lichenan
  carbon-13 nuclear magnetic resonance spectroscopy and chemical structure, 156, 157
  methylated, partial hydrolysis, 72, 73
  nuclear magnetic resonance spectroscopy, conformations and relaxation characteristics, 161
Lipopolysaccharides, 1, 2
  immune response, 315, 316
  Smith degradation, 89
Locust bean gum, structure and interactions, 268
Locust bean gum, viscosity and solution concentration, 262
Lossen rearrangement, 115

Lymphocytes
  B, humoral response, 296–298
  T, cell-mediated response, 296–298
D-Lyxose, Fischer convention, 5, 6

M

Macrophages, immune response, 296–298
Maltose
  conformational analysis, 200–203
  nomenclature, 9
  structures, 14
Mannans
  nuclear magnetic resonance spectroscopy, 165, 166
    and chemical structure, 154–156
  primary structure, 225
  yeast, acetolysis, 64, 65
D-Mannose, Fischer convention, 5, 6
D-Mannose, structure, 9, 10
D-Mannuronic acid
  Fischer convention, 7
  nomenclature, 5
  structure, 9, 11
Mass spectrometry
  of methylated sugar derivations, 52–56
  nomenclature, 76
  of oligosaccharide derivatives, 73, 75–81
Meningococcal polysaccharides, nuclear magnetic spectroscopy, conformations and relaxation characteristics, 162
Mercaptolysis, polysaccharide, 67
Mesquite gum, oligosaccharides from degraded polysaccharides, 58, 59
Methanolysis, polysaccharide, 67, 68
Methylation
  polysaccharide, procedures, 49–51
  structural information, 45–56
Methyl α-D-glucofuranoside, structure, 9, 13
Methyl β-D-glucofuranoside, structure, 9, 13
Methyl α-D-glucopyranoside, structure, 9, 13
Methyl β-D-glucopyranoside, structure, 9, 13
(4-*O*-Methyl)-D-glucurono-D-xylan, shorthand nomenclature, 16

Micrococcus lysodeikticus, polysaccharide, deamination cleavage, 122
Molecular weight, polysaccharide, measurement, 217–219
Monosaccharide, building blocks, 196, 197
Muramic acid, structure, 9, 10
Myeloma, 302

N

Neoagarose, 70
Neoagarotetraose, 70
Neuraminic acid, structure, 9, 11
Nitrocellulose, carbon-13 nuclear magnetic resonance spectroscopy, 169
Nomenclature
mass spectral, 76
polysaccharide, basic chemistry of constituent sugars, 5–18
polysaccharide, shorthand, 15, 16
Nuclear magnetic resonance spectroscopy
applications, 169–172
chemical shifts, 146–149
chemical shift standards, solvents, 142
conformational analysis of carbohydrate chains, 214–216
polysaccharide, 134–172
anomeric configuration in multiresidue types, 158–160
characteristics, 145–151
and chemical structure, 151–158
conformations and relaxation characteristics, 161–166, 214
conformations of individual residues, 160
methods and instrumentation, 135–145
sensitivity enhancement, quantitative measurements, 142, 143

O

Oligosaccharides
fractionation and characterization from degraded polysaccharides, 57–62
nuclear magnetic resonance spectroscopy, 166
Optical rotatory dispersion, conformational analysis of carbohydrate chains, 206–214

Osmotic pressure, incompatible polymers, 276
Oxidation, *see also* Lead tetraacetate; Periodate oxidation
anodic, of uronic acids, 117
Oxidative hydrolysis, polysaccharide, 68
Oxirane ring, epoxide formation in polysaccharide from *Ulva lactuca,* 95, 97
Oxolane ring, formation in polysaccharides, 95–97

P

Pectin
calcium-induced gelation, 234, 235
carbohydrate chain interruptions, 263, 264
Peptidoglycans, 1, 2
chain conformations, 228
Periodate oxidation, carbohydrate, 81–86
Phosphorodiester linkages, 112
Plant cell walls, cellulose structure, ribbon sequences, 226, 227
Polyadenylic acid, nuclear magnetic resonance spectroscopy, conformations, 162, 163
Poly(L-guluronate), egg box model, 233
Polymer entanglement, 219
Polymers, carbohydrate-containing, 1–4
Polysaccharides, *see also* Carbohydrate chains; Copolysaccharides; Heteropolysaccharides; Homopolysaccharides
acetolysis, 63
algal, network formations, 266
oxolane and oxirane formation, 95–97
Barry degradation, 89
branched, immune response, 315
chemical characterization and structure determination, 35–130
composition, sugar components and removable substituents, 37–43
conformations, relaxation characteristics, 160–166
and conjugates, 1–4
degradative aspects, 86–89
degraded, fractionation and characterization, 57–62
depolymerization by hydrolysis and related reactions, 56–81

derivatives, nuclear magnetic resonance spectroscopy, 166–169
desulfation, 98
disorder-order transition kinetics, 216, 217
fractionation, 19–34
functional group modifications of sugar residues, 9, 12
functions, 3
hydrodynamic and rheological techniques, 219–222
immunology, 4, 291–330
infrared-Raman spectroscopy, 173–175
  functional group detection and configuration, 176–180
isolation, 19–34
literature, 17, 18
methylation, 45–56
mixed interactions with proteins, 276–278
molecular size determination, 43, 44
molecular weight and particle size, 217–219
nuclear magnetic resonance spectroscopy, 151–158
periodate oxidation, 83–86
permethylated, characterization and analysis, 51–56
  partial depolymerization, 71–73
protective, 4
removable substituents, location, 97–100
ribbon sequences, in crustaceans, insects, bacteria, and fungi, 227, 228
selective substitution and structural modification, 94, 95
shorthand forms, 15, 16
spectroscopy, 133–193
stoichiometry of cation binding, 222, 223
structural modifications, 89–100
structure, 3, 4
  anomeric configuration in multiresidue type, 158–160
  conformation of individual residues, 160
  determination problems, 44, 45
  nomenclature, 14–16
  primary or covalent, 4
  sequences and chemical heterogeneity, 152–158
virulence factors, 321

Polyuridylic acid, nuclear magnetic resonance spectroscopy, conformations, 162, 163
Polyuronates
  conformational analysis by circular dichroism, 211–214
  primary structures, 225
  ribbon sequences and cation binding, 228–235
  stoichiometry of cation binding, 222, 223
Polyuronides, infrared-Raman spectroscopy, 178
Porphyran, structure, 96
Precipitin reaction, antibody, 305, 306
Proteins, polysaccharide mixed interactions, 276–278
Proteoglycans, 1, 2
  conformational behavior, 248, 254
Proton nuclear magnetic resonance spectroscopy, polysaccharide, 135–140
Pullulan, structure, 21
Pyruvic acid ketals
  as antigenic determinant groups, 317
  location of, 99–100

R

Radioimmunoassay, antibody, 309, 310
Raman spectroscopy, see Infrared-Raman spectroscopy
Relaxation, see Nucleic magnetic resonance spectroscopy, Polysaccharide conformations
L-Rhamnose, structure, 9, 10
Rhizobium trifollii, capsular polysaccharide, methylation and hydrolysis, 99, 100
D-Ribose, Fischer convention, 5, 6
D-Ribose, structure, 9, 10
Ring size, determination in polysaccharides, 73, 74

S

Saccharinic acid-forming reactions, in depolymerization of polysaccharides, 102
Scleroglucan, structure, 22
Serotransferrin, acetolysis, 65
Shigella dysenteriae type 1, lipopolysaccharide, deamination cleavage, 119, 121

Smith degradation, polysaccharide, 86–89
Solvents
  nuclear magnetic resonance spectros-
    copy, 142
  for polysaccharide extraction, 26–28
Spectroscopy
  comparative evaluation of methods, 184–
    186
  polysaccharide, 133–193
Starch
  inclusion complexes, 281
  nuclear magnetic resonance spectros-
    copy, 166
Streptococcus pneumoniae type 1, capsu-
    lar polysaccharide, deamination cleav-
    age, 121
Streptococcus Pneumoniae type 2, poly-
    saccharide, base-catalyzed degradation
    of sulfone derivative, 108
Streptococcus pneumoniae type 6, deg-
    radation of specific substance, 112
Sugars, reducing, structure, 5
Sulfate hemiesters, 97–98
Svensson oxidation alkaline degradation,
    polysaccharide, 111, 112

T

D-Talose, Fischer convention, 5, 6
T cell, independence, 299, 300
Teichoic acids, 1, 2
  capsular, depolymerization, 112
D-Threose, Fischer convention, 5, 6
Tragacanthic acid, acetolysis, 65, 66
Trifluoroacetolysis, of glycoconjugates and
    oligosaccharides, 66, 67

U

Ulva lactuca polysaccharide epoxide
    formation, 97

Uronic acid residues, epimerization, 25
Uronic acids, reduction and fragmentation,
    90–94

V

Vaccines, protective, capsular polysac-
    charides, 321–324
Vacuum ultraviolet circular dichroism,
    conformational analysis of carbohy-
    drate chains, 208, 209
van der Waals repulsion, polysaccharide
    conformations, 199
Virulence factors, capsular polysaccha-
    rides, 321
Viscosity, carbohydrate polymers, 220,
    255–263

X

Xanthan
  higher-order interactions, 272–276
  repeating units, 239
  structure, 245–247
Xanthan gum, methylated, partial hydroly-
    sis, 72
X-ray fiber diffraction, carbohydrate
    chains, 204–206
Xylanase, in polysaccharide hydrolysis,
    69, 70
D-Xylose, Fischer convention, 5, 6
D-Xylose, structure, 9, 10

Y

Yeast mannans, *see* Mannans
Yeasts, molecular architecture, 296

Molecular Biology

An International Series of Monographs and Textbooks

Editors

BERNARD HORECKER

Roche Institute of Molecular Biology
Nutley, New Jersey

NATHAN O. KAPLAN

Department of Chemistry
University of California
At San Diego
La Jolla, California

JULIUS MARMUR

Department of Biochemistry
Albert Einstein College of Medicine
Yeshiva University
Bronx, New York

HAROLD A. SCHERAGA

Department of Chemistry
Cornell University
Ithaca, New York

HAROLD A. SCHERAGA. Protein Structure. 1961

STUART A. RICE AND MITSURU NAGASAWA. Polyelectrolyte Solutions: A Theoretical Introduction, *with a contribution by Herbert Morawetz.* 1961

SIDNEY UDENFRIEND. Fluorescence Assay in Biology and Medicine. Volume I—1962. Volume II—1969

J. HERBERT TAYLOR (Editor). Molecular Genetics. Part I—1963. Part II—1967. Part III—Chromosome Structure—1979

ARTHUR VEIS. The Macromolecular Chemistry of Gelatin. 1964

M. JOLY. A Physico-chemical Approach to the Denaturation of Proteins. 1965

SYDNEY J. LEACH (Editor). Physical Principles and Techniques of Protein Chemistry. Part A—1969. Part B—1970. Part C—1973

KENDRIC C. SMITH AND PHILIP C. HANAWALT. Molecular Photobiology: Inactivation and Recovery. 1969

RONALD BENTLEY. Molecular Asymmetry in Biology. Volume I—1969. Volume II—1970

JACINTO STEINHARDT AND JACQUELINE A. REYNOLDS. Multiple Equilibria in Protein. 1969

DOUGLAS POLAND AND HAROLD A. SCHERAGA. Theory of Helix-Coil Transitions in Biopolymers. 1970

JOHN R. CANN. Interacting Macromolecules: The Theory and Practice of Their Electrophoresis, Ultracentrifugation, and Chromatography. 1970

WALTER W. WAINIO. The Mammalian Mitochondrial Respiratory Chain. 1970

LAWRENCE I. ROTHFIELD (Editor). Structure and Function of Biological Membranes. 1971

ALAN G. WALTON AND JOHN BLACKWELL. Biopolymers. 1973

WALTER LOVENBERG (Editor). Iron-Sulfur Proteins. Volume I, Biological Properties—1973. Volume II, Molecular Properties—1973. Volume III, Structure and Metabolic Mechanisms—1977

A. J. HOPFINGER. Conformational Properties of Macromolecules. 1973

R. D. B. FRASER AND T. P. MACRAE. Conformation in Fibrous Proteins. 1973

OSAMU HAYAISHI (Editor). Molecular Mechanisms of Oxygen Activation. 1974

FUMIO OOSAWA AND SHO ASAKURA. Thermodynamics of the Polymerization of Protein. 1975

LAWRENCE J. BERLINER (Editor). Spin Labeling: Theory and Applications. Volume I, 1976. Volume II, 1978

T. BLUNDELL AND L. JOHNSON. Protein Crystallography. 1976

HERBERT WEISSBACH AND SIDNEY PESTKA (Editors). Molecular Mechanisms of Protein Biosynthesis. 1977

TERRANCE LEIGHTON AND WILLIAM F. LOOMIS, JR. (Editors). The Molecular Genetics of Development: An Introduction to Recent Research on Experimental Systems. 1980

ROBERT B. FREEDMAN AND HILARY C. HAWKINS (Editors). The Enzymology of Post-Translational Modification of Proteins, Volume 1. 1980

WAI YIU CHEUNG (Editor). Calcium and Cell Function, Volume I: Calmodulin. 1980. Volume II. 1982

OLEG JARDETZKY and G. C. K. ROBERTS. NMR in Molecular Biology. 1981

DAVID A. DUBNAU (Editor). The Molecular Biology of the Bacilli, Volume I: *Bacillus subtilis*. 1982

GORDON G. HAMMES. Enzyme Catalysis and Regulation. 1982

GUNTER KAHL and JOSEF S. SCHELL (Editors). Molecular Biology of Plant Tumors. 1982

P. R. CAREY. Biochemical Applications of Raman and Resonance Raman Spectroscopies. 1982

OSAMU HAYAISHI and KUNIHIRO UEDA (Editors). ADP-Ribosylation Reactions. 1982

G. O. ASPINALL. The Polysaccharides, Volume 1. 1982

In preparation

CHARIS GHELIS and JEANNINE YON. Protein Folding. 1982

WAI YIU CHEUNG (Editor). Calcium and Cell Function, Volume III. 1983

ALFRED STRACHER (Editor). Muscle and Non-Muscle Motility, Volume 1. 1982